Studies in Classification, Data Analysis, and Knowledge Organization

Studies in Classification, Data Analysis, and Knowledge Organization is a book series which offers constant and up-to-date information on the most recent developments and methods in the fields of statistical data analysis, exploratory statistics, classification and clustering, handling of information and ordering of knowledge. It covers a broad scope of theoretical, methodological as well as application-oriented articles, surveys and discussions from an international authorship and includes fields like computational statistics, pattern recognition, biological taxonomy, DNA and genome analysis, marketing, finance and other areas in economics, databases and the internet. A major purpose is to show the intimate interplay between various, seemingly unrelated domains and to foster the cooperation between mathematicians, statisticians, computer scientists and practitioners by offering well-based and innovative solutions to urgent problems of practice.

Antonio D'Ambrosio · Mark de Rooij ·
Kim De Roover · Carmela Iorio ·
Michele La Rocca
Editors

Supervised and Unsupervised Statistical Data Analysis

Springer

Editors
Antonio D'Ambrosio
Department of Economics and Statistics
University of Naples Federico II
Naples, Italy

Mark de Rooij
Institute of Psychology
Leiden University
Leiden, The Netherlands

Kim De Roover
Research Group of Quantitative Psychology
and Individual Differences
KU Leuven
Leuven, Belgium

Carmela Iorio
Department of Economic and Statistics
University of Naples Federico II
Naples, Italy

Michele La Rocca
Department of Economics and Statistics
University of Salerno
Fisciano, Italy

ISSN 1431-8814 ISSN 2198-3321 (electronic)
Studies in Classification, Data Analysis, and Knowledge Organization
ISBN 978-3-032-03041-2 ISBN 978-3-032-03042-9 (eBook)
https://doi.org/10.1007/978-3-032-03042-9

This Springer imprint is published by the registered company Springer Nature Switzerland AG
The registered company address is: Gewerbestrasse 11, 6330 Cham, Switzerland

Preface

This book contains peer-reviewed contributions from the 15th Scientific Meeting of the Classification and Data Analysis Group (CLADAG) of the Italian Statistical Society (SIS), held in Naples, Italy, from 8 to 10 September 2025. This edition marks a historic milestone: for the first time, CLADAG has hosted a full joint scientific meeting with another classification society—the Vereniging voor Ordinatie en Classificatie (VOC), the Dutch-Flemish classification society.

Over the years, CLADAG has organized other joint scientific events with classification societies, such as the Société Francophone de Classification (SFC) in 2008, the German Classification Society (GfKl) in 2010, and the Japanese Classification Society (JCS) in 2012. However, those meetings were not part of the official CLADAG biennial cycle and were instead smaller, intermediate events. The 2025 edition stands out as the first time CLADAG and VOC have co-organized an official conference of both societies. The CLADAG-VOC 2025 conference brings together two long-standing traditions in statistical science. Since its founding, CLADAG has been Italy's leading platform for advancing methodological research in multivariate statistics, particularly in classification and data analysis. A member of the International Federation of Classification Societies (IFCS) since 1985, CLADAG has promoted scientific progress through biennial conferences hosted across Italy—from Pescara in 1997 to Salerno in 2023—as well as through specialized schools and collaborative initiatives with other IFCS members.

This joint meeting takes on special significance through the collaboration with VOC, which has cultivated an equally distinguished tradition since its establishment in 1989. Also, a founding member of the IFCS, VOC has consistently supported innovative research in data analysis, classification, clustering methodology, and ordination, both through its annual meetings and international collaborations. The intellectual synergy created by the joint CLADAG-VOC meeting elevates this volume beyond standard conference proceedings. It exemplifies how two societies can connect theoretical innovation with practical application. The selected papers demonstrate how international collaboration can generate novel solutions to long-standing challenges in statistical data analysis and address the emerging needs of an increasingly complex data landscape. The Scientific Committee of CLADAG-VOC 2025 designed the keynote sessions to provide fresh perspectives on the current state of research in clustering, classification, and data analysis. The scientific program is particularly rich, featuring five keynote lectures, 42 invited sessions sponsored by members of the scientific committee, and, at the time of writing, 14 contributed sessions.

We wish to express our deepest gratitude to all participants—authors, reviewers, session chairs, speakers, and committee members—whose dedication made this joint meeting possible. Special thanks goes to the Italian Statistical Society (SIS) and the Dutch-Flemish Classification Society (VOC) for their institutional support, to the University of Naples Federico II and its Department of Economics and Statistics for both institutional and financial backing, and to Springer for its continued collaboration in disseminating research. We hope that this volume will serve as a testament to the value of international collaboration and inspire future breakthroughs at the forefront of statistical science. This volume captures the intellectual vitality and collaborative spirit that characterized the meeting, showcasing how the convergence of distinct statistical traditions can enrich data analysis. The contributions offer new insights into the theoretical and practical challenges of supervised and unsupervised learning, highlighting the remarkable breadth of contemporary statistical research while maintaining methodological rigour.

Innovative approaches to statistical modelling—addressing spatial dependencies and circular data structures—are presented alongside significant advances in interpretable machine learning that reconcile statistical precision with algorithmic transparency. Particularly noteworthy is the volume's treatment of complex data structures, including novel methods for network analysis, high-dimensional clustering, temporal pattern recognition, optimization techniques. What distinguishes this volume is its seamless integration of methodological innovation and practical relevance. The applications span diverse domains, including the social sciences and biomedical engineering, each demonstrating the effective translation of statistical theory into real-world impact. The 30 selected papers collectively push the boundaries of the field: reconciling traditional statistical approaches with modern computational demands, developing new tools for interpreting increasingly complex data, and illustrating how well-crafted methodologies can address substantive questions across multiple disciplines. This interdisciplinary character reflects the unique value of international collaborations like the CLADAG-VOC partnership, and we are truly grateful for it.

As editors, we hope this volume will serve both as a snapshot of current innovations at the intersection of classification and data analysis, and as an inspiration for future research and collaboration.

June 2025

Antonio D'Ambrosio
Mark de Rooij
Kim De Roover
Carmela Iorio
Michele La Rocca

Contents

Measuring Discrimination in Decision-Making Algorithms: An Approach Based on Causal Inference

Francesco Pauli and Roberta Pappadà(✉)

Dipartimento di Scienze Economiche, Aziendali, Matematiche e Statistiche 'B. de Finetti', Università degli Studi di Trieste, Trieste, Italy
francesco.pauli@deams.units.it, rpappada@units.it

Abstract. Machine learning algorithms are routinely used for business decisions that may directly affect individuals in various contexts, such as credit scoring, employment, and criminal justice. When such algorithms are used in the decision process, their behavior concerning discrimination depends on the information it is given, and discrimination may occur unconsciously or explicitly based on sensitive attributes. Statistical tools and methods are then required to handle such potential biases. We propose to exploit the Coarsened Exact Matching (CEM) algorithm to measure discrimination against a protected group to be used in data pre-processing for discrimination detection and removal. Experiments are conducted to test the proposed methodology on real data and a comparison with related work is also discussed.

Keywords: Discrimination · Machine learning · Coarsened Exact Matching · Sensitive attribute

1 Introduction

Machine learning procedures are nowadays exploited to support decisions in many contexts, such as credit scoring, criminal justice, job hiring, which affect individuals and the whole society. A crucial issue in the growing use of such procedures is the potential discrimination that can lead to an unjust treatment of particular groups: treating a person or a group depending on some sensitive attribute (s.a., S) such as race (skin color), sex, religious orientation, etc.

A human may discriminate either because of irrational prejudice induced by ignorance and stereotypes or based on statistical generalization: lacking specific information on an individual, the prevalent characteristics in the sensitive attribute category to which the individual belongs are assigned. For example, in the United States, without information on education, it can be assumed that a black person has a relatively low level since, on average, unequal learning opportunities based on social status often penalize black people in the country [1]. As discussed in [2], the definition of fairness, that is, the absence of discrimination in the outcome of an automated decision-making system, is not univocal

A. D'Ambrosio et al. (Eds.): CLADAG-VOC 2025, SCDAKO, pp. 1–11, 2025.
https://doi.org/10.1007/978-3-032-03042-9_1

and depends on the specific context. A major categorization distinguishes group level fairness, when we are concerned with achieving parity across predefined groups, and individual fairness, when we aim at guarantying that individuals that are similar but for the s.a. are treated similarly.

Moreover, discrimination against individuals or groups based on sensitive or protected characteristics can arise from various sources, pertaining to the algorithm, the data, or the population itself. As such, several approaches to deal with unfairness and discrimination have been proposed in various contexts: fairness through unawareness aims at achieving fairness by ignoring the s.a. in the decision process whereas fairness through intervention involves recognizing the presence of bias and introducing appropriate corrections either to the data, the algorithm or the results (pre-processing, in-processing, or post-processing interventions); see, e.g., [3] and the references therein. In this work, we consider a possible strategy to achieve individual fairness that can be used as a pre-processing technique to determine whether the data contains discrimination, or as a post-processing intervention method to establish whether and to what extent the algorithm produces discriminatory results.

When a statistical or machine learning algorithm is used in the decision process, its behavior regarding discrimination depends on the information it is given. In particular, if the sensitive attribute is available to the algorithm (i.e., it is included in the learning data and can be used for predictions), it can discriminate either because the data it is taught contain irrational prejudice or because the sensitive attribute is associated with an unobserved attribute that is relevant for the prediction of the outcome of interest (Y).

If the algorithm is constrained not to use the s.a. (S) in the final rule, i.e., it is excluded from the data from which the algorithm learns, discrimination could still happen either because a variable related to the outcome is also related to S or because a variable included in the data that is unrelated to the outcome is related to S, which is associated with the outcome (either directly or mediated). Whether or not the situations described above are instances of undesired discrimination depends on how the concept is defined from a legal or ethical standpoint.

In this work, we investigate conditional equal treatment that is, assuming that S and Y are dichotomous, $Y = 1$ is the desired outcome and $S = 1$ "belonging to a protected category" (i.e., possessing the s.a.), it is required that

$$P(Y = 1|S = 1, \mathbf{X} = \mathbf{x}) = P(Y = 1|S = 0, \mathbf{X} = \mathbf{x}),$$

where $\mathbf{X} = (X_1, \ldots, X_p)$ is the set of p attributes for n individuals in the dataset. Hence, the kind of discrimination we refer to occurs when the collective possessing the s.a. experiences the desired outcome less frequently than those not possessing the s.a. ($S = 0$) but similar in other respects.

In this context, data pre-processing techniques aim to establish whether and to what extent the available data are discriminatory. If discrimination is detected, data are modified to make them discrimination-free and then used with a standard algorithm (see, e.g., [4–9]). Such methods hence require a suitable strategy for discrimination discovery from data, which consists in the actual discovery of discriminatory situations in historical decision records. Comparing a protected unit's outcome with the outcome of unprotected units that are similar with respect to the available attributes is a reasonable way to detect whether the unit has been discriminated against (conditionally). Along this direction, [9] propose using the difference in the rate of unfavorable outcomes in units with $S = 1$ and units with $S = 0$, i.e., non-sensitive ones, by exploiting a variant of the k-nearest neighbor (k-NN) classification.

Our proposal is to measure discrimination against the units in a protected category via a sequential procedure based on Coarsened Exact Matching (CEM), a technique from causal inference that aims to estimate the causal effect of a treatment on an outcome [10]. Units labeled as treated will be individuals belonging to a protected category and (conditional) discrimination is detected by comparing the outcome of a protected unit with the outcome of unprotected units that have similar values on the observed variables other than S, after preliminary coarsening of the continuous ones.

The paper is organized as follows. In Sect. 2, we present the framework and introduce the proposed CEM-based discrimination measure. An application to real data is illustrated in Sect. 3, which also considers the comparison with the approach proposed in [9]. Section 4 presents some final remarks.

2 Measuring Discrimination via CEM

Let $\mathcal{S}$ denote the set of all units having the s.a. ($S = 1$). To measure the potential discrimination against unit $i \in \mathcal{S}$ one possible strategy is to compare it to its neighbors using a suitable distance, as done in [9].

Our approach to measuring the potential discrimination against unit $i \in \mathcal{S}$ is based on comparing the outcome y_i with the outcome of the units matched with i by the CEM technique, without relying on a specific distance measure. The CEM algorithm matches a treated unit to those untreated units that share the same values on the available (continuous and categorical) attributes, possibly after coarsening the continuous ones. Moreover, the CEM algorithm discards those units that can not be matched.

Our objective is to detect whether unit $i \in \mathcal{S}$ has experienced a different outcome because it possesses the s.a.. Let $\bar{y}_i^{(S=0)}$ be the relative frequency of positive outcomes among units matched with i and not possessing the s.a., that is, assuming that a non empty subset C_i of similar units can be found,

$$\bar{y}_i^{(S=0)} = \#\{j|y_j = 1, s_j = 0, j \in C_i\}/\#\{j|s_j = 0, j \in C_i\},$$

with $j \neq i$, where $\#G$ is the cardinality of a set G. A measure of discrimination for unit i can be expressed in terms of the difference

$$D_i = y_i - \bar{y}_i^{(S=0)},$$

taking negative values when the unit i is discriminated against and positive values when the unit i is favored ($D_i \in [-1, 1]$). However, not all units are matched in a standard implementation of the CEM algorithm, making it unsuitable for our purpose: after coarsening, a larger contingency table is created, if a cell contains both individuals with and without the s.a., we can compare their outcomes; however, some cells will contain only individuals having the s.a., which can not be compared to anything. To overcome such limitation, we propose to apply the CEM technique sequentially: rather than splitting the sample using all covariates at once, we add variables sequentially so the cells become smaller at each step. We compare each individual having the s.a. to the individuals in the same cell as long as there are non s.a. individuals available; if this is not the case, the last available D is accepted. Specifically, the procedure consists of the following steps:

Step 0 For unit i such that $s_i = 1$, set $D_i = D_i^{(0)}$, with

$$D_i^{(0)} = y_i - \bar{y} = y_i - \hat{P}\{Y = 1|S = 0\} = y_i - \frac{\#\{j|y_j = 1, s_j = 0\}}{\#\{j|s_j = 0\}},$$

that is, no matching is performed.

Step 1 Let $\bar{S}$ denote the set of all units not possessing the s.a. (that is, for $j \in \bar{S}$, $s_j = 0$). Variable X_1 is considered in matching, and let $C_i^{(1)}$ be the set of units matched with unit i based on X_1. Then, if $C_i^{(1)} \cap \bar{S} \neq \emptyset$,

$$D_i^{(1)} = y_i - \bar{y}_i^{(S=0)} = y_i - \frac{\#\{j|y_j = 1, s_j = 0, j \in C_i^{(1)}\}}{\#\{j|s_j = 0, j \in C_i^{(1)}\}},$$

otherwise, set $D_i^{(1)} = D_i^{(0)}$.

Step 2 The attributes X_1 and X_2 are considered in matching, leading to the set $C_i^{(2)}$ of matched units based on both X_1 and X_2. Then, if $C_i^{(2)}$ is non-empty and $C_i^{(2)} \cap \bar{S} \neq \emptyset$, set

$$D_i^{(2)} = y_i - \frac{\#\{j|y_j = 1, s_j = 0, j \in C_i^{(2)}\}}{\#\{j|s_j = 0, j \in C_i^{(2)}\}},$$

otherwise, set $D_i^{(2)} = D_i^{(1)}$.

Step k The attributes $X_1, \ldots, X_k$ are considered in matching, leading to the set $C_i^{(k)}$ of matched units based on all k variables. Then, if $C_i^{(k)}$ is non-empty and $C_i^{(k)} \cap \bar{S} \neq \emptyset$, set

$$D_i^{(k)} = y_i - \frac{\#\{j|y_j = 1, s_j = 0, j \in C_i^{(k)}\}}{\#\{j|s_j = 0, j \in C_i^{(k)}\}},$$

otherwise, set $D_i^{(k)} = D_i^{(k-1)}$.

The procedure is continued until the variables $X_1, \ldots, X_p$ have been sequentially included and the value $D_i^{(p)}$ is obtained by using all variables or a subset of them.

Clearly, the final discrimination score $D_i^{(p)}$ depends on the order in which the attributes are used. Hence, the sequential CEM procedure is repeated over M random permutations $i_1, \ldots, i_p$ of indices $1, \ldots, p$. Finally, for each unit $i \in \mathcal{S}$, the discrimination score is computed as the average

$$D_i = \frac{1}{M} \sum_{i_1,\ldots,i_p} D_i^{(i_1,\ldots,i_p)}, \tag{1}$$

where $D_i^{(i_1,\ldots,i_p)} = D_i^{(p)}$ is obtained through Step 0–p with the variables in order $X_{i1}, \ldots, X_{ip}$.

3 Experiments Based on the Adult Dataset

We conduct experiments for discrimination detection and removal using the Adult dataset [11]. The Adult dataset consists of 48,842 records (which reduce to 45,222 observations after removing missing values) with both categorical and continuous attributes containing information on age, education, nationality, marital status, relationship status, occupation, work classification, gender, race, working hours per week, capital loss and capital gain. This is a relatively small dataset, in particular in terms of number of covariates. Our proposed method is, in principle, applicable to datasets of any dimension (with obvious increase of computational time); the sequential nature of our implementation of CEM in fact mitigates the sparsity issue: if by using a particular set of covariates there are not enough available observations to compare a units with other units, than that combination is discarded for that particular unit.

The income level is the decision attribute expressing a binary condition on the range of income per year in Thousand Dollars: lower or equal to \$50K and higher than \$50K. $Y = 1$ is the desired outcome of having an income greater than \$50,000 (corresponding to the 28.76% of units). For the sensitive attribute, we consider "being non-white" ($S = 1$), to investigate the possible presence of social discrimination against non-white people.

As a preliminary step required by the CEM algorithm, we first choose the coarsening for both categorical and continuous attributes, that is, we adopt meaningful breaks for the numerical attributes to create the coarsening and, when appropriate, we aggregate the existing levels for the categorical ones to avoid categories with too few observations (see Table 1). In doing so, we introduce an element of arbitrariness in the procedure whose robustness to this choice is to be investigated in future work. Finally, the dataset is split into learning (30,162 units) and test (15,060 units) subsamples for the analysis.

In order to evaluate the performance of the proposed method we compare the distribution of the D-scores calculated on the original dataset, on an artificial version of the dataset where no bias is present (that is, it has been eliminated) and on artificial versions where more bias is present. If the proposed measure is sensitive to the presence of discrimination, differences in the left tail should be seen. We consider two alternative scenarios:

Table 1. Name, original type and number of bins or levels for the numerical and categorical attributes, respectively, used in the CEM procedure for the Adult dataset.

Attribute	Type	n. bins/levels
`age`	numerical	13
`workclass`	categorical	6
`education.num`	numerical	8
`marital.status`	categorical	5
`occupation`	categorical	14
`relationship`	categorical	5
`sex`	categorical	2
`capital.gain`	numerical	5
`capital.loss`	numerical	4
`hours.per.week`	numerical	12
`native.country`	categorical	7

Scenario (1) data are modified to build a discrimination-free dataset;
Scenario (2) data are modified to introduce discrimination.

To simulate Scenario (1) where no discrimination is present, we modify the records of all units via a random permutation on the sensitive attribute. Note that, in such case, we do not alter the outcome Y in the dataset.

To simulate Scenario (2), we modify a subset of units in the original data by adding discrimination against the protected category or in favor of the unprotected group. To this aim, we proceed as follows:

(a) Y is changed from positive to negative ($Y = 1 \rightarrow Y = 0$) for a set of randomly selected units belonging to the protected category ($S = 1$);
(b) Y is changed from negative to positive ($Y = 0 \rightarrow Y = 1$) for a set of randomly selected units not belonging to the protected category ($S = 0$).

Once discrimination has been removed or added, we compute the discrimination scores via the iterated sequential CEM algorithm described in Sect. 2, where the number of repetitions is decided based on the convergence of the D_i measures. The implementation of the CEM procedure is carried out via the package `cem` of the R statistical software [12].

For comparison purposes, we consider the measure introduced in [9] based on k-NN, which exploits a similar strategy in that the authors compare each individual with the other units that are similar based on the available non-sensitive attributes. Specifically, the authors introduce a measure computed, for each unit $i \in \mathcal{S}$, as the difference $p_1 - p_2$, where p_1 (respectively, p_2) is the proportion of units in the set of k-nearest neighbors of unit i having $S = 1$ (respectively, $S = 0$), whose outcome is the same of unit i. Such difference represents the bias of the decision for i due to membership to the protected group, with a positive value indicating discrimination against i if $y_i = 0$, and discrimination in favor of the protected category, otherwise. Hence, an individual is labeled as discriminated if there is a significant difference of treatment among its neighbors belonging to a protected-by-law group and its neighbors not belonging to it.

To compare it with our proposal, we express such measure as

$$\delta_i = \frac{1}{k}\#\{j|j \neq i, x_j \in U_i^{k,1}, y_j = 0\} - \frac{1}{k}\#\{j|j \neq i, x_j \in U_i^{k,0}, y_j = 0\}, \quad (2)$$

$\forall i \in \mathcal{S}$, where $U_i^{k,s}$ denotes the set of the k-nearest neighbors of x_i for which $S = s$ ($s \in \{0,1\}$), and the distance between units is a Gower's type distance.

Figure 1 shows the Q-Q plots for comparing the proposed discrimination measure computed via the sequential CEM for the original Adult data against the modified data with discrimination removed or added. Here, we add discrimination against the protected category by changing the outcome for 5% of units possessing (respectively, not possessing) the s.a. whose outcome is the desired one, i.e., $Y = 1$ (respectively, $Y = 0$). Figure 1 also displays the Q-Q plots for the discrimination score δ computed as in Eq. (2), where $k = 32$. As can be seen, the different behavior of D when discrimination is removed or added is clearly spotted for negative values, which indicate discrimination against the protected category. Moreover, the measure D appears more sensitive to detect discrimination than the measure δ, for which positive values identify discriminated units. This may be due to the fact that the k-NN approach is always based on groups of fixed size k, while very small (CEM) strata may contribute to the values of D.

Since our focus is on discrimination against a protected group, we test the discrimination measure D and compare it with the measure δ on the Adult dataset when discrimination is added based on Correct Prediction Ratio (CPR), True Positive Ratio (TPR), and False Negative Ratio (FNR). Specifically, the outcome of a proportion v of units in the protected category is changed from positive to negative–thus introducing discrimination against the protected category, with $v \in \{2.5, 5, 10\}$; then, the discrimination scores are obtained via Eq. (1) using M repetitions of the sequential CEM algorithm. The ratios $\frac{CPR-CEM}{CPR-KNN}$, $\frac{TPR-CEM}{TPR-KNN}$, and $\frac{FNR-CEM}{FNR-KNN}$ are computed based on specific quantiles of the distribution of the discrimination measure D. Figure 2 shows the results for the Adult data based on various quantiles of D for discrimination detection and $M = 20$ repetitions. The proposed measure shows an overall satisfactory performance and seems to outperform the measure based on k-NN in the considered scenarios.

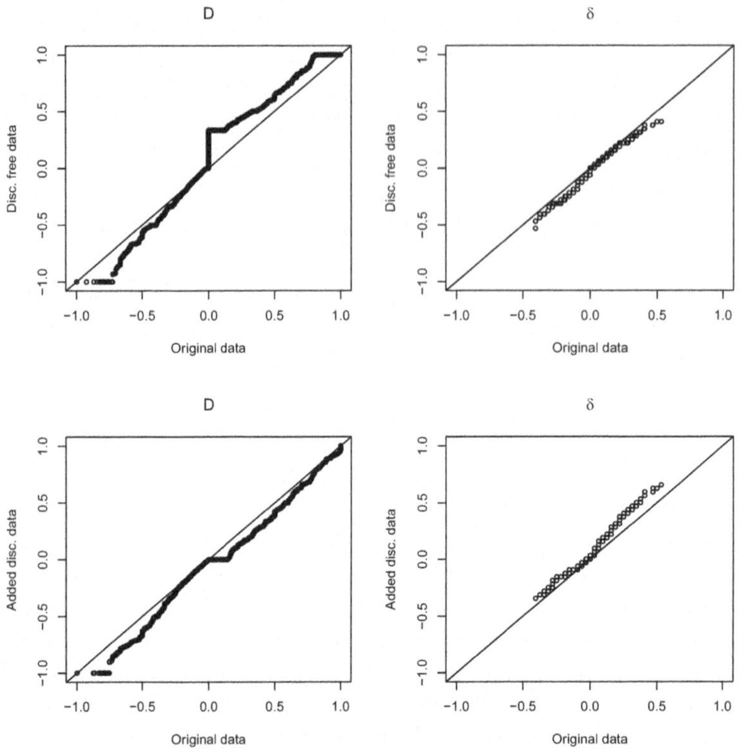

Fig. 1. Q-Q plots for the proposed discrimination measure D and the alternative measure δ based on the original data and the modified data with discrimination removed (Scenario (1), first row) or increased by modifying the outcome for 5% of units having the s.a and 5% of units not possessing the s.a (Scenario (2), second row).

4 Conclusions and Future Work

In the framework of discrimination detection in automated-decision algorithms, we propose a sequential procedure based on Coarsened Exact Matching, a matching method from causal inference that is fast computationally even with very large datasets. The proposed approach allows us to obtain a discrimination score for each unit in the dataset by pursuing a requirement of conditional equal treatment, thus seeking for those units that are matched with respect to non-sensitive attributes in the CEM strata (at worst, it is based on the rate of positive outcome for the whole non-sensitive sample).

Compared to existing proposals considering similarity between units, the approach based on sequential CEM has the advantage to match the units with respect to the coarsened variables, rather than relying on an arbitrary distance measure. Notably, by defining appropriate coarsened versions of the continuous variables (and possibly convenient recoding of the nominal ones), our method

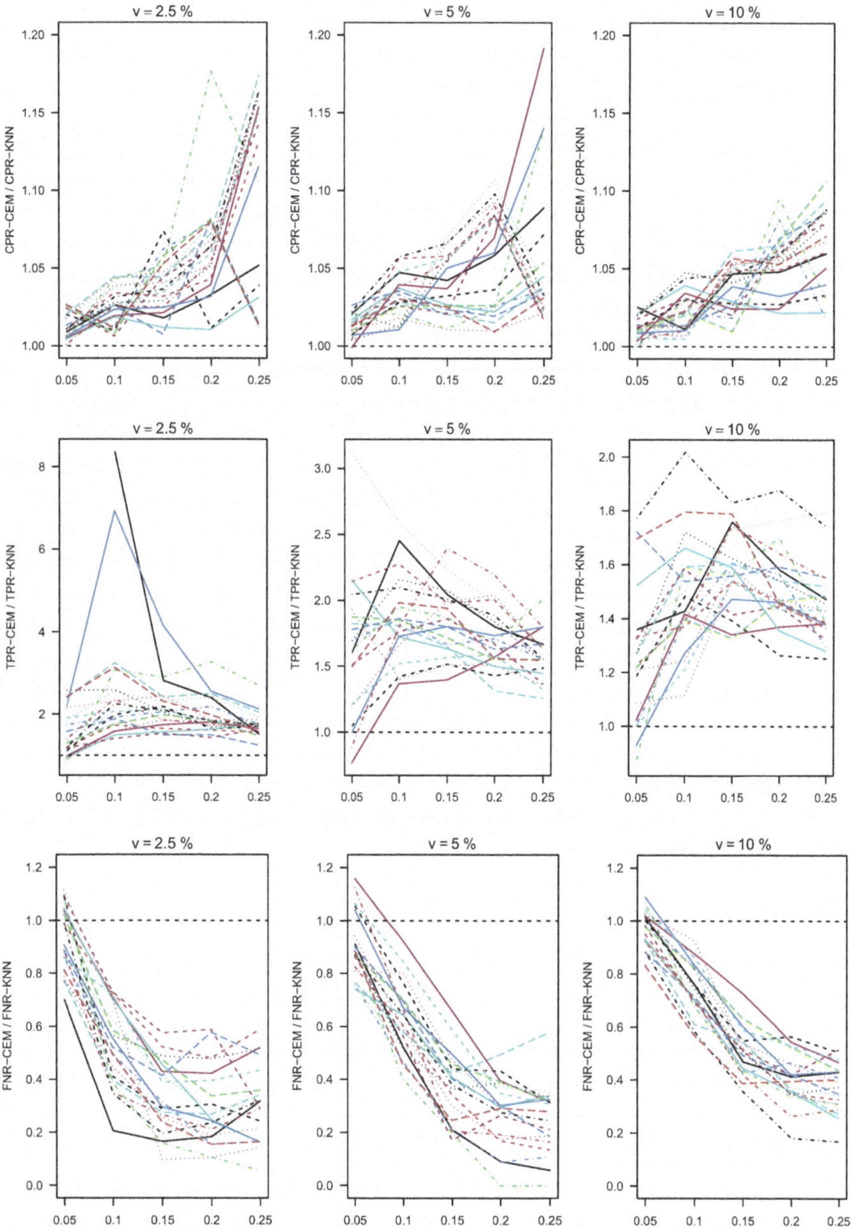

Fig. 2. Comparison of discrimination measures D and δ for Adult data in terms of Correct prediction ratio (CPR-CEM/CPR-KNN, top row), True Positive Ratio (TPR-CEM/TPR-KNN, middle row), and False Negative Ratio (FNR-CEM/FNR-KNN, bottom row) based on $M = 20$ repetitions (lines in each panel) and various thresholds reported on the x-axis. Discrimination is introduced in the original data by modifying the outcome for 2.5%, 5%, 10% of sensitive units, respectively (columns from left to right).

enforces the concept of similarity among the units. On the other hand, we acknowledge that using a distance measure can ease the implementation and do not require choosing a coarsening strategy.

Experiments using the Adult dataset show the effectiveness of our proposal in detecting and measuring discrimination against a protected group, yielding promising results in terms of discrimination discovery performance. Once discrimination is detected, a discrimination-free dataset can be obtained by changing the outcome of the units with discrimination above a certain threshold, which can be chosen by balancing false positives and false negatives as deemed appropriate for the specific application.

It is worth pointing out that the current work opens several challenges and issues that need to be further investigated. Specifically, the CEM procedure requires coarsening of all continuous variables. In this work, we used manually chosen breaks, and we acknowledge that the performance of the procedure may be sensitive to this particular choice. A sensitivity analysis may be put in place to assess the robustness of the method to the choice of the breaks; however, given the degrees of freedom in the choice of the breaks, it would be difficult to design a study that may encompass a sufficient number of different coarsening strategies. An alternative strategy (which we plan to explore in future work) would be to perform a random coarsening at each iteration of the algorithm, whenever a quantitative variable is selected, so that the final discrimination measure will not depend on a particular choice of the breaks. Furthermore, in order to properly address the issue of algorithmic fairness, it is of paramount importance to intersect a plurality of s.a.; although in principle our procedure can be extended to allow for more than one s.a., this will be exacerbate sparsity. Future work will explore whether the sequential procedure is apt to mitigate such an issue.

References

1. Romei, A., Ruggieri, S.: A multidisciplinary survey on discrimination analysis. Knowl. Eng. Rev. **29**(5), 582–638 (2014). https://doi.org/10.1017/S0269888913000039
2. Gottard, A.: Uncertainty & fairness metrics. In: Chelli, F.M., Ciommi, M., Ingrassia, S., Mariani, F., Recchioni, M.C. (eds.) Book of the Short Papers - SIS 2023 (SEAS IN), pp. 180–185. Pearson (2023)
3. Caton, S., Haas, C.: Fairness in machine learning: a survey. ACM Comput. Surv. **56**(7), 1–38 (2024). https://doi.org/10.1145/3616865
4. Kamiran, F., Calders, T.: Data preprocessing techniques for classification without discrimination. Knowl. Inf. Syst. **33**(1), 1–33 (2012). https://doi.org/10.1007/s10115-011-0463-8
5. Žliobaite, I., Kamiran, F., Calders, T.: Handling conditional discrimination. In: 2011 IEEE 11th International Conference on Data Mining, pp. 992–1001. IEEE (2011). https://doi.org/10.1109/ICDM.2011.72
6. Pedreschi, D., Ruggieri, S., Turini, F.: Measuring discrimination in socially-sensitive decision records. In: Proceedings of the 2009 SIAM International Conference on Data Mining (SDM), Philadelphia, pp. 581–592. SIAM (2009). https://doi.org/10.1137/1.9781611972795.50

7. Mancuhan, K., Clifton, C.: Combating discrimination using Bayesian networks. Artif. Intell. Law **22**(2), 211–238 (2014). https://doi.org/10.1007/s10506-014-9156-4
8. Feldman, M., Friedler, S.A., Moeller, J., Scheidegger, C., Venkatasubramanian, S.: Certifying and removing disparate impact. In: Proceedings of the 21st ACM SIGKDD Conference on Knowledge Discovery and Data Mining, pp. 259–268. ACM, New York (2015). https://doi.org/10.1145/2783258.2783311
9. Luong, B.T., Ruggieri, S., Turini, F.: k-NN as an implementation of situation testing for discrimination discovery and prevention. In: Proceedings of the 17th ACM SIGKDD International Conference on Knowledge Discovery and Data Mining, pp. 502–510. ACM, New York (2011). https://doi.org/10.1145/2020408.2020488
10. Iacus, S.M., King, G., Porro, G.: Multivariate matching methods that are monotonic imbalance bounding. J. Am. Stat. Assoc. **106**(493), 345–361 (2011). https://doi.org/10.1198/jasa.2011.tm09599
11. Lichman, M.: UCI machine learning repository (2013). http://archive.ics.uci.edu/ml
12. Iacus, S., King, G., Porro, G.: CEM: software for coarsened exact matching. J. Stat. Softw. **30**(13), 1–27 (2009). https://doi.org/10.18637/jss.v030.i09

Leveraging Social Network Analysis for Semantic Differential Scale: An Application to Survey Data

Ilaria Primerano[1(✉)] and Maria Carmela Catone[2]

[1] National Research Council of Italy, Institute for Research on Population and Social Policies, Corso S. Vincenzo Ferreri, 12, Fisciano, SA, Italy
ilaria.primerano@cnr.it

[2] Department of Political and Social Studies, University of Salerno, Via Giovanni Paolo II, 132, Fisciano, SA, Italy
mcatone@unisa.it

Abstract. Advancing methodological strategies for analyzing subjective perceptions is crucial in social research, particularly when dealing with large-scale digital transitions. This study proposes a quantitative approach that leverages Social Network Analysis (SNA) methods to extract a network of adjectives from Semantic Differential scales. The approach is applied to data collected from an online survey realized during the pandemic among undergraduate students at a Southern Italian university, providing insights into the dimensional structure of their evaluations. The results highlight a positive perception of distance learning services, particularly in terms of activities conducted, use of digital platforms, online interactions, and self-study attitudes. Additionally, the study demonstrates that the perception of these services varies depending on students' prior digital skills.

Keywords: Semantic Differential · Social Network Analysis · Students' perception · Distance learning · Survey Responses

1 Introduction

The objective of this paper is to apply an innovative statistical approach to analyze students' perceptions of distance learning, with a particular focus on the emotional connotations associated with these perceptions. In order to achieve this, an integrated strategy of analysis combining Semantic Differential scales [1] by means of Social Network Analysis (SNA) methods [2] has been proposed to gain a peculiar network of adjectives [3] Specifically, SNA centrality measures [4] allowed us to identify the set of adjectives characterizing the most cohesive component of this network, i.e. the adjectives that mainly contribute to the semantic definition of distance learning process with respect to the dimensions investigated. This methodology provides insights into the underlying emotional aspects of students' evaluations, offering a richer understanding of how they

A. D'Ambrosio et al. (Eds.): CLADAG-VOC 2025, SCDAKO, pp. 12–24, 2025.
https://doi.org/10.1007/978-3-032-03042-9_2

perceive the quality of distance learning services. To validate this innovative methodology, we present a case study using real-life data collected from students' assessments of their distance learning experiences. The data were gathered through an online survey administered to students enrolled in a bachelor's degree course at a Southern Italian university during pandemic in the second term of the 2021–2022 academic year. The use of SNA for semantic differential scale enables a comprehensive exploration of students' emotional reactions to distance learning services, providing valuable feedback for policymakers aiming to enhance the quality of these services.

The paper is structured as follows: Sect. 2 presents the main analytical approaches used to explore university students' perceptions of distance learning; Sect. 3 describes the survey design; Sect. 4 outlines the methodology, detailing the integration of Semantic Differential data into an adjective network; Sect. 5 discusses the results; and Sect. 6 offers concluding remarks.

2 Exploring University Students' Perceptions of Distance Learning: Analytical Approaches

In recent years, the use of Information Communication Technology (ICT) has reshaped education in many ways, contributing to the development of different approaches in the teaching and learning processes. This shift became even more pronounced during the pandemic, which allowed educational activities to continue by temporarily moving them online. These transformation from face-to-face to online education, has become a key focus of scholarly inquiry.

Researchers have employed various methodological approaches to analyze the implications of this abrupt shift from the students' perspective [5–9]. These studies, which have utilized both qualitative and quantitative methods, aim to assess multiple dimensions of the online learning experience, such as student engagement, satisfaction, and the perceived effectiveness of digital tools and platforms. By analyzing students' feedback, scholars have sought to uncover patterns and trends, identifying factors that influence their evaluations of remote education and offering valuable insights into the strengths and challenges of this new learning environment.

A significant body of research has focused on students' perceptions and evaluations of their online learning experiences during the pandemic, addressing a range of interconnected aspects such as the services offered, the tools and resources supporting the learning process, the perceived usefulness of e-learning platforms, and interactions with peers and instructors. Specifically, Tejedor et al. (2021) carried out a comparative exploratory study in three countries (Spain, Italy and Ecuador), based on a research survey, in order to investigate teachers and students perceptions toward the university experience during the pandemic, in terms of evaluation of tools, materials, and digital platforms [10]. The main results indicate the preference of students for a mixture of learning resources, a greater role for podcasts and a typology of assessment tests different from the traditional exams. Fuady et al. (2021), through a survey and descriptive statistical analysis, explored how students' behaviors in using online platforms were influenced by the perceived ease of use and usefulness of those platforms [11]. Similarly, Nikou and Maslov (2021) applied structural equation modeling techniques to show that perceived challenges and COVID-19 awareness directly impacted students' intention to use ICT tools, with these effects mediated by the perceived usefulness and ease of use of e-learning systems [12]. Ramirez-Hurtado et al. (2021), by means of a Structural Equation Model, identified some priority areas on which to focus to improve students satisfaction, such as the interaction among peers, the level of concentration during online lessons, and the usefulness of the virtual system [13].

In addition, the students' perceptions of distance learning has also been investigated taking into account the impact of learner behaviors and attitudes towards online education [14]. For instance, Maphosa (2021) carried out a quantitative study aimed at exploring the perceptions of university students regarding the deployment of e-learning during the COVID-19 pandemic and the factors that affected usage. Using a multiple regression analysis, the main results showed that performance expectancy and facilitating conditions had a positive impact on the students' behavioral intentions to use an e-learning platform [15]. Baber (2020) building regression models found that interaction in the online class, students' motivation, course structure, and instructor knowledge and facilitation are important determinants impacting on the students' perceived learning outcomes and satisfaction [16]. Analyzing data collected through an online questionnaire and using different statistical tests (chi-squared, Kruskal-Wallis Mann-Whitney tests), Szopinski and Bachnik (2022) showed that students engagement, the change in frequency of attending online classes, and the favorite modality of study (online or offline) impacts on the evaluation of online studying [17].

Another important issue in the analysis of students' perceptions of the distance learning regards their level of digital skills [18]. According to Sá and Serpa (2020), fostering digital skills during the pandemic was a key issue, especially among those students who had limited access to digital technologies and lacked IT skills [19]. Through a descriptive-inferential quantitative methodology, Garcìa Prieto et al. (2022) analyzed the impact of university students' digital competence on their academic performance [20]. Tejedor et al. (2020), using an exploratory-correlational approach, carried out comparative research into three higher education institutions of different countries, aimed at understanding the student's evaluation of a teacher's digital skills, and digital sources for learning [21]. The main results indicated an overall need to reinforce communication, teaching, and digital competences. Also, Akcil and Bastas (2020) carried out a correlational survey research that demonstrated that there is a positive relationship between digital citizenship behaviors and e-learning attitudes [22].

Building upon this framework, this paper employs a quantitative methodology to investigate real-life data, such as the university students' perceptions of distance learning activities. Inspired by work on semantic connectivity in symbolic networks [23], which emphasized the role of network structures in shaping meaning, our study employs SNA methods applied to a set of adjectives derived from Semantic Differential scales. This approach allows for the identification of emotional dimensions in students' evaluations of their distance learning experiences. The analysis reveals distinct perceptions of distance learning that vary according to students' prior digital skills. In particular, we investigate whether students' previous familiarity with digital technologies and their ability to use technological tools shape their perception of distance learning. The findings highlight how different levels of digital competence influence students' emotional responses and overall evaluation of their remote learning experience.

3 Survey Design

In this work, we analyzed data on students' real-life experiences, collected during the 2021–2022 academic year through an online questionnaire, aimed at investigating various aspects of their university experience during the pandemic [24,25]. The survey questionnaire was administered to students enrolled in a bachelor's degree program at a Southern Italian university. For the purpose of this paper, we focused on specific sections of the questionnaire for analysis, namely: perceptions of distance learning, the frequency and ease of use of digital tools, and digital skills.

In particular, concerning students' assessment of distance learning, ad-hoc Semantic Differential scales were used to explore students' perceptions and emotional connotations. Drawing on the theoretical framework of [1], the Semantic Differential scale specifically designed for this study were adapted through a preliminary conceptual analysis of the dimensions most relevant to the context of distance learning and aligned with the characteristics of the target population. A set of nine pairs of adjectives, measured on a seven-point bipolar scale and grouped into the Evaluation, Potency, and Activity (EPA) dimensions, was identified and applied to analyze various aspects related to distance learning, such as the activities carried out, the use of digital platforms, online interactions, and self-study attitudes. Furthermore, an equal number of adjective pairs was selected for each EPA dimension:

- **Evaluation**: [unpleasant–pleasant], [unclear–clear], [ugly–nice]
- **Potency**: [cumbersome–light], [troublesome–simple], [difficult–easy]
- **Activity**: [rigorous–permissive], [useless–useful], [demanding–undemanding]

As regards the frequency and ease of use of digital tools, the following items, measured on a 5-point ordinal scale, were considered: chat, video conferencing, collaborative files, recorded audio/video files, and forums. Finally, students' level of digital competence was operationalized through several items, measured on a 7-point scale (with 1 = completely unskilled, and 7 = completely skilled), based on the competence areas outlined by the European Commission in the European Digital Competence Framework for Citizens (EU DigComp 2.1). The eight items considered are: 1) saving information via cloud platforms; 2) creating and managing content with collaborative tools; 3) interacting on social networks or other online communities; 4) using programming languages; 5) using office automation programs; 6) checking the configuration and operating status of technology devices; 7) using software systems for the security and protection of personal data; and 8) avoiding online threats and fraud.

4 The Methodological Strategy of Analysis

In this study, we adopt an useful methodology introduced in [3] to analyze Semantic Differential data. This approach leverages SNA techniques to detect semantic similarities among adjectives used in the Semantic Differential task, treating them as relational ties.

The analysis produces a weighted, undirected one-mode network, where nodes represent adjectives, links reflect co-occurrences, and weights are based on respondents' scores. It follows that in such a network two adjectives are linked if they have been chosen by a large number of respondents with high scores.

Constructing this network requires transforming the raw Semantic Differential data, including double coding and adjacency matrix construction, key steps in SNA.

Formally, let **X** be the individuals by adjectives matrix holding the scores given by N individuals on all the p pairs of adjectives used in the Semantic Differential scale. The use of doubling coding expands this matrix, assigning positive scores to the dominant adjective in each bipolar scale while setting its antonym to 0. This transforms the matrix **X** into an affiliation matrix. The product **X'X** gives rise to a weighted adjacency matrix **W**, holding the interconnections between all the adjectives obtained by the cross-product of the scores assigned by all respondents to any pair of adjectives.

Since **W** could generate a complete graph, a normalization procedure follows, ensuring weights fall within [0, 1].

The scaled adjacency matrix **W*** is obtained by dividing the w_{ij} weights by the theoretical maximum value, i.e., the squared highest score of the Semantic Differential scale times the number of respondents. Here, $w{*}_{ii} = 1$ if all respondents assigned the highest score to an adjective; $w{*}_{ij} = 1$ if all respondents rated the same pair of adjectives with the highest score.

To identify the most important relationships, the maximum difference criterion (MDC) [3] determines the optimal threshold by looking at the maximum difference between the number of components in two successive percentiles. This process highlights the key adjectives defining the assessed dimensions. SNA centrality measures and visualization tools further clarify the role and importance of adjectives within the network.

This procedure has been applied to real data collected through the survey described in the Sect. 3, which explores students' perceptions of the distance learning experience according to their level of digital skills.

5 Main Results

5.1 Students' Profiles

The survey was completed by 188 students enrolled in the second and third years of a bachelor's degree program, representing nearly 70% of the total student population. Specifically, 66.5% of the students are in their second year, while 33.5% are in their third year. The gender composition of the respondents shows a significant prevalence of females, accounting for 91.5%, compared to 8.5% male students, which is typical for humanistic and social sciences degree programs. The average age of respondents is 23.2 years, with a standard deviation of 5.9, influenced by a small group of students aged between 50 and 54 years. Additionally, 93.6% of the respondents are full-time students, while the remaining 6.4% are part-time students. Regarding their residential status, 12.8% of students moved from their place of origin to be near the university, 15.4% are off-site students who did not relocate, 47.9% are commuting students, and 23.9% are on-site students.

Focusing on the central sections of the questionnaire regarding digital and technological aspects, such as prior digital skills and the frequency and ease of use of distance learning platform tools, it is evident that students did not encounter significant difficulties in using these tools. In particular, they reported ease in using the chat feature, participating in video lectures, and downloading shared learning materials, which are also the most frequently used tools. Regarding prior digital skills, the responses show that students possess high proficiency in interacting on social networks, sharing files on the cloud, and using the Office suite. However, lower levels of digital competence are observed in the use of collaborative files and programming languages.

These items were analyzed using a hierarchical Cluster Analysis to identify homogeneous student profiles based on their digital expertise, including digital skills, frequency, and ease of use of learning platform tools. Specifically, this technique was performed to identify distinct groups of students based on their digital expertise. This procedure revealed the presence of three clusters:

- Students with a low level of digital skills, frequency of use, and ease of use of the tools and resources available on online platforms.
- Students with a medium level of digital skills, frequency of use, and ease of use of the tools and resources available on online platforms.
- Students with a high level of digital skills, frequency of use, and ease of use of the tools and resources available on online platforms.

These clusters are analyzed in relation to the networks of adjectives derived from the Semantic Differential scales, allowing for an exploration of how students' perceptions vary according to the cluster they belong to.

5.2 Students' Perception of Distance Learning

The results of the Semantic Differential scales, based on students' ratings for each pair of adjectives across the dimensions under analysis, show that students predominantly attributed high scores to adjectives with positive connotations in all dimensions. In particular, students clearly recognized the usefulness of the distance learning activities. Low scores (<5, on average) are associated with negative connotation adjectives, such as *cumbersome*, *rigorous*, and *demanding*, indicating feelings of cognitive overload, stress and reduced clarity.

This pattern becomes even clearer in Fig. 1, which displays the results for each cluster. In fact, the high or low scores assigned to each adjective across the clusters provide valuable insights into the affective or connotative meaning of each dimension under analysis. In general, a similar structure emerges across all dimensions when considering the scores given by respondents to the adjectives. However, when examining the scores given by students in the different clusters across the four dimensions, noticeable differences emerge. Specifically, students' perceptions appear to be influenced by their level of digital expertise.

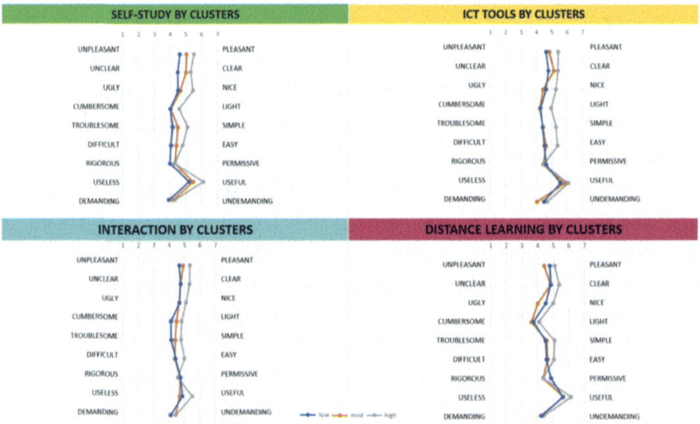

Fig. 1. Line plot of the Semantic differential tasks of the four dimensions by clusters of students

Specifically, the cluster of highly skilled students perceives all aspects of distance learning positively, as illustrated by the gray line, with high scores assigned to positive meaning adjectives. In contrast, the cluster of low-skilled students, represented by the blue line, assigned lower scores to nearly all adjective pairs in the Semantic Differential scales, reflecting a somewhat negative perception. According to these results, these students seem to have experienced distance learning as a complex activity and mentally fatiguing, suggesting a struggle to keep up with digital demands and feelings of marginalization. Finally, the definitions of the four dimensions provided by the cluster of medium-skilled students tend to reflect more balanced evaluations, occasionally interspersed with lower scores on specific adjective pairs. In other words, the experience of distance learning during the pandemic appears to reveal a spectrum of emotional responses - ranging from feeling of uncertainty and difficulty - largely shaped by students' level of digital competence. Differences between student profiles with respect to perceptions of distance learning emerge in greater detail through the application of the proposed analysis by looking at the networks of adjectives obtained.

5.3 The Adjectival Networks of Students' Perception

Following the analytical procedure above described, students scores are organized into as many affiliation matrices $X_{188,24}$, respondents by adjectives, as the dimensions being analyzed. The resulting normalized adjacency matrices $W{*}_{24,24}$, adjectives by adjectives, obtained by re-scaling the matrix W for the theoretical maximum value (i.e., $188 * 3^2 = 1692$) hold normalized weighted scores in [0,1], for each pair of adjectives. By carrying out the MDC procedure to set the optimal threshold on each normalized matrix, the prominent adjectives in the definition of the analyzed dimensions emerges. The analysis of such graphs in the scope of SNA permits to investigate the roles and the positions of the adjectives in the network.

The networks obtained for all the dimensions by considering the three students' profiles are presented. Figures 2, 3, and 4 show, respectively, the networks of adjectives derived from the scores assigned by the low, medium and high level digital skilled students. For all the networks, a common structure emerges.

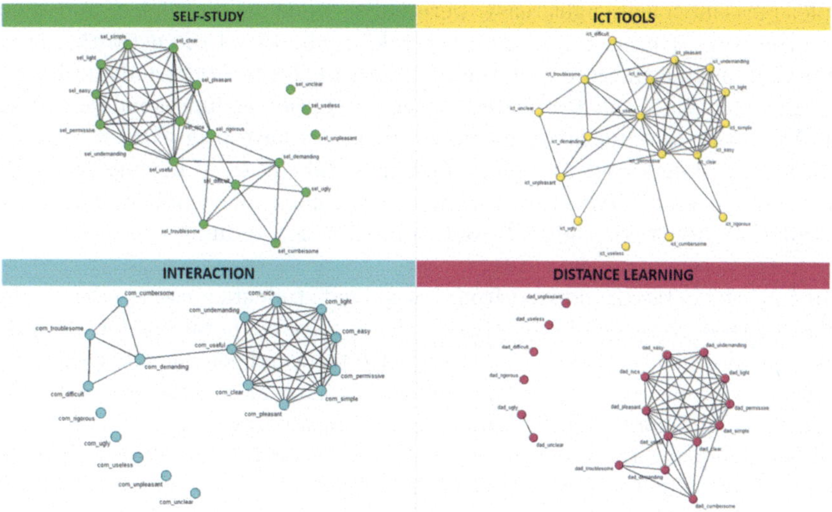

Fig. 2. Network of adjectives derived from the Semantic Differential tasks for the four dimensions according to low level skilled students' scores

There is a core made up of well-connected adjectives which mainly contribute to the definition of the dimensions being analyzed. These adjectives highlight the positive connotation of the four dimensions for all the students. So, the most

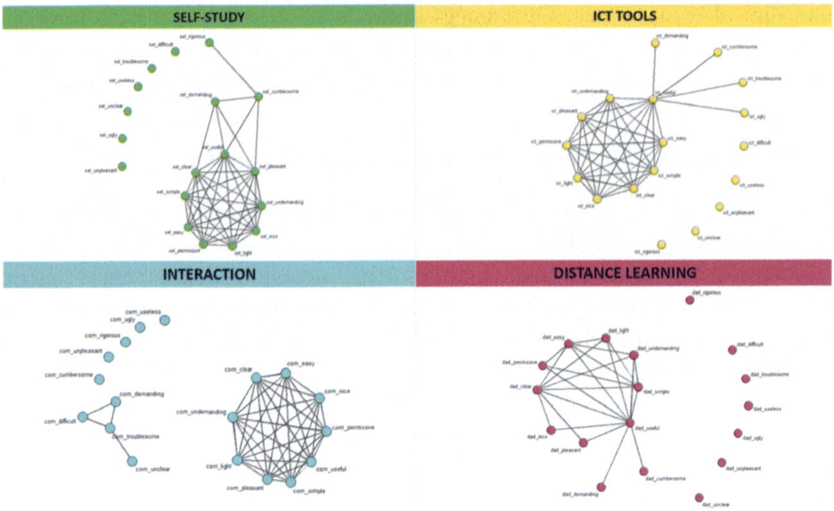

Fig. 3. Network of adjectives derived from the Semantic differential tasks for the four dimensions according to medium level skilled students' scores

chosen adjectives with the highest scores lying in the cores of the networks are: simple, pleasant, clear, light, easy, permissive, undemanding, nice, and useful. The latter adjective, useful, plays a key role in almost all the networks. It acts as a bridge, linking the core with the periphery of the networks and emphasizing a positive emotional valence related to the perception of functionality and support. Moreover, some negative meaning adjectives have been chosen regardless the students' skills profile, emphasizing their difficulties in dealing with these dimensions. The adjectives are: rigorous, demanding and cumbersome. This is particularly evident for the ICT tools and distance learning dimensions, with some differences among the clusters.

In fact, the networks derived from low skilled students answers (Fig. 2) show a more marked presence of adjectives with negative connotations for all dimensions. This prevalence of emotionally negative terms reflects an experience characterized by insecurity, difficulty, maybe linked to low confidence with their digital skills and emotional discomfort in the learning process.

When analyzing the group of students with a medium level of digital competence, a lower presence of negatively connoted adjectives emerges across the four dimensions considered, compared to the previous group. This reduction is noticeable in evaluations related to self-study and the use of ICT tools, suggesting a more positive - or at least less critical - perception of these aspects within the context of learning during the pandemic. This shift can be interpreted as an increase in emotional confidence and perceived effectiveness, which contribute to reducing anxiety and discomfort associated with technology use.

By contrast, the network of adjectives derived from high-level students' scores are mainly characterized by the presence of positive meaning adjectives, as shown

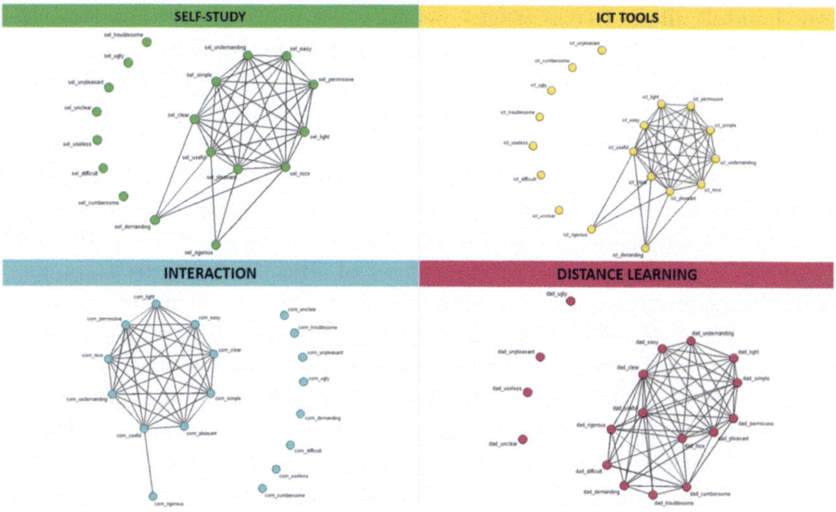

Fig. 4. Network of adjectives derived from the Semantic differential tasks for the four dimensions according to high level skilled students' scores

in Fig. 4. These students tended to describe distance learning in terms of accessibility, efficiency and autonomy. Furthermore, looking at the distance learning activities dimension for high-level students, it emerges that the greatest component of the network is also characterized by the presence of negative meaning adjectives. This suggest that even digitally skilled students experienced a form of ambivalence - valuing the flexibility of online education while recognizing its cognitive and relational limitations, generating mixed feeling of satisfaction and difficulty. The perception of distance learning, therefore, emerges as a hybrid terrain, shaped by individual competencies and structural constraints.

6 Conclusion

In this paper, we have applied a SNA approach to examine Semantic Differential data, offering a useful methodology for analyzing survey responses based on students' emotional and perceptual reactions to distance learning during the pandemic. According to the main results, adjectives with positive connotations were used by students in all clusters to define the four concepts related to distance learning activities (networks cores). At the same time, the edges linking the adjectives in the core of the networks with those in the peripheral area, mainly with a negative connotation, highlight some peculiar characteristics of the dimensions under study for the different clusters. This method gives a comprehensive understanding of the relationships of the emerging networks of adjectives by identifying those that have greater importance in the final definition of the phenomenon under analysis. This approach boasts a twofold advantage. On the one hand, it allows to construct a network of adjectives based on the perception about a service assessed by a set of respondents, from the co-occurrences observed among the adjectives in the Semantic Differential. On the other hand, it allows us to consider people's reaction towards a service in terms of the ratings they give on the bipolar scales. Furthermore, its visualization, which consists of a graph structure showing the overall perspective of the respondents provides important information about their perception. The emerging graph of adjectives shows different important features in the assessment of a service: i) the importance of each adjective, ii) the prominent set of adjectives used by respondents, and iii) the positive/negative connotation of the emotional meaning given by respondents. One strength of this method is its capability to reveal complex patterns in perception by mapping how adjectives relate to each other, rather than treating responses as isolated points. Among its limitations, the approach may be sensitive to the quality of the initial adjective selection and can be influenced by respondent biases, which should be carefully managed in future applications. Future directions could include expanding the application of this approach to larger and more diverse student populations to test its generalizability. Further development of automated tools for real-time semantic network analysis could facilitate rapid assessment and monitoring of student perceptions. Additionally, integrating this method with other quantitative data sources, such as learning analytics or performance metrics, may offer a more comprehensive view of the factors influencing distance learning experiences. Concluding, the use of Semantic Differential scale, together with its network analysis, are found to be a useful tool for assessing a learning experience since they provide decision makers with important implications directly attributable to the emotions and perceptions on the part of its actual users.

References

1. Osgood, C.E., Suci, G.J., Tannenbaum, P.H.: The Measurement of Meaning. University of Illinois Press, Urbana, IL (1957)
2. Wasserman, S., Faust, K.: Social Network Analysis: Methods and Applications. Cambridge University Press, Cambridge (1994)
3. Giordano, G., Primerano, I.: The use of network analysis to handle semantic differential data. Qual. Quant. **52**(3), 1173–1192 (2018)
4. Freeman, L.C.: Centrality in social networks conceptual clarification. Soc. Netw. **1**(3), 215–239 (1978)
5. Aloulou, M., Grati, R.: An e-learning system in a higher education institution in the UAE during the Covid-19 pandemic: a students' perspective. Glob. J. Eng. Educ. **24**(1) (2022)
6. Cicha, K., Rizun, M., Rutecka, P., Strzelecki, A.: COVID-19 and higher education: first-year students' expectations toward distance learning. Sustainability **13**(4), 1889 (2021)
7. Hervás-Gómez, C., Díaz-Noguera, M.D., De la Calle-Cabrera, A.M., Guijarro-Cordobés, O.: Perceptions of university students towards digital transformation during the pandemic. Educ. Sci. **11**(11), 738 (2021)

8. Giovannella, C.: Effect induced by the covid-19 pandemic on students' perception about technologies and distance learning. In: Mealha, Ó., Rehm, M., Rebedea, T. (eds.) Ludic, Co-design and Tools Supporting Smart Learning Ecosystems and Smart Education. SIST, vol. 197, pp. 105–116. Springer, Singapore (2021). https://doi.org/10.1007/978-981-15-7383-5_9
9. Hodges, C., Moore, S., Lockee, B., Trust, T., Bond, A.: The difference between emergency remote teaching and online learning. EDUCAUSE Rev. **27** (2020)
10. Tejedor, S., Cervi, L., Pérez-Escoda, A., Tusa, F., Parola, A.: Higher education response in the time of coronavirus: perceptions of teachers and students, and open innovation. J. Open Innov. Technol. Mark. **7**(1), 43 (2021)
11. Fuady, I., Sutarjo, M.A.S., Ernawati, E.: Analysis of students' perceptions of online learning media during the Covid-19 pandemic (Study of e-learning media: Zoom, Google Meet, Google Classroom, and LMS). Randwick Int. Soc. Sci. J. **2**(1), 51–56 (2021)
12. Nikou, S., Maslov, I.: An analysis of students' perspectives on e-learning participation–the case of COVID-19 pandemic. Int. J. Inf. Learn. Technol. **38**(3), 299–315 (2021)
13. Ramírez-Hurtado, J.M., Hernández-Díaz, A.G., López-Sánchez, A.D., Pérez-León, V.E.: Measuring online teaching service quality in higher education in the covid-19 environment. Int. J. Environ. Res. Public Health **18**(5), 2403 (2021)
14. Boca, G.D.: Factors influencing students' behavior and attitude towards online education during COVID-19. Sustainability **13**(13), 7469 (2021)
15. Maphosa, V.: Factors influencing student's perceptions towards e-learning adoption during COVID-19 pandemic: a developing country context. Eur. J. Interact. Multimedia Educ. **2**(2), e02109 (2021)
16. Baber, H.: Determinants of students' perceived learning outcome and satisfaction in online learning during the pandemic of COVID-19. J. Educ. E-Learn. Res. **7**(3), 285–292 (2020)
17. Szopiński, T., Bachnik, K.: Student evaluation of online learning during the COVID-19 pandemic. Technol. Forecast. Soc. Chang. **174**, 121203 (2022)
18. Arce, D.D., Illescas, E.L.: Competencias digitales en el contexto COVID 19: una mirada desde la educación. Revista Innova Educación **3**(1), 120–150 (2021)
19. Sá, M.J., Serpa, S.: COVID-19 and the promotion of digital competences in education. Univ. J. Educ. Res. **8**(10), 4520–4528 (2020)
20. García Prieto, F.J., López Aguilar, D., Delgado García, M.: Competencia digital del alumnado universitario y rendimiento académico en tiempos de COVID-19. Pixel-Bit (2022)
21. Tejedor, S., Cervi, L., Pérez-Escoda, A., Jumbo, F.T.: Digital literacy and higher education during COVID-19 lockdown: Spain, Italy, and Ecuador. Publications **8**(4), 48 (2020)
22. Akcil, U., Bastas, M.: Examination of university students' attitudes towards e-learning during the COVID-19 pandemic process and the relationship of digital citizenship. Contemp. Educ. Technol. **13**(1), ep291 (2020)
23. Carley, K.M., Kaufer, D.S.: Semantic connectivity: an approach for analyzing symbols in semantic networks. Commun. Theory **3**(3), 183–213 (1993)

24. Primerano, I., Catone, M.C., Giordano, G., Vitale, M.P.: Assessing undergraduate students' perceptions of distance learning during the COVID-19 pandemic. In: Lombardo, R., Camminatiello, I., Simonacci, V. (eds.) Book of Short Papers 10th International Conference IES 2022 Innovation and Society 5.0: Statistical and Economic Methodologies for Quality Assessment, pp. 263–268 (2022)
25. Primerano, I., Giordano, G., Vitale, M.P.: Exploring factors affecting the evaluation of online learning services: evidence from a social science bachelor's degree. In: Bini, M., Balzanella, A., Masserini, L., Verde, R. (eds.) SIS 2022. Springer Proceedings in Mathematics & Statistics, vol. 467, pp. 179–189. Springer, Cham (2024). https://doi.org/10.1007/978-3-031-65699-6_20

Extending the Boosted-Oriented Probabilistic Clustering to the Unit Hypersphere: A Textual Data Perspective

Rebecca Riviecchio[1(✉)] and Roberta Siciliano[2]

[1] Department of Physics, University of Naples Federico II, Naples, Italy
rebecca.riviecchio@unina.it

[2] Department of Electrical Engineering and Information Technology, University of Naples Federico II, Naples, Italy
roberta.siciliano@unina.it

Abstract. Clustering techniques play a crucial role in uncovering patterns and structures within complex datasets. In this study, we extend the boosted-oriented probabilistic clustering algorithm, originally proposed for time series data, to address the unique challenges posed by directional data. This approach is applied to a textual dataset to categorize documents based on their word content. The results highlight the effectiveness of this technique in organizing high-dimensional, sparse data, offering a novel perspective for clustering in challenging domains.

Keywords: Boosting · Clustering · Text Mining · Directional data

1 Introduction

Clustering methods have garnered growing interest in the literature over the years, aiming to divide datasets into groups where objects exhibit similar characteristics. The partitioning process is shaped by how units are assigned to centers. This results in two main types of clustering: crisp (hard), where each unit is exclusively assigned to one cluster, and fuzzy (soft), where the rigid assignment is relaxed, allowing units to simultaneously belong to multiple clusters based on degrees of membership. Fuzzy clustering, which enables the representation of overlapping groups, offers a more flexible and realistic model compared to traditional partitional methods. Two prominent clustering approaches have significantly influenced research in these areas: k-means [11,13,22,24], which is based on minimizing distances between objects and cluster centers, and fuzzy c-means [5,9], which also aims to minimize distances but incorporates a soft membership function. In fuzzy c-means, the degree of fuzziness is determined by a parameter known as the *fuzzifier*, which controls how much overlap or "softness" is allowed between clusters.

Taking into account the importance of clustering, several applications have been proposed for various fields of statistics, such as directional data. Directions,

A. D'Ambrosio et al. (Eds.): CLADAG-VOC 2025, SCDAKO, pp. 25–36, 2025.
https://doi.org/10.1007/978-3-032-03042-9_3

which represent orientations in space, require a different approach compared to standard data. Given the lack of magnitude, these can be interpreted as points on a unit circle in two dimensions or on a unit sphere in three dimensions [19]. Therefore, they are often referred to as circular or spherical data. Analyzing directional data presents several unique challenges, such as the absence of a reference direction, the ambiguity in defining rotation and the lack of a natural ordering. This type of data appears in various fields, like meteorology, where wind direction is analyzed, physics, in the study of sound waves, and psychology, where mental maps illustrate how individuals perceive their surroundings [25]. Furthermore, directional clustering has been used further to uncover hidden concepts within text datasets, and document similarity can be measured by examining the relative frequencies of words found in each document [8]. As a result, clustering algorithms process sparse and high-dimensional matrices, where the number of shared terms between documents is low relative to the total vocabulary of the entire collection. When dealing with vectors containing nonnegative elements that sum up to one, such as the relative frequencies of words in texts, these data can be treated as compositional. As noted by Stephens [31], by applying the square root transformation to each element of these vectors, the data can be represented as directional data on a unit hypersphere of $(p-1)$ dimensions. Therefore, L^2 normalization reduces biases caused by document length [30] and shifts the focus to the direction of the vectors rather than their magnitudes [2]. When treating textual data as vectors on a hyperspherical unit space, a suitable measure of similarity should be computed to capture homogeneity among documents. Cosine similarity is the most widely implemented, given its intuitive interpretation and its easy computation for sparse vectors [8]. Formally, let $\mathbf{x}$ and $\mathbf{y}$ be two vectors defined on the unit hypersphere S^{p-1} and $\theta(\mathbf{x}, \mathbf{y})$ be the angle between them. The cosine similarity is defined as:

$$\cos(\theta(\mathbf{x}, \mathbf{y})) = \frac{\mathbf{x}^T\mathbf{y}}{||\mathbf{x}||\,||\mathbf{y}||} = \mathbf{x}^T\mathbf{y}$$

where $\mathbf{x}^T\mathbf{y}$ is the inner product between the two vectors and $||\cdot||$ is the Euclidean norm. While cosine similarity measures how aligned two vectors are, cosine distance provides a complementary perspective by quantifying dissimilarity and it is formulated as:

$$d(\mathbf{x}, \mathbf{y}) = 1 - \cos(\theta(\mathbf{x}, \mathbf{y}))$$

Unlike Euclidean distance, cosine distance is insensitive to the magnitudes of vectors and focuses solely on their orientation, making it well-suited for the analysis of directional data. Strehl et al. [32] have demonstrated that cosine distance outperforms traditional Euclidean distance in clustering tasks involving sparse matrices, especially in textual data applications.

Section 2 provides an overview of several existing methods for clustering directional data. In Sect. 3, we recall the underlying concept of the boosted-oriented probabilistic clustering algorithm, detailing its extension to effectively handle hyperspherical data. Section 4 presents a real-world application, offering valuable insights and discussions. Finally, Sect. 5 concludes the study with key remarks and considerations for future research directions.

2 An Overview of Clustering Methods for Directional Data

The foundational reference in directional statistics is the book by Mardia et al. [25], which offers comprehensive methods specifically designed for handling directional data, setting it apart from traditional statistical approaches. This is followed by the more recent work of Ley and Verdebout [21], which presents modern advancements in the field. Recent developments in this area are discussed in the article by Pewsey and García-Portugués [28]. Clustering in hyperspherical spaces has gained increasing attention in applications such as text mining and gene expression analysis, where data points are effectively represented as directions. Despite its relevance, clustering techniques for directional data remain relatively underexplored in statistical research. This section provides an overview of well-established and recent methodologies for clustering on the unit hypersphere, highlighting both traditional approaches and innovative techniques in this area. Existing approaches can be broadly categorized into discriminative (non-parametric) and generative (parametric) models.

Among non-parametric methods, the most widely used algorithm is spherical k-means, introduced by Dhillon and Modha [8]. This method employs cosine similarity to partition data, with cluster centers interpreted as concept vectors that provide compact summaries of the data. Banerjee and Ghosh [3] introduced frequency-sensitive spherical k-means, a variation designed to mitigate convergence issues associated with poor local optima. Another non-parametric approach is FCM4DD, developed by Kesemen et al. [20], which extends fuzzy c-means to directional data without assuming an underlying distribution. Benjamin et al. [23] introduced PCM direction (PCM-D) algorithm for clustering directional data which is not necessarily based on mixture distribution models. Oba et al. [26] formulated mean-shift clustering based on the kernel density estimation for directional data Additionally, DBMCA, proposed by Pandolfo and D'Ambrosio [27], utilizes angular depth functions to perform clustering in high-dimensional, sparse spaces.

On the parametric side, Banerjee et al. [2] proposed a generative mixture model based on the Expectation-Maximization (EM) algorithm, utilizing the von Mises-Fisher (movMF) distribution, a natural choice for data constrained to the unit hypersphere. Their findings revealed that spherical k-means can be interpreted as a special case of von Mises-Fisher mixtures under restrictive assumptions and in its variation the centroids of the mixture components are estimated following the hard movMF approach. The von Mises-Fisher distribution is a key parametric model used to represent directional data. However, these models can be sensitive to the presence of mild outliers or clusters with heavier tails than the von Mises-Fisher distribution can handle. To address these limitations, a mixture of contaminated von Mises-Fisher distributions is introduced by Zhang et al. [33]. Also other distributions are used to model clustering on directional data. Rodríguez et al. [29] proposed a new approach for clustering circular data using a Bayesian mixture model framework. Agiomyrgiannakis et al. [1] present a method for modeling dispersion phases using wrapped Gaussian mixture models (WGMMs) tailored for circular data.

3 Extending Boosted-Oriented Probabilistic Clustering to the Unit Hypersphere

Iorio et al. [18] introduced boosted-oriented probabilistic clustering, a novel unsupervised method for time series analysis that utilizes the boosting paradigm, originally designed for supervised learning, in an unsupervised setting. In this work, we propose extending this algorithm to a different type of data, such as textual data.

The boosting approach [12] relies on a sequential learning process that iteratively reduces errors. Specifically, resampling assigns higher weights to misclassified instances, thereby improving the model's performance in regions where it exhibits weaknesses. Unlike the boosting approach, in the unsupervised context, the weight of an object in the resampling process increases with its membership probability to a given cluster. Let X be a $N \times p$ data set containing N objects $\{x_1, x_2, ..., x_N\}$ defined in S^{p-1} and K be the number of clusters chosen a-priori. Let $P_{i,k} = p(x_i, C_k)$ be the probability that the $i-th$ object belongs to the $k-th$ cluster, and let $d_{i,k} = d(x_i, c_k)$ represent the distance between the $i-th$ unit and the $k - th$ center. To define unit-cluster membership probabilities, the relation between probabilities and distances proposed in the Probabilistic Distance (PD) clustering approach [4] is employed:

$$P_{i,k} d_{i,k} = constant(i) \tag{1}$$

The Eq. 1 states that the product between the distance and the probability is a constant depending only on x_i. Therefore, a data point has a higher probability of belonging to a cluster if it is closer to its cluster center. In particular, the constant in 1 remains independent of cluster k. Consequently, the equation for the probabilities can be obtained as follows:

$$P_{i,k} = \frac{\prod_{j \neq k} d_{i,j}}{\sum_{k=1}^{K} \prod_{j \neq k} d_{i,j}}. \tag{2}$$

Furthermore, in this approach, a well-defined loss function can be constructed without requiring a fuzzifier parameter. Since the probabilities quantified by the Eq. 2 sum to one across clusters, the authors [18] introduce the badness of clustering index (BC), which is defined as:

$$BC = \frac{1}{N} \sum_{i=1}^{N} \left(\prod_{k=1}^{K} P_{i,k} \right) K^K \tag{3}$$

When a data point has a probability of one for belonging to a specific cluster, the probabilities for all other clusters are zero, making $\prod_{k=1}^{K} P_{i,k}$ equal to zero. In the case of perfect uncertainty, a data point has an equal membership probability of K^{-1} for each cluster. Thus, according to Eq. 3, the certainty of the clustering procedure is ensured when the value of this index is low.

From the Eq. 3 it is possible to define the loss function as:

$$\beta = \sum_{i=1}^{N} \left(\prod_{k=1}^{K} P_{i,k} \right) K^K \tag{4}$$

Consequently, following the boosting paradigm, the loss function in Eq. 4 allows for the construction of weights for the observations, which are then used in the resampling process. The weight of the $i-th$ object to the $k-th$ cluster is given by:

$$w_{i,k} = \beta^{\gamma_{i,k} \Gamma_{i,k}}$$

where $\gamma_{i,k} = d_{i,k}/max_{k=1}^{K} d_{i,k}$ represents the contribution of the $i-th$ unit to the $k-th$ cluster and $\Gamma_{i,k}$ is a $N \times K$ indicator matrix containing 1 if $P_{i,k} > P_{i,h}(k,h = 1,...,K, k \neq h)$ and -1 otherwise. Then, the weights are first normalized in this way:

$$w^{\cdot}_{i,k} = \frac{w_{i,k}}{\sum_{h=1}^{K} w_{i,k}}$$

and the final weights for the resampling procedure are obtained as:

$$W_{i,k} = \frac{w^{\cdot}_{i,k}}{\sum_{i=1}^{N} w^{\cdot}_{i,k}}. \tag{5}$$

To estimate cluster centers, the original study [18] proposes computations specifically designed for time series data. Here, we adapt this approach to accommodate the characteristics of directional data.

The algorithm randomly selects at each iteration a set O^k of N objects from X with replacement considering $W_{i,k}$. The weights of the selected units are then further multiplied by the absolute frequencies of their extraction during the resampling process, resulting in $w_{i,k}^{boost}$. Then, the weights are normalized in this way for each cluster:

$$W_{i,k}^{boost} = \frac{w_{i,k}^{boost}}{\sum_{i=1}^{N} w_{i,k}^{boost}}. \tag{6}$$

Let w_k^b be the $N \times 1$ vector containing the boosted weights defined in Eq. 6 for the $k-th$ cluster. The weighted mean vector m_k for each cluster can be computed as follows:

$$m_k = (w_k^b)^T X. \tag{7}$$

The mean vectors m_K are subsequently normalized using the L^2 norm to ensure a unit vector length, obtaining the clusters centers c_k^* that can be formulated as:

$$c_k^* = \frac{m_k}{\|m_k\|}. \tag{8}$$

These steps are repeated until the maximum number of iterations is reached. The pseudocode of the algorithm is provided in Box 1. Regarding the choice of the distance measure, since the goal is to partition objects located on a unit high-dimensional sphere, cosine distance emerges as the most appropriate option.

Algorithm 1 Boosted-oriented probabilistic clustering for directional data

Input: X
Initialize: K = number of clusters; $c_k, k = 1, \ldots, K$ random cluster centers; it = number of iterations; d = a distance measure.
for $iter = 1 : it$ **do**
  -compute the $N \times K$ distance matrix D
  -compute the $N \times K$ membership probabilities matrix P as in Eq. 2
  -compute the BC index as in Eq. 3
  -compute β as in Eq. 4
  -compute the $N \times K$ matrix W containing the weights as in Eq. 5
  for $k = 1 : K$ **do**
    -extract a sample O^k with replacement from X taking into account W
    -compute the $N \times 1$ boosted weights vector w_k^b as in Eq. 6
    -compute the mean vector m_k as in Eq. 7
    -compute the cluster center c_k^* as in Eq. 8
  end for
end for
Output: c_k^* = final cluster centers; P = membership probabilities matrix; BC = badness of clustering index.

4 Real-World Text Clustering: A Practical Insight

In this section, the 're0' dataset (https://github.com/ntyaan/some-datasets/blob/master/datasets_cluto_20171213/CLUTO/re0/sparse_re0.txt), sourced from the internal repository of the CLUTO software, is used to demonstrate an application of clustering high-dimensional text data using the boosted-oriented approach, which was fully implemented in R. These documents were published on the Reuters newswire and manually categorized by Reuters Ltd. employees. This dataset captures the occurrence of 2886 unique terms across a collection of 1504 individual documents, which have already been preprocessed and classified into 13 categories, as shown in Table 1. The approach involves representing word frequencies as directional data by applying L^2 normalization to scale them to a unit length. This transformation ensures that documents of varying lengths become comparable. Moreover, mapping text data onto the unit hypersphere is particularly advantageous for handling sparsity, a common characteristic of textual datasets. We applied the proposed algorithm, detailed in Sect. 3, using cosine distance and setting the maximum number of iterations to 100. The fuzzy silhouette method [6] was employed to determine the optimal number of clusters within the range of 2 to 15, with the resulting values shown in Fig. 1. The higher the values of this index, the better the partitions in fuzzy logic. Based on the results presented in Fig. 1, the optimal number of clusters is 15, which closely aligns with the number of classes identified by the personnel.

Table 1. Relative frequency distribution of document categories.

Class	Proportion	Class	Proportion	Class	Proportion	Class	Proportion
bop	0.0253	cpi	0.0399	gnp	0.0532	housing	0.0106
interest	0.1456	ipi	0.0246	jobs	0.0259	lei	0.0073
money	0.4043	reserves	0.0279	retail	0.0133	trade	0.2121
wpi	0.0100						

Setting the number of clusters to 13 and the iterations to 100, the BC index is 0.6397, suggesting a moderate overlap among the document groups. To compare the obtained clusters with the a priori classifications, a suitable external validation measure is required to allow comparison between soft and hard clusters. The normalized degree of concordance (NDC) [17] between a fuzzy and a crisp partition or between two fuzzy partitions, as well as its chance-adjusted counterpart, the adjusted concordance index (ACI) [10], were calculated using the R package `ConsRankClass` [7]. The NDC and ACI values, 0.3195 and 0.0061 respectively, suggest a notable divergence between the clusters identified by the algorithm and the predefined document categories. However, it is crucial to emphasize that clustering, as an unsupervised method, does not aim to replicate pre-established labels as in supervised learning. Instead, clustering can serve as a valuable tool for understanding the relationships between objects, offering insights into their similarities, even when these belong to different categories.

The same comparison is carried out using the movMF and spherical k-means algorithms, implemented using the R packages `movMF` [16] and `skmeans` [14]. To allow for fuzzy partitions, the extended version of spherical k-means, as described by Hornik et al. [15], is used with a low fuzziness parameter set to $m = 1.2$, in line with the settings adopted in previous experiments [15]. Regarding the degree of concordance between spherical k-means and the a priori categorization, the NDC and ACI values are 0.6405 and 0.0736, respectively. Concerning the movMF algorithm, the NDC and ACI values are 0.7688 and 0.1921, respectively, further indicating a discrepancy between the obtained clusters and the pre-established document classification.

Furthermore, the pairwise comparisons between the membership matrices obtained using the aforementioned methods are presented in the Table 2. The spherical k-means and movMF evaluation exhibit a high NDC value of 0.7489, but the ACI is notably low at 0.2378, suggesting that the observed degree of concordance is likely due to chance. Meanwhile, for the proposed version of boosted-oriented probabilistic clustering, both the NDC and ACI results reveal a clear

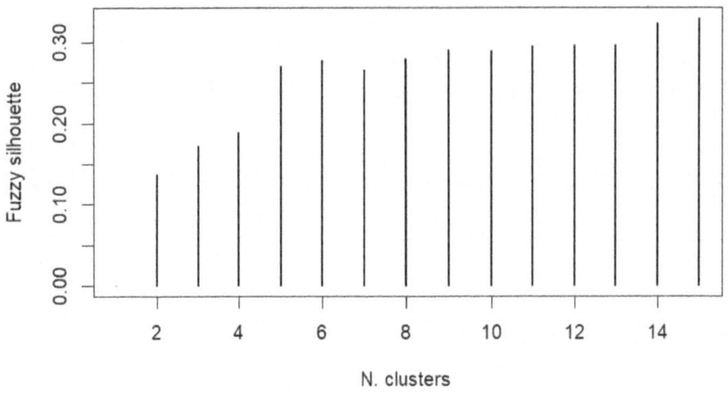

Fig. 1. Fuzzy silhouette values. Higher fuzzy silhouette values indicate better partitions.

distinction in how these methods partition documents within the hyperspherical space.

Table 2. Pairwise comparisons of the clusters based on the NDC and ACI indices.

	movMF spherical k-means	boosted-oriented spherical k-means	movMF boosted-oriented
NDC	0.7489	0.4360	0.2342
ACI	0.2378	0.0063	0.0200

We can also compare the centers obtained by the boosted-oriented algorithm with the normalized mean vectors conditioned on the a priori categories. The heatmap illustrating the cosine similarities between the centers matrix and the normalized mean vectors of the classes is shown in Fig. 2. As depicted in the figure, many categories exhibit strong similarities with specific cluster centers, including: bop, gnp, interest, money, reserves, and trade. The remaining categories and the fifth center have similar orientation in the space, but the latter shows the strongest similarity with the cpi class.

Additionally, to enhance cluster interpretability, six partitions were selected based on fuzzy silhouette values, with a maximum of 100 iterations. When applying clustering techniques, the cluster center vectors encapsulate the information that guided the partitioning process, summarizing how documents are associated with each cluster based on a certain probability. These vectors can be analyzed to identify the topics to which the documents are most closely related, by examining the words with the highest values within each center. The results can help steer the research towards texts that are most aligned with the intended knowledge goals, allowing individuals to focus on documents that best represent the topics of interest. The top 7 most relevant words for each cluster center are presented in Table 3. The top words for clusters 2 and 3 indicate a strong association with the Japanese and English markets, respectively, while cluster 6 is primarily concerned with foreign trade mechanisms. The most relevant terms of cluster 5, on the other hand, is linked to scenarios of dynamic growth, whereas those of the remaining two clusters are associated with the banking and financial market contexts.

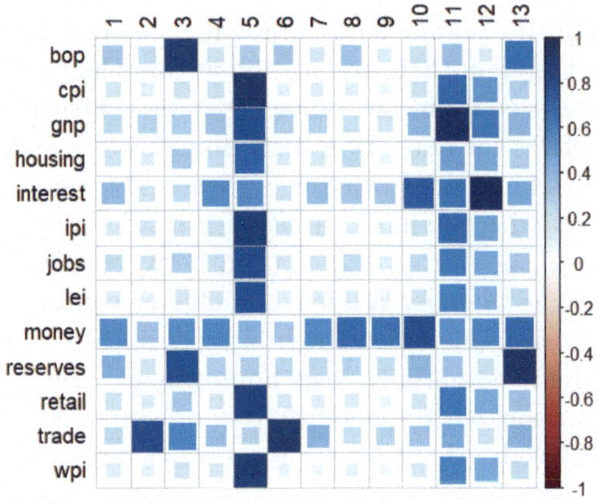

Fig. 2. Heatmap of cosine similarities between the normalized mean vectors of the classes and the cluster centers.

Table 3. Top 7 words for each cluster center.

	word 1	word 2	word 3	word 4	word 5	word 6	word 7
cluster 1	billion	dlr	fed	mln	reserv	week	bank
cluster 2	japan	trade	japanes	offici	market	industri	yen
cluster 3	mln	stg	bank	market	monei	bill	england
cluster 4	bank	rate	dollar	pct	market	currenc	exchang
cluster 5	pct	februari	januari	rate	rise	rose	price
cluster 6	trade	countri	export	dlr	deficit	billion	foreign

5 Conclusions

In this study, an extension of the boosted-oriented algorithm for probabilistic clustering, originally introduced for time series data, has been proposed and implemented for directional data. This algorithm, which is inherently driven by the philosophy of boosting, is capable of creating clusters predominantly influenced by the most representative observations. In fact, unlike traditional boosting logic, this approach prioritizes the most impactful areas, meaning that objects presenting a higher probability of belonging to a cluster are assigned greater weight to define the center vector. By using the same approach as PD clustering, it is possible to obtain a membership probability matrix whose degree of softness does not depend on the fuzzifier parameter.

This approach was applied to a textual dataset, where the documents were normalized to unit length to remove any influence of their size. By doing so, the distances were solely based on their orientation within the unit hypersphere, utilizing the widely recognized cosine distance measure, which is considered one of the most effective metrics for clustering high-dimensional, sparse datasets. The proposed algorithm has uncovered valuable insights. By comparing the membership probability matrix with the hard partition from the a priori document classification, we can assess the degree of deviation between the algorithm's output and the manually assigned labels. While some cluster centers align closely with specific a priori categories, one cluster exhibits significant similarity with multiple classes, indicating a potentially high homogeneity among those document types. A brief comparison with the movMF and extended spherical k-means methods is also presented. All of these approaches provide different findings compared to the a priori of documents, and furthermore, there is no redundancy between these algorithms. Additionally, the most relevant words were utilized to enhance the interpretation of the algorithm's results.

Future research could explore the application of this algorithm to other datasets. A potential improvement could also involve adapting the algorithm to dynamically update the cluster centers during iterations, while taking into account the influence of previous centers.

References

1. Agiomyrgiannakis, Y., Stylianou, Y.: Wrapped gaussian mixture models for modeling and high-rate quantization of phase data of speech. IEEE Trans. Audio Speech Lang. Process. **17**(4), 775–786 (2009)
2. Banerjee, A., Dhillon, I.S., Ghosh, J., Sra, S.: Clustering on the unit hypersphere using von mises-fisher distributions. J. Mach. Learn. Res. **6**, 1345–1382 (2005)
3. Banerjee, A., Ghosh, J.: Frequency sensitive competitive learning for clustering on high-dimensional hyperspheres. In: Proceedings of the 2002 International Joint Conference on Neural Networks. IJCNN'02 (Cat. No. 02CH37290), vol. 2, p. 1590–1595 (2002)
4. Ben-Israel, A., Iyigun, C.: Probabilistic d-clustering. J. Classif. **25**, 5–26 (2008)
5. Bezdek, J.C.: Objective function clustering. In: Pattern Recognition with Fuzzy Objective Function Algorithms, pp. 43–93. Springer, Boston (1981)
6. Campello, R.J.G.B., Hruschka, E.R.: A fuzzy extension of the silhouette width criterion for cluster analysis. Fuzzy Sets Syst. **157**(21), 2858–2875 (2006)
7. D'Ambrosio, A.: Consrankclass: classification and clustering of preference rankings. R package version 1.0.1 (2021)

8. Dhillon, I.S., Modha, D.S.: Concept decompositions for large sparse text data using clustering. Mach. Learn. **42**, 143–175 (2004)
9. Dunn, J.C.: A fuzzy relative of the isodata process and its use in detecting compact well-separated clusters. J. Cybern. **3**(3), 32–57 (1973)
10. D'Ambrosio, A., Amodio, S., Iorio, C., Pandolfo, G., Siciliano, R.: Adjusted concordance index: an extension of the adjusted rand index to fuzzy partitions. J. Classif. **38**(1), 112–128 (2021)
11. Forgy, E.W.: Cluster analysis of multivariate data: efficiency vs interpretability of classifications. Biometrics **21**, 768–769 (1965)
12. Freund, Y., Schapire, R.E.: A decision-theoretic generalization of on-line learning and an application to boosting. J. Comput. Syst. Sci. **55**(1), 119–139 (1997)
13. Hartigan, J.A., Wong, M.A.: Algorithm as 136: a k-means clustering algorithm. Appl. Stat. **28**, 100–108 (1979)
14. Hornik, K., Feinerer, I., Kober, M.: Skmeans: Spherical k-Means Clustering. R package version 0.2-18 (2024)
15. Hornik, K., Feinerer, I., Kober, M., Buchta, C.: Spherical *k*-means clustering. J. Stat. Softw. **50**(10), 1–22 (2012)
16. Hornik, K., Grün, B.: movMF: Mixtures of von Mises-Fisher Distributions. R package version 0.2-9 (2024)
17. Hullermeier, E., Rifqi, M., Henzgen, S., Senge, R.: Comparing fuzzy partitions: a generalization of the rand index and related measures. IEEE Trans. Fuzzy Syst. **20**(3), 546–556 (2012)
18. Iorio, C., Frasso, G., D'Ambrosio, A., Siciliano, R.: Boosted-oriented probabilistic smoothing-spline clustering of series. Stat. Methods Appl. **32**(4), 1123–1140 (2023)
19. Jammalamadaka, S.R., Sengupta, A., Sengupta, A.: Topics in Circular Statistics. Series on Multivariate Analysis. World Scientific (2001)
20. Kesemen, O., Tezel, Ö., Özkul, E.: Fuzzy C-means clustering algorithm for directional data (FCM4DD). Expert Syst. Appl. **58**, 76–82 (2016)
21. Ley, C., Verdebout, T.: Modern Directional Statistics. Chapman and Hall/CRC, Florida (2017)
22. Lloyd, S.P.: Least squares quantization in PCM. IEEE Trans. Inf. Theory **28**, 128–137 (1982). Technical Note, Bell Laboratories. (Originally published in 1957)
23. Benjamin, J.B.M., Hussain, I., Yang, M.-S.: Possiblistic C-means clustering on directional data. In: 2019 12th International Congress on Image and Signal Processing, BioMedical Engineering and Informatics (CISP-BMEI), pp. 1–6 (2019)
24. MacQueen, J.: Some methods for classification and analysis of multivariate observations. In: Le Cam, L.M., Neyman, J. (eds.) Proceedings of the Fifth Berkeley Symposium on Mathematical Statistics and Probability, vol. 1, pp. 281–297, Berkeley, CA. University of California Press (1967)
25. Mardia, K.V., Jupp, P.E.: Directional Statistics. Wiley (2009)
26. Oba, S., Kato, K., Ishii, S.: Multi-scale clustering for gene expression profiling data. In: Fifth IEEE Symposium on Bioinformatics and Bioengineering (BIBE 2005), New York, pp. 210–217. IEEE (2005)
27. Pandolfo, G., D'ambrosio, A.: Clustering directional data through depth functions. Comput. Stat. **38**, 1487–1506 (2022)
28. Pewsey, A., García-Portugués, E.: Recent advances in directional statistics. TEST **30**(1), 1–58 (2021). https://doi.org/10.1007/s11749-021-00759-x
29. Rodríguez, C.E., Núñez-Antonio, G., Escarela, G.: A Bayesian mixture model for clustering circular data. Comput. Stat. Data Anal. **143**, 106842 (2020)

30. Singhal, A., Buckley, C., Mitra, M.: Pivoted document length normalization. In: Proceedings of the 19th Annual International ACM SIGIR Conference on Research and Development in Information Retrieval, pp. 21–29. Association for Computing Machinery (1996)
31. Stephens, M.A.: Use of the von mises distribution to analyse continuous proportions. Biometrika **69**(1), 197–203 (1982)
32. Strehl, A., Ghosh, J., Mooney, R.: Impact of similarity measures on web-page clustering. In: Proceedings of the AAAI Workshop on Artificial Intelligence for Web Search, pp. 58–64. AAAI Press (2000)
33. Zhang, Y., Melnykov, V., Melnykov, I.: On model-based clustering of directional data with heavy tails. J. Classif. **40**, 527–551 (2023)

Understanding ESG Scores Through Network Analysis: A Study Using Graph Neural Networks

Brian Daniel Bernhardt[1], Chiara Marciano[1](✉), Sara Gigli[1], Lucia Maddalena[2], and Mario Rosario Guarracino[1]

[1] University of Cassino and Southern Lazio, 03043 Cassino, Italy
{briandaniel.bernhardt,chiara.marciano,sara.gigli, mario.guarracino}@unicas.it
[2] National Research Council, 80131 Naples, Italy
lucia.maddalena@cnr.it

Abstract. Graph neural networks (GNNs) are powerful tools for analyzing graph-structured data and have numerous applications. In this study, GNNs are used to explore the connections between a set of Italian companies, focusing on how sharing decision-makers influence their Environmental, Social, and Governance (ESG) scores. The research compares various GNN models, including GraphSAGE, Graph Attention Networks, and Graph Neural Additive Networks, with traditional classification techniques such as kNN, Random Forest, Decision Trees, and Naïve Bayes. The results demonstrate that GNNs provide a higher node classification accuracy and suggest that more central nodes tend to have higher ESG scores, indicating that companies with greater network connectivity may provide higher ESG performance. This suggests that sharing decision-makers could be a strategic tool to enhance efforts in sustainability and social responsibility.

Keywords: ESG Perception Index · Decision-Makers · Graph Neural Network · Interpretability

1 Introduction

Globalization and market evolution have made companies increasingly interconnected, requiring them to collaborate to remain competitive and dynamic. Cooperation between companies not only facilitates access to new ideas but also generates positive effects in relation to sustainability and social responsibility. Business interactions can take various forms, ranging from direct connections, such as corporate shareholdings or partnerships, to indirect ones, as in cases where multiple companies share board members, the so-called decision-makers (DMs). Indeed, DMs play a strategic role in shaping corporate policies, influencing not only business objectives but also the impact of companies on environmental, social, and governance (ESG) issues, which have become a key source of competitive advantage [23].

A. D'Ambrosio et al. (Eds.): CLADAG-VOC 2025, SCDAKO, pp. 37–47, 2025.
https://doi.org/10.1007/978-3-032-03042-9_4

Various studies have explored the effects that interconnections among companies can have on business performance. Previous studies have indeed documented a positive association between shared leadership and team or firm performance, as well as firm value [3,11]. Ponomareva's study conducted on Italian firms shows that co-CEO structures can harm ESG performance, particularly for family businesses [22]. In contrast, Hoje's study focuses on business groups in Korea and shows that they tend to disclose higher ESG scores [12].

Since the results on this topic are still conflicting, our study aims to contribute to the existing literature. First, this study enriches the literature investigating the effect of shared governance, particularly the sharing of decision-makers, on the ESG performance of some listed Italian companies. Our goal is to predict ESG ratings using financial economic variables and information related to decision-makers, using machine learning models that support multi-class classification. In accordance with the literature [1,4,6], we selected four models based on traditional machine learning, namely Random Forest, Decision Tree, k-Nearest Neighbors (kNN), and Naïve Bayes classifiers. These models have demonstrated good performance in different financial classification problems. In addition to these, we considered Graph Neural Networks (GNNs), which have emerged as the predominant approach for learning over graph-structured data in various application domains [9,16,19,24], selecting three models: GraphSAGE (Graph Sample and Aggregation) [10], GAT (Graph Attention Networks) [20], and the recently proposed GNAN (Graph Neural Additive Networks) [2]. The first two are state-of-the-art, while the latter was selected specifically because it is a promising method capable of ensuring interpretability, which is one of the main limitations of traditional GNN-based methods. The motivation behind selecting multiple models is to identify the best model for ESG rating prediction that can lead to reliable results, which could then be extended to other fields of research.

The main contents of this paper are as follows. Section 2 reviews related work to provide context for the current advancements in GNNs for the ESG topic. Section 3 describes the dataset, variables, and methodology used. Section 4 shows our empirical results. Section 5 discusses the results and draws the conclusion.

2 Related Work

ESG factors are defined as the "environmental, social or governance matters that may have a positive or negative impact on the financial performance or solvency of an entity, sovereign or individual" [8]. The topic of ESG is currently widely studied in the literature. In recent years, there has been a significant increase in publications on the subject, with a notable rise in the number of articles and research dedicated to ESG. A particularly relevant area of research concerns the use of Artificial Intelligence (AI) techniques for evaluating ESG performance. To identify and measure the ESG impacts of their activities, organizations increasingly resort to data analysis techniques that can support them in decision-making processes, aiming to improve their ESG performance. These techniques can be used not only for analytical activities but also to predict possible scenarios or opportunities that can enhance ESG performance. In the literature, there are different works. The study by Chowdhury et al. [4] explores how ESG ratings are influenced by firm-specific and macroeconomic variables, employing traditional machine learning algorithms. The findings indicate that the Random Forest classifier provides the highest accuracy in ESG ratings among the algorithms tested. A recent review on ESG and AI [14] has highlighted that the application of advanced AI techniques, such as graph neural networks, has been limited in this context. However, a recent study by Liu et al. [15] investigates the impact of ESG factors on the industrial chain using GNNs. The researchers introduce a cross-attention module to quantify the effects of ESG shocks by analyzing the interaction between ESG impacts and the characteristics of industrial sectors. Experimental results on two real datasets demonstrate that ESG shocks significantly influence the extension of industrial value and provide insights into how such shocks impact the performance of the industrial chain. Here, we compare traditional classification methods with some GNN techniques in the context of ESG performance. This comparison aims to highlight the strengths and limitations of each approach, providing a deeper understanding of the potential of GNNs in analyzing corporate sustainability practices. In particular, our study focuses on the advancements of GraphSAGE, GAT, and GNAN. GraphSAGE is an inductive framework designed for generating node embeddings in large graphs [10]. Unlike traditional transductive methods that require the entire graph during training, GraphSAGE learns a function to generate embeddings by sampling and aggregating features from a node's local neighborhood. GAT utilizes an attention mechanism to update node representations in a graph [20]. The importance of each neighboring node is determined by its features and proximity. The model first applies a linear transformation to node features, followed by the computation of attention coefficients, which are normalized using softmax to aggregate information from neighboring nodes. To enhance expressiveness, GAT employs multi-head attention, combining multiple attention mechanisms in parallel. This architecture is particularly effective for complex graphs and tasks such as classification and link prediction. GNAN extends the concept of Generalized Additive Models to graph data, enabling interpretable and high-performance predictions without the complexity and opacity found in other deep learning models [2]. It is a neural network model designed for graph processing that generates node representations through the learning of distance functions and feature shape functions. Each node is represented as a weighted combination of the features of its neighboring nodes, where the influence of each node depends on its distance from the reference node. GAT, GraphSAGE, and GNAN are of interest because they show state-of-the-art performance in various graph-based tasks. In the context of these models, our work proposes a comparison between models in order to highlight how the performance of the models can improve.

3 Methodology

The adopted dataset contains information about 180 major Italian companies across various sectors (automotive, manufacturing, etc.), each provided with the following information

- the ESG Perception Index, henceforth referred to as ESG index. This index, created by Reputation Manager [7], monitors the perception of the sustainability of companies, indicating the aggregate six-monthly trend of the ESG reputation of the major companies on the Italian market. It can take a value between 0 and 100;
- the names of the decision-makers (DM), i.e., people who cover one of the following roles: Member of the Board, President of the Board of Directors, General Manager, Chief Executive Officer, Chairman;
- some of the main financial-economic indicators, ranging from the Coverage Index to the Structure Margin, briefly described in Table 1.

Based on this information, we created an undirected weighted network $\mathcal{G}$, where each node u represents a company and its features are the financial-economic indicators related to the company. Nodes are labeled based on the value of the ESG Index; specifically, we divided the companies into three classes: class A with ESG Index between 51 and 100, class B with ESG Index between 26 and 50, and class C with ESG Index between 0 and 25. An edge $e(u, v)$ connects two nodes u and v if the two companies share at least one DM; the associated weight is given by the Jaccard Index $J(u, v)$, which is the number of DM's shared by the two companies divided by the total number of their DM's:

$$J(u, v) = \frac{|DM_u \cap DM_v|}{|DM_u \cup DM_v|},$$

where DM_u indicates the DM's of company u. The obtained network $\mathcal{G}$ is shown in Fig. 1, where nodes are colored according to their labels (green for class A, yellow for class B, and red for class C), and their size reflects their degree centrality (in the following simply referred to as centrality), i.e., the number of edges; the width of each edge reflects its weight.

For the experiments, we considered the giant component $\overline{\mathcal{G}}$ of $\mathcal{G}$, i.e., the connected component that contains the most significant fraction of the entire network's nodes, which comprises 138 nodes (20 in class A, 75 in class B, and 43 in class C) and 346 edges. Given this graph, we formulate a node classification problem, the aim of which is to build a model that predicts the nodes' class label. For this purpose, we adopted various classifiers, including four traditional machine learning classifiers (Random Forest, Decision Trees, k-NNN, and Naïve Bayes) and three models based on GNNs (GraphSAGE, GAT, and GNAN). The data was split into 60% for training, 20% for validation, and 20% for test using a fixed seed for reproducibility. Each GNN model was trained for 100 epochs using Adam's optimization [13], with the fixed parameter values shown in Table 2. Results were aggregated over 10 runs, and model performance was evaluated using Accuracy, Balanced Accuracy, and F1 Score.

Table 1. Brief descriptions of the financial-economic indicators adopted as node features

Features	Category	Description
Number of employees	Dimension	Number of people working for the company
Coverage index	Coverage	Measures the ability to meet financial obligations, often related to debt
Asset coverage ratio	Coverage	Assesses the capacity to finance fixed assets with equity and debt
Debt Ratio	Debt ratio	Measures the proportion of total debt to total assets
Financial charges	Financial	Costs incurred for interest on loans and other debt
Financial position	Financial	Financial obligations that the company has towards third parties
Debt Equity ratio	Financial	Proportion of borrowed capital relative to own capital
Debt/EBITDA ratio	Financial	Measures ability to repay its debt using earnings
Liquidity index	Liquidity	Measure of a company's liquidity
Current ratio	Liquidity	Measures the liquidity
EBITDA	Redditivity	Earnings Before Interest, Taxes, Depreciation and Amortisation
EBITDA/Sales	Redditivity	Measure of operating income generated per unit of turnover
ROA	Redditivity	Measures how efficiently a company uses its assets to generate profits
ROE	Redditivity	Measures a company's profitability in relation to its equity
Consumption Margin	Redditivity	Difference between the available income and the consumer expenditure
Investment capital rotation	Rotation	Efficiency in utilizing the invested capital to generate revenue or profit
Working capital rotation	Rotation	Difference between current assets and current liabilities
Treasury Margin	Structure	Difference between liquid assets and short-term financial obligations
Structure Margin	Structure	Difference between net worth and fixed assets

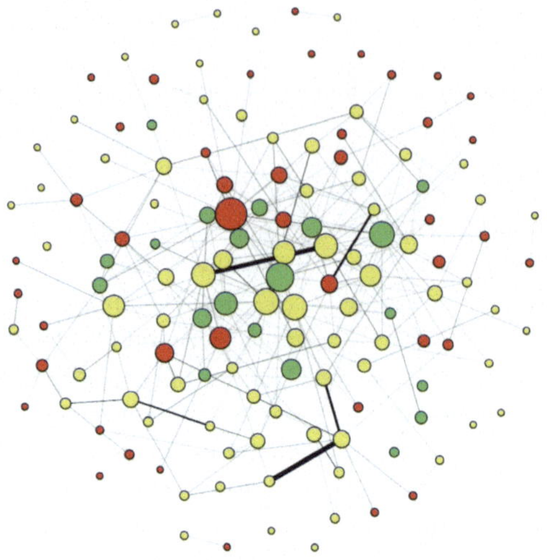

Fig. 1. The network $\mathcal{G}$ constructed by the adopted dataset. Nodes are colored according to their labels (green for class A, yellow for class B, and red for class C), and their size reflects their centrality (i.e., the number of edges); the width of each edge reflects its weight. Plot obtained using the igraph package [5] from the R software

Table 2. Hyperparameters used for training the GNN models

Hyperparameter	Value
Weight Decay	0
Dropout	0.2
Learning Rate	0.05
Hidden Channels	30
Number of Layers	2
Number of Convolutional Layers	2
Epochs	100

Additionally, to analyze whether there are significant differences in network centrality among companies with different ESG classifications, we performed an ANOVA test using the centrality measure. This test helped us to check if the average centrality differs significantly across ESG classes. After performing ANOVA on our dataset, we conducted a Tukey HSD (Honestly Significant Difference) test for post-hoc pairwise comparisons of ESG classes to determine which specific groups exhibit significant differences in network centrality.

4 Experiments

The experiments were conducted on a MacBook Pro, featuring an Apple M3 chip with 8 cores and 8 GB of unified memory. Model training and evaluation were implemented in Python 3.9.21, leveraging PyTorch [17], Deep Graph Library [21], and Scikit-learn [18] for deep learning and machine learning computations. Network centrality analysis and graph visualizations were carried out in R 2024.04.2+764, using the igraph package [5] for structural analysis.

The comparison of centrality across the ESG classes, as shown in the box plots of Fig. 2, revealed statistically significant differences among the three groups.

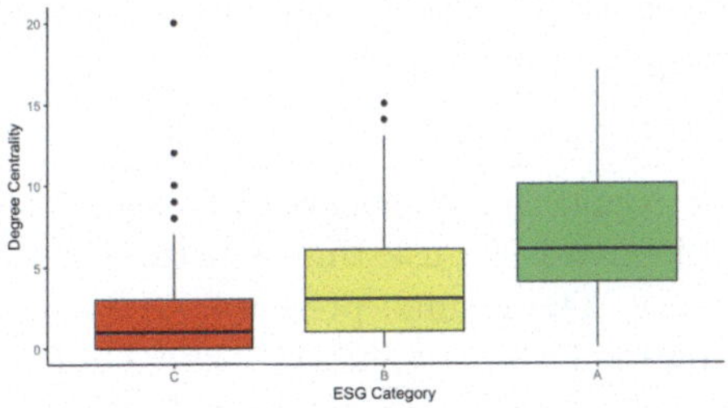

Fig. 2. Distribution of centrality across the ESG classes

The ANOVA test confirmed the presence of variations among the classes, and the post-hoc Tukey HSD test further identified that at least the ESG class A exhibits a significantly different centrality compared to the other two classes, as shown in Table 3. Indeed, these results show significant differences in centrality between ESG classes. Specifically, the comparisons between class A and both classes B and C show statistically significant differences, as indicated by the confidence intervals that do not contain zero (see these bounds in Fig. 3). This suggests that companies from class A have a significantly higher centrality compared to those in other classes. In contrast, the difference between classes B and C is not statistically significant, as its confidence interval includes zero.

Table 3. Tukey HSD test results for centrality across the ESG classes

Comparison	Mean Diff.	Lower Bound	Upper Bound	Adjusted p-value
B-C	1.2664	−0.2169	2.7496	0.1108
A-C	4.7143	2.4093	7.0193	8.6e-6
A-B	3.4479	1.2442	5.6517	8.4e-4

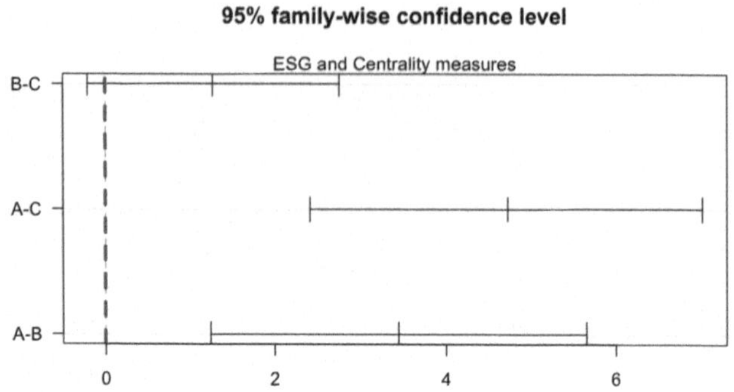

Fig. 3. Confidence intervals for pairwise comparisons of centrality across ESG classes using the Tukey HSD test

Given that network connectivity appears to be associated with ESG classes in our sample, our objective is to use this structural information to improve classification performance. The results, presented in Table 4, demonstrate that GNN-based models, particularly GraphSAGE and GAT, outperform conventional classifiers such as Random Forest, Decision Trees, kNN, and Naïve Bayes. GNNs achieve higher accuracy, balanced accuracy, and F1 Scores, indicating their ability to integrate both node features and network topology for ESG classification.

Table 4. Performance comparison of different classification models

Model	Accuracy	Balanced Accuracy	F1 Score
GAT	57.93%	51.44%	56.20%
GraphSAGE	57.24%	51.06%	53.67%
GNAN	56.21%	44.56%	51.14%
Random Forest	53.10%	45.11%	46.44%
Decision Trees	52.07%	47.22%	50.87%
kNN	48.28%	32.22%	40.23%
Naïve Bayes	34.48%	30.59%	30.59%

As a final remark, the execution times for each run (100 epochs) of GNN-based methods were about 0.14 s for GraphSAGE, 0.21 s for GAT, and 14 s for GNAN. Times for one execution of the traditional machine learning methods were about 0.64 s for Random Forest, 0.05 s for Decision Trees, 0.04 for kNN, and 0.03 for Naïve Bayes.

5 Conclusions

This study indicates that company connectivity, defined by shared decision-makers, has a meaningful relationship with ESG classification. The ANOVA and Tukey HSD tests confirmed that firms with higher ESG Perception Index scores tend to have significantly higher degree centrality in the network, demonstrating that firms with more interconnections might have an advantage in achieving better ESG performance. Given this network structure, we adopted GNN models to improve ESG classification performance. The results indicate that GNN models (GraphSage, GAT, and GNAN) generally outperform traditional machine learning classifiers such as Random Forest, Decision Trees, kNN, and Naïve Bayes, suggesting that incorporating both node attributes and network topology leads to better classification outcomes. These results highlight the potential of GNNs as effective ways for analyzing corporate connections and could encourage further research into the role of network structures in financial and sustainability studies. However, it is well known that one of the main limitations to the widespread adoption of GNNs is their lack of interpretability, to the point of being considered "black-box" models. Research in this field is continuously evolving, with several post-hoc models now available in the literature to enhance the explainability of GNNs. In this context, the GNAN method emerges as a promising solution, as it provides interpretability at both global and local levels. Although GNAN performs worse than other GNN models, this should not be surprising, as greater interpretability inevitably entails a slight decrease in performance. Therefore, in the future, we intend to conduct an in-depth study of the GNAN model to further evaluate its potential.

Acknowledgements. The work of M.R.G. and C.M. was partially supported by the PNRR MUR M4C2 Investment 1.5, project Rome Technopole PE00000013, Spoke 8 FP8. This work has also been carried out within the activities of L.M. and M.R.G. as members of the CDS-group at ICAR-CNR, the INdAM Research group GNCS, and the ICAR-CNR INdAM Research Unit.

References

1. Abdullah, M.: The implication of machine learning for financial solvency prediction: an empirical analysis on public listed companies of Bangladesh. J. Asian Bus. Econ. Stud. **28**(4), 303–320 (2021). https://doi.org/10.1108/JABES-11-2020-0128
2. Bechler-Speicher, M., Globerson, A., Gilad-Bachrach, R.: The intelligible and effective graph neural additive networks (2024). https://arxiv.org/abs/2406.01317
3. Carson, J., Tesluk, P., Marrone, J.: Shared leadership in teams: an investigation of antecedent conditions and performance. Acad. Manag. J. **50**, 1217–1234 (2007). https://doi.org/10.5465/amj.2007.20159921
4. Chowdhury, M., Abdullah, M., Azad, M., Sulong, Z., Islam, M.: Environmental, social and governance (ESG) rating prediction using machine learning approaches. Ann. Oper. Res. **1**(1), 1–12 (2023). https://doi.org/10.1007/s10479-023-05633-7

5. Csardi, G., Nepusz, T.: The igraph software package for complex network research. Int. J. Complex Syst. **1695** (2006). https://igraph.org
6. D'Amato, V., D'Ecclesia, R., Levantesi, S.: ESG score prediction through random forest algorithm. Comput. Manag. Sci. **19** (2022). https://doi.org/10.1007/s10287-021-00419-3
7. ESG: ESG global perception index (2025). https://www.esgreputation.it/esg-global-perception-index
8. European Banking Authority: On management and supervision of ESG risks for credit institutions and investment firms (2021). https://www.eba.europa.eu/sites/default/files/document_library/Publications/Reports/2021/1015656/EBA%20Report%20on%20ESG%20risks%20management%20and%20supervision.pdf. EBA Report, EBA/REP/2021/18
9. Giordano, M., Falbo, E., Maddalena, L., Piccirillo, M., Granata, I.: Untangling the context-specificity of essential genes by means of machine learning: a constructive experience. Biomolecules **14**(1) (2024). https://doi.org/10.3390/biom14010018. https://www.mdpi.com/2218-273X/14/1/18
10. Hamilton, W., Ying, R., Leskovec, J.: Inductive representation learning on large graphs. Curran Associates, Inc. (2017). https://doi.org/10.48550/arXiv.1706.02216
11. Hmieleski, K., Baron, R.: Shared authentic leadership and new venture performance. J. Manag. **38**, 1476–1499 (2012). https://doi.org/10.1177/0149206311415419
12. Hoje, J., Hongmin, C., Hakjoon, S.: Shared governance and ESG rating: evidence from Korea. Corp. Soc. Responsib. Environ. Manag. **32**(1), 1–15 (2025). https://doi.org/10.1002/csr.3104
13. Kingma, D.P., Ba, J.: Adam: a method for stochastic optimization (2017). https://arxiv.org/abs/1412.6980
14. Lim, T.: Environmental, social, and governance (ESG) and artificial intelligence in finance: state-of-the-art and research takeaways. Artif. Intell. Rev. **57** (2024). https://doi.org/10.1007/s10462-024-10708-3
15. Liu, B., He, J., Li, Z., Huang, X., Zhang, X., Yin, G.: Interpret ESG rating's impact on the industrial chain using graph neural networks. In: Proceedings of the Thirty-Second International Joint Conference on Artificial Intelligence (IJCAI-23), pp. 6076–6084. International Joint Conferences on Artificial Intelligence Organization (2023). https://doi.org/10.24963/ijcai.2023/674
16. Manzo, M., Giordano, M., Maddalena, L., Guarracino, M.R., Granata, I.: Novel data science methodologies for essential genes identification based on network analysis, pp. 117–145. Springer, Cham (2023). https://doi.org/10.1007/978-3-031-24453-7_7
17. Paszke, A., et al.: Pytorch: an imperative style, high-performance deep learning library. In: Advances in Neural Information Processing Systems 32, pp. 8024–8035. Curran Associates, Inc. (2019). http://papers.neurips.cc/paper/9015-pytorch-an-imperative-style-high-performance-deep-learning-library.pdf
18. Pedregosa, F., et al.: Scikit-learn: machine learning in Python. J. Mach. Learn. Res. **12**, 2825–2830 (2011)
19. Veličković, P.: Everything is connected: graph neural networks. Curr. Opin. Struct. Biol. **79**, 102538 (2023). https://doi.org/10.1016/j.sbi.2023.102538. https://www.sciencedirect.com/science/article/pii/S0959440X2300012X
20. Veličković, P., Cucurull, G., Casanova, A., Romero, A., Liò, P., Bengio, Y.: Graph attention networks (2018). https://arxiv.org/abs/1710.10903
21. Wang, M., et al.: Deep graph library: a graph-centric, highly-performant package for graph neural networks (2020). https://arxiv.org/abs/1909.01315

22. Yuliya, P., Francesco, P., Domenico, C., Marc, G.: The ties that bind or those that tear us apart? Co-CEO constellations and ESG performance in family firms. SSRN Electron. J. (2024). https://doi.org/10.2139/ssrn.4754036
23. Zaccone, M., Pedrini, M.: Sostenibilità e innovazione: dalla visione all'azione. Giappichelli Editore (2024)
24. Zhou, J., et al.: Graph neural networks: a review of methods and applications. AI Open **1**, 57–81 (2020). https://doi.org/10.1016/j.aiopen.2021.01.001. https://www.sciencedirect.com/science/article/pii/S2666651021000012

Innovative Applications of Supervised Learning in Addressing Missing Data: A Case Study on Social Surveys

Simona Cafieri(✉), Francesco Pugliese, and Mauro Sodani

ISTAT, Rome, Italy
{cafieri,frpuglie}@istat.it, sodani@istst.it

Abstract. In today's interconnected and dynamic global landscape, addressing key challenges such as public health, environmental sustainability, and social inequality demands coordinated, data-informed policy responses. Addressing these issues requires coordinated and evidence-based policy interventions, for which national statistical offices (NSOs) play a crucial role by providing high-quality and timely data. However, the presence of missing or incomplete information—whether arising from surveys or administrative sources—represents a persistent threat to the validity and reliability of statistical outputs. This study investigates the potential of advanced imputation techniques based on supervised machine learning (ML) and deep learning (DL) approaches, which fall within the broader domain of Artificial Intelligence (AI), to improve the handling of missing data in official statistics. A comparative analysis is conducted using real-world microdata from the Italian National Institute of Statistics (Istat), specifically from the Multipurpose Survey on Households, which is characterized by a non-negligible incidence of item nonresponse. Preliminary findings indicate that AI-driven imputation strategies outperform traditional statistical methods in terms of accuracy and robustness, particularly in complex social datasets. The results contribute to the growing body of literature advocating for the integration of modern computational tools within the framework of official statistics, with the aim of enhancing data quality on critical societal issues.

1 Introduction

The rise of artificial intelligence (AI) is having a significant impact on official statistics. AI methods provide solutions to data incompleteness and support informed decision-making (Sun et al. 2023). National Statistical Institutes have developed frameworks to support policy decisions, particularly in relation to environmental sustainability, health and social inequalities (Rigo 2022)[1].

[1] In this paper, we adopt a standard hierarchical definition of terms: *Artificial Intelligence (AI)* refers to the broad field of systems capable of performing tasks that normally require human intelligence. Within AI, *Machine Learning (ML)* includes algorithms that learn patterns from data. *Deep Learning (DL)* is a specialized subfield of ML that uses multi-layered neural networks. We follow the hierarchy: DL $\subset$ ML $\subset$ AI.

A. D'Ambrosio et al. (Eds.): CLADAG-VOC 2025, SCDAKO, pp. 48–58, 2025.
https://doi.org/10.1007/978-3-032-03042-9_5

In Italy, these issues are of particular importance as they form the basis of the BES (balanced and sustainable well-being) indicators that underpin the government's economic and financial planning document (Istat 2024). It is worth noting that a significant proportion of these indicators are based on survey data, which are prone to inaccuracy and unreliability if incomplete or missing responses are not addressed. The application of AI is a promising solution to the problem of missing data in surveys. Such methods can be used to predict and impute missing values, thereby improving the overall quality of statistical datasets. Our goal is not simple matrix completion, which focuses solely on predicting missing entries, but rather statistical imputation, aiming to preserve the overall distribution and structural integrity of the dataset. This is confirmed by comparing not only prediction errors but also descriptive statistics and distributional alignment between original and imputed data. Traditional techniques such as mean or median imputation often introduce bias, whereas AI-based methods can provide more accurate and impartial estimates. Several machine learning and deep learning models can be employed to impute missing data in official statistics. A substantial body of research has already been conducted to apply these methods, with positive results (De Fausti et al. 2023). This paper explores how AI can improve the quality of official statistics, focusing on the potential of machine learning (ML) and deep learning (DL) to improve the accuracy and reliability of health and environmental data.

2 Related Works

According to the literature, many deep learning and machine learning algorithms have been considered for 'data imputation'. Support vector machines (SVMs), a supervised learning method that identifies optimal hyperplanes for class separation in high-dimensional spaces, have been used to deal with missing values (Honghai et al. 2005). In addition, decision trees, which divide data into branches for decision-making, are evaluated for interpretability and used for data imputation (Nikfalazar 2020).

XGBoost (Extreme Gradient Boosting) improves model performance by combining predictions from multiple estimators and is known for its efficiency (Mitchell 2017; Rusdah & Murfi 2020). The k-Nearest Neighbours (KNN) algorithm classifies data points based on their nearest neighbours and is popular for missing value imputation (Guo et al. 2003.

Linear regression models relationships between variables and is often used for predictive analysis (Montgomery et al. 2021). Random forest (RF) constructs multiple decision trees to improve model accuracy (Breiman 2001) and has been used to impute missing values (Tang & Ishwaran 2017). Long short-term memory (LSTM) networks capture long-term dependencies in sequential data and have been used to impute missing values (Hochreiter & Schmidhuber 1997). Gated recurrent units (GRUs) provide a simplified architecture for a similar task (Dey & Salem 2017.

3 Methods

3.1 Dataset

One of the main sources of social and household health data in Italy is the Aspects of Daily Life (AVQ) survey, carried out annually by Istat (2022). AVQ represents an integral component of a unified system of social surveys. Indeed, collecting data is indispensable for understanding the daily lives of individuals and households. The survey provides information on the habits of citizens and the problems they face in everyday life through interviews with a sample of 20,000 households, representing approximately 50,000 individuals. Since 2018, the survey has been carried out using a sequential CAWI/PAPI mixed-mode technique. The survey investigates a range of social aspects, including education, employment, family and social life, leisure time, political and social participation, health, lifestyles, access to services and other factors relevant to the study of quality of life. These topics are investigated from a social perspective, with particular consideration given to behaviors, motivations and opinions as key elements in the definition of social information. The survey is included in the National Statistics Plan, which collates the statistical investigations that are required for the Country. However, it is not uncommon for questionnaires to be incomplete, which can affect the precision and dependability of the resulting data.

To address this issue, we designed and implemented an imputation experiment by using the AVQ dataset from the 2021 survey comprising 735 variables.

3.2 Preprocessing

The presence of missing values in the dataset, frequently represented as blanks or NaN, is incompatible with scikit-learn estimators, which require all values to be numeric and significant. A fundamental approach is the complete case analysis, whereby rows (dropNA) or columns with missing values (list-wise deletion) are excluded. Nevertheless, this may result in a significant reduction of the available information. An effective strategy is to impute missing data by inferring it from the available data. Conversely, techniques such as the use of central tendency measures (mean, median, etc.) can be employed. This approach appears to be relatively straightforward and robust. However, there is a risk of underestimating or overestimating the true values, which could introduce bias into the resulting estimates. This phenomenon occurs when an algorithm produces results that are systematically biased due to incorrect assumptions, which are typically present in the data set or in the machine learning process.

3.3 Imputation Strategy

In this study, we used missing data imputation techniques known as 'regression imputation'. Essentially, this method estimates missing values using a regressor (e.g. support vector regressor or random forest regressor), with the missing variable as the target and the other variables as inputs. Regression imputation is divided into 'deterministic' and 'stochastic'. The main difference between these two approaches is how the missing values are estimated and how uncertainty is taken into account. In deterministic regression imputation, missing values are estimated using a deterministic relationship between the variables. The trained regression model is used to predict the missing values for incomplete observations. The predicted value is used directly as an estimate for the missing value, hence the term deterministic. Deterministic imputation does not take into account the uncertainty associated with the estimate, so the predicted values are always the same for a given combination of input values. In contrast, stochastic regression imputation incorporates the uncertainty of the estimate into the imputation process. A stochastic noise term is added to the prediction. This noise term can be generated using the distribution of residual errors from the regression model. For example, if the regression model has a residual variance of sigma squared, noise can be added extracting it from a normal distribution with a mean of 0 and a variance of sigma squared, resulting in a more realistic estimate. However, in our work we only used deterministic regression imputation, with the intention of exploring stochastic imputation adapted to machine learning and deep learning models in the future. As a final observation, we cannot directly impute regression values before preprocessing. The input predictor variables also contain missing data themselves, which would cause issues for the machine learning and deep learning models, as the libraries we used (i.e., scikit-learn or Keras) do not accept null values. Therefore, we imputed all input variables with missing values using a method called "Simple Random Imputation," which involves replacing the missing value with a random value. This method, although simple, introduces randomness that may bias subsequent supervised learning. We recognize that imputing predictors in this way can influence the final model's performance. While this initial study relies on Simple Random Imputation (SRI) for predictor variables due to library constraints, we acknowledge that this method introduces random noise that can propagate bias through the supervised models.

To mitigate this, we implemented basic quality control measures, including outlier filtering and validation of domain-consistent values. In the context of official statistics, Multiple Imputation (MI) is widely regarded as a gold standard approach for handling missing data, particularly due to its ability to account for uncertainty by generating multiple plausible datasets and combining results through Rubin's rules. While this study focuses on deterministic imputation via supervised learning models, we acknowledge that MI offers a statistically rigorous alternative, especially in policy-sensitive domains. However, we opted not to implement MI at this stage for several reasons. Firstly, our primary goal was to explore the potential of machine and deep learning models in high-dimensional imputation tasks, where traditional MI methods often face computational challenges and convergence issues. Secondly, implementing MI with modern ML/DL models requires specialized adaptations—such as Bayesian neural networks or multiple draws from posterior predictive distributions—which are outside the current scope. That said, MI will serve as a benchmark for future comparisons, and we plan to include a full evaluation of its performance alongside both deterministic and stochastic supervised models in subsequent work. Moreover, we are currently experimenting **with** iterative imputation strategies, such as chained equations and model-based refinement, where the values imputed in one step are recursively improved in subsequent iterations.

These enhancements, although not applied in the current version, are planned for future implementations to reduce the randomness and improve the accuracy of the imputed predictors. It is crucial to emphasize that the dataset for the subsequent machine learning models must adhere to specific rules and requirements to ensure optimal final imputation. Missing data should be random or at least analysable, for example, it should be Missing Completely at Random (MCAR) or Missing at Random (MAR). In the event that the data is missing not at random (MNAR), the process of imputation becomes more complex. Regarding the percentage of missing data, datasets exhibiting a markedly high percentage of missing values (>50%) may potentially compromise the accuracy of the imputation models. It is optimal for the percentage of missing data to fall within the range of 12% to 50%. In the case of datasets with a missing data percentage lower than 12%, it would be preferable to utilize simpler imputation methods, such as setting the missing values to the mean of the variable or other constant values. Moreover, a significant proportion of the variables must remain complete, as these data points are indispensable for reconstructing the relationships between variables and predicting the missing ones. In particular, when employing deep learning models, it is crucial to ensure that the dataset is sufficiently large to enable effective training of the model, as these models typically require a significant number of samples to generalise effectively. Conversely, for a traditional machine learning model, the inclusion of a greater number of useful variables facilitates a more accurate prediction of the missing values. Additionally, the dataset must be structured in a way that meets the statistical production requirements. For example, existing values must be consistent with domain rules. In the case of a variable representing age, for instance, negative values or values exceeding a realistic threshold should be avoided. Imputation models must be capable of accommodating both qualitative variables, which should be accurately encoded using one-hot or ordinal encoding, and quantitative variables, which should be standardized or normalized in accordance with the models employed. The imputation process may be distorted by extreme outliers, particularly in the case of traditional models such as KNN or regression. Outliers should be identified and managed, either through transformations or exclusion. Variables must have a stable and meaningful distribution, as this enables the models to accurately capture the patterns. Variables with highly imbalanced distributions may reduce the effectiveness of the imputation. Finally, for models such as linear regression or logistic regression, it is essential to reduce multicollinearity (strong correlation between independent variables), as this could negatively affect the imputation. This can be addressed using techniques such as the Variance Inflation Factor (VIF). In the literature, this method is more efficient than other methods of replacing predictor variables with a zero, the mean, etc. (Kalton & Kish 1984).

3.4 Model Training

In this work, we trained all the traditional machine learning models described in the previous section: SVM, DT, RF, XGBoost, KNN, and the most recent deep learning models: MLP, LSTM, GRU, CONV1D. The objective of the training was to create models for the imputation of the following health-related variables: The variables of interest were body mass index (BMI) for individuals aged 18 and over. The same models have been trained for the imputation of additional environmental variables, namely SODPOAP (resident satisfaction with household waste collection services).

3.5 Evaluation Metrics

The ML and DL models were evaluated using a range of metrics that are appropriate for regression problems. The Root Mean Squared Error (RMSE), the Mean Absolute Percentage Error (MAPE), the Mean Absolute Error (MAE) and the R2 Score were employed for the assessment of the models. We observed that even though the RMSE yields absolute values, it was an adequate metric for comparing the performance of the different models and for their ranking. Accordingly, the trained models were ordered in descending order of RMSE, with the most effective model identified as the one with the lowest error. Furthermore, models that demonstrated minimal overfitting, as evidenced by a minimal discrepancy between the RMSE values for the training and test sets, were deemed the most optimal. Prior to the commencement of the training phase, a train-test split was conducted, with the test set comprising 20% of the entire dataset. No pre-processing was applied to the traditional machine learning models, given that the data is entirely numeric and lacks qualitative variables. For the deep learning models based on neural networks, all numerical inputs were normalized between 0 and 1. The training phase used Python's Hyperopt framework to optimize hyperparameters for scikit-learn (ML) and Keras (DL) models. Given the strong impact of initial hyperparameters, optimization was integrated directly into each model's training, adding computation time but ensuring optimal model selection, as Hyperopt returns the best model similarly to RandomizedSearch and GridSearch. Hyperopt is more efficient than these methods, using the Tree-structured Parzen Estimator (TPE) and Bayesian inference. TPE classifies parameter sets as "good" or "bad," focusing sampling on "good" regions. Iteratively, Hyperopt refines probability estimates, narrowing the search and optimizing models faster. All results here were obtained using Hyperopt. Relative training times were compared using KNN as a baseline (set to 1x). GRU showed the longest training time (approximately 4767x), while traditional models such as DT and LR were among the fastest (<15x). These values were obtained on a system with Intel Core i7 CPU, 16 GB RAM, and no GPU acceleration.

4 Results

Upon completion of the training phase, it was observed that the combination of the most effective models varied depending on the variable being imputed. This finding may be related to the No-Free-Lunch Theorem (ADAM, et al. 2019), which sets a theoretical limit in machine learning. The No-Free-Lunch Theorem postulates that no single optimal machine learning model exists for every task. Consequently, the strength of our method lies in its capacity to identify the optimal model for each variable (for each variable, the task is completely distinct) to be imputed, which is meticulously selected from a vast array of models. About the health-related variable "BMI", the results for all the models are presented in Table 1.

It can be seen that the Deep Learning model is the most effective. Long Short-Term Memory (LSTM) model. Table 2 presents a comparison of the descriptive statistics (mean, standard deviation, and quartiles) calculated for the original variable prior to imputation and the imputed variable. This preliminary assessment indicates that the imputed distribution is not markedly disparate from the original distribution.

Table 1. Table of metrics of the models trained to predict the variable "BMI".

MODEL	Training RMSE	Test RMSE
LSTM	0.7489	0.7546
GRU	0.7423	0.7551
CONV1D	0.7637	0.7854
MLP	0.7866	0.8122
SVM	0.8524	0.8823
KNN	0.6954	0.7949
LR	0.6400	0.7936
XG Boost	0.3898	0.7318
RF	0.2666	0.7240
DT	0.0000	1.0143

Table 2. Table of comparisons between descriptive statistics of BMI and Imputed BMI (DetBMI).

MODEL	Mean	STD	MIN	25%	50%	75%	MAX
BMI	2.5695	0.7413	1	2	2	3	4
DetBMI	2.4296	0.7683	1	2	2	3	4

Figure 1 provides a comparison between the original distributions (omitting the nulls in the BMI column) and the distribution with imputed data (the BMI column without nulls plus imputed values).

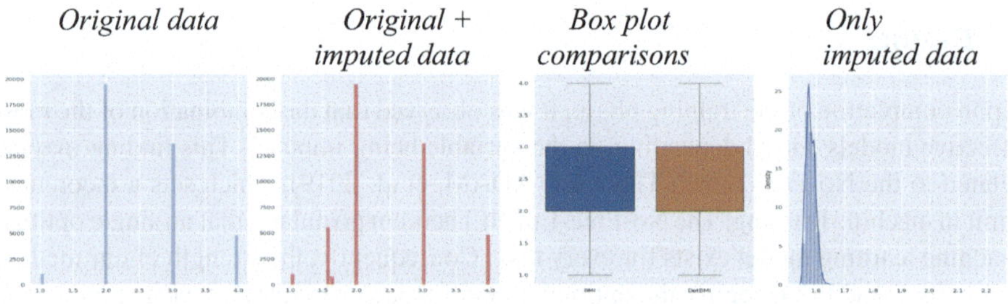

Fig. 1. Comparisons among distributions charts for the variable BMI.

Furthermore, an analysis of the box plots of both distributions and the distribution of only the imputed values that replace the nulls is provided. As can be observed, the two distributions are similar, although a slight margin of error is to be expected at this stage. Figure 2 presents a further comparison between the mean and standard deviation of the imputed data and the original data. Furthermore, a comparison between the original and imputed univariate and cumulative distributions is presented. The distributions are highly similar, which reinforces the assertion that AI models can markedly enhance the handling of missing data.

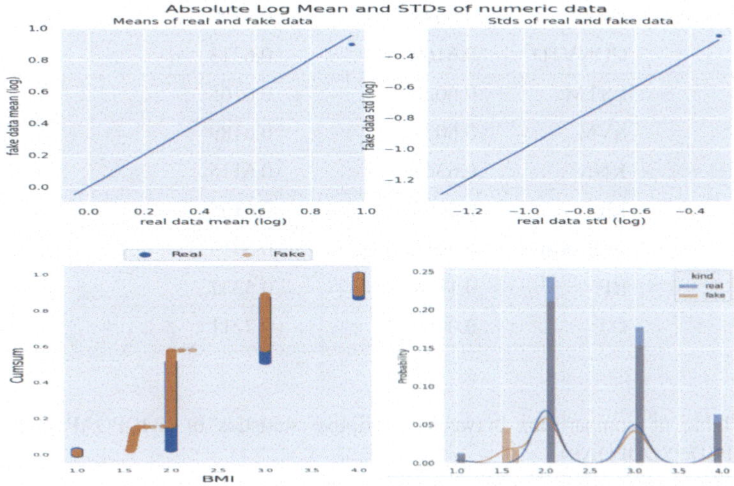

Fig. 2. Comparisons among means, standard deviations, cumulate and univariate distributions charts

With regard to the environmental variable SODPOAP, Table 3 provides a comparison of the metrics for all models, indicating that the Multi-Layer Perceptron (MLP) is the optimal model in this context. The most recent models exhibit a proclivity for overfitting, as evidenced by the markedly lower error rate on the training set in comparison to the test set. It is also noteworthy that, as anticipated, deep learning models demonstrate superior performance compared to machine learning models on these high-dimensional imputation datasets. Indeed, deep learning models consistently rank among the top performers, irrespective of whether they are recurrent or not. This suggests that the longitudinal aspect of the data does not influence the models' performance in this particular dataset. In comparison to traditional machine learning models, simpler models such as SVM and LR appear to demonstrate superior performance, whereas more sophisticated ensemble models like XGBoost and RF tend to exhibit a higher propensity for overfitting.

In Table 4, we present comparisons between descriptive statistics.

Figure 3 presents a comparison of the original and imputed distributions of SODPOAP, while Fig. 4 provides a comparison of the means, standard deviations, univariate, and cumulative distributions. As with the previous results, excellent results are obtained. However, it can be observed that the outcome and behaviour of the models change depending on the difficulty level of the variable. The principal advantage of our methodology is the construction of a bespoke model for each variable to be imputed. The results demonstrate that deep learning models, such as long short-term memory (LSTM) and gated recurrent unit (GRU) networks, exhibit high performance in terms of root mean square error (RMSE) on every task, indicating their suitability for handling sequential

Table 3. Table of metrics of the models trained to predict the variable "SODPOAP".

MODEL	Training RMSE	Test RMSE
MLP	0.6843	0.6970
GRU	0.6048	0.6190
CONV1D	0.6169	0.6313
LSTM	0.6020	0.6165
SVM	0.6031	0.6189
KNN	0.5347	0.6115
LR	0.5093	0.6195
XGBoost	0.2915	0.5315
RF	0.1964	0.5337
DT	0.0000	0.7211

Table 4. Table of comparisons between descriptive statistics of SODPOAP and Imputed SODPOAP (DetSODPOAP).

MODEL	Mean	STD	MIN	25%	50%	75%	MAX
SODPOAP	1.8563	0.6260	1	1	2	2	4
DetSODPOAP	1.7666	0.5532	1	1.5	2	2	4

data. Random Forest and XGBoost also demonstrated satisfactory performance, however, they tended to over-fit, rendering them unsuitable for the task of the imputation of missing values. The support vector machine models demonstrated reliable performance in imputation, although they exhibited slightly higher root mean square error (RMSE) compared to the top-performing models.

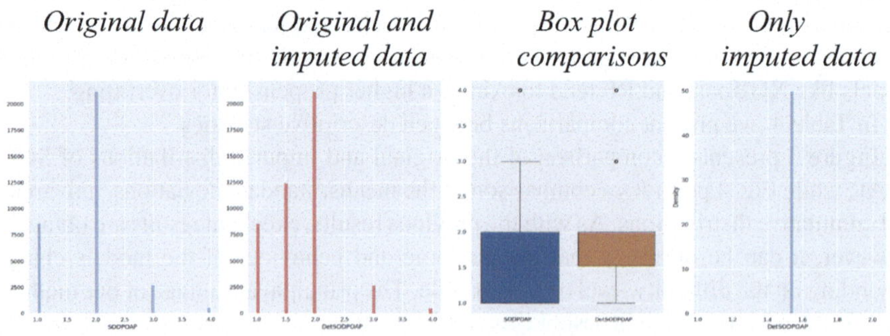

Fig. 3. Comparisons among distributions charts for the variable SODPOAP.

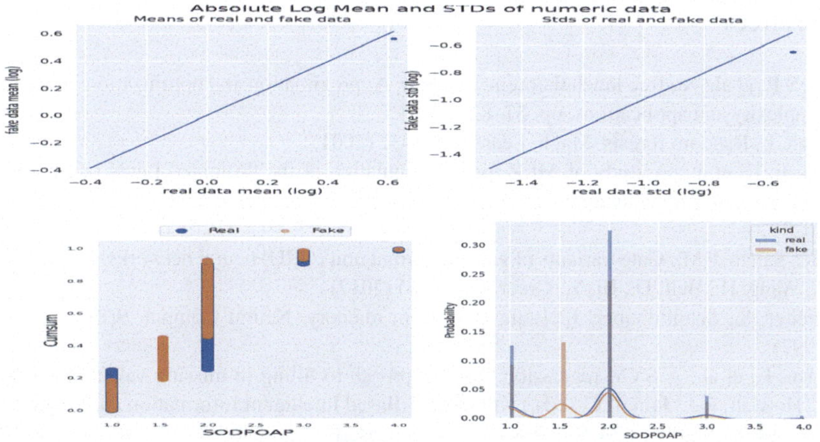

Fig. 4. Comparisons among means, standard deviations, cumulate and univariate distributions charts

5 Conclusion

The examination of health and environmental statistics offers a promising avenue for improving the quality and reliability of data. It should be noted that this research was conducted on a single dataset; however, it could have been performed on multiple datasets. However, we encourage you to consider a few points. The dataset is not a simple one; rather, it is essential for Istat's production needs. The objective was to concentrate on this particular dataset, comprising over 700 variables, to ascertain the limits of AI capabilities in the context of this specific imputation problem. Furthermore, the No-Free-Lunch theorem indicates that no optimal model or AI exists that can perform equally well on all tasks and datasets. All such applications are domain-specific or task-specific, and thus dataset-oriented. Consequently, even if the AI had been tested on all imputation datasets, it would not be possible to guarantee that it would always work. It can be stated with certainty that AI-based imputation methods can effectively address the issue of missing data, thereby enhancing the overall integrity of statistical surveys. However, it is essential to note that the most suitable model must be carefully designed and implemented for each imputation problem, as there is currently no general AI (AGI) that can be applied to all types of problems and datasets. A comprehensive analysis allows for the selection of the optimal model, the imputation of missing data using this model, and the subsequent evaluation of the quality of the imputed data. Furthermore, while this study focuses on deterministic imputation methods, we acknowledge the importance of Multiple Imputation (MI) as a standard benchmark in official statistics. Future work will include a comparative evaluation with MI and the exploration of stochastic approaches. While the current approach uses simple random imputation for predictors, future work will incorporate more sophisticated iterative techniques to refine these variables and reduce the introduction of noise into the models.

References

Adam, S.P., et al.: No free lunch theorem: a review. Approximation and optimization: algorithms, complexity and applications, pp. 57–82 (2019)

Breiman, L. Random forests. Mach. Learn. **45**, 5–32 (2001)

De Fausti, F., et al.: A study of MLP for the imputation of the "Attained Level of Education" in base register of individuals. In: Workshop on Methodologies for Official Statistics, p. 69 (2023)

Dey, R., Salem, F.M.: Gate-variants of gated recurrent unit (GRU) neural networks. In: 2017 Guo, G., Wang, H., Bell, D., Bi, Y., Greer, K. (2003) (2017)

Hochreiter, S., Schmidhuber, J.: Long short-term memory. Neural Comput. **9**(8), 1735–1780 (1997)

Honghai, F., et al.: A SVM regression based approach to filling in missing values. In: Khosla, R., Howlett, R.J., Jain, L.C. (eds.) Knowledge-Based Intelligent Information and Engineering Systems. KES 2005. LNCS, vol 3683. Springer, Berlin, Heidelberg. https://doi.org/10.1007/11553939_83

Istat. Indagine Aspetti della vita quotidiana 2021 (2022)

Istat. Rapporto BES 2023 (2024)

Kalton, G., Kish, L.: Some efficient random imputation methods. Commun. Statist.-Theory Meth. **13**(16), 1919–1939 (1984)

Mitchell, R., Frank, E.: Accelerating the XGBoost algorithm using GPU computing. PeerJ Comput. Sci. **3**, e127 (2017)

Montgomery, D.C., Peck, E.A., Vining, G.G.: Introduction to linear regression analysis. John Wiley & Sons (2021)

Nikfalazar, S., Yeh, C.H., Bedingfield, S., Khorshidi, H.A.: Missing data imputation using decision trees and fuzzy clustering with iterative learning. Knowl. Inf. Syst. **62**, 2419–2437 (2020)

Rigo, A.: Programmazione e innovazione: il percorso verso l'efficienza interna delle Pubbliche Amministrazioni (2022)

Sun, Y, et al.: Deep learning versus conventional methods for missing data imputation: a review and comparative study. Expert Syst. **227**, 120201 (2023)

Tang, F., Ishwaran, H.: Random forest missing data algorithms. Statist. Anal. Data Min.: ASA Data Sci. J. **10**(6), 363-377 (2017)

The Relevance of Information in Changing the Structure of Consumer Preferences: A Pre-post Sensory Experiment on Seven Olive Oils

Marco Cardillo(✉), Alfonso Piscitelli, and Raffaele Sacchi

Department of Agricultural Sciences, University of Naples Federico II, Naples, Italy
marco.cardillo@unina.it

Abstract. This paper describes the design and the results of a pre-post sensory evaluation experiment aimed at evaluating the possible effects of the seminar on sensory perceptions of olive oil and its impact on consumers in terms of preferences. The experiment was designed to control for the effects of possible spontaneous maturation of consumers due to pre-seminar measurement of their preferences for the same olive oils evaluated (i.e., post vs pre measurement). The seminar, lasting about one hour, was held after the first tasting session and aimed to inform participants about the commercial classification of olive oils, their olfactory and flavour characteristics and problems, their aftertaste, and how to identify qualified and/or fresh oil (e.g., EVOO). After one week, the second tasting session involved 98 assessors who had participated in the first taste session and the seminar. The seminar was able to affect the structure of individuals' preferences. In particular, the evaluation of quality of the oils of uninformed consumers appears to be influenced by the reduction of information asymmetry.

Keywords: Consensus Ranking · Paired sample · consumer preferences · Sensory Experiment

1 Introduction

The globalization of markets has made it possible to offer consumers a wide variety of choice even among olive oils. Understanding the mechanisms on which consumer choice is based is fundamental in marketing campaigns.

Perceived quality refers the consumers' value judgments or perceptions of quality [20,25]. The concept of perceived quality can be classified in two groups of factors that allow to the consumer to evaluate the products: intrinsic attributes and extrinsic attributes [15,21]. The intrinsic attributes are related to the physical aspects of a product (e.g., colour, flavour, form and appearance); on the other hand, extrinsic attributes are related to the product, but not in the physical part of this one (e.g., brand name, price, country of origin, packaging). Consumers

A. D'Ambrosio et al. (Eds.): CLADAG-VOC 2025, SCDAKO, pp. 59–69, 2025.
https://doi.org/10.1007/978-3-032-03042-9_6

tend to use both intrinsic and extrinsic attributes simultaneously when choosing a food product [2,3,22,24]. The relative importance of different attributes of olive oil helps consumers to determine their preference for a hypothetical product [11,14,23]. Physical characteristics such as colour, taste, and flavour play an important role in consumers' perception [9,17]. This paper describes the design and the results of a pre-post experiment aimed at evaluating the possible effects of a short seminar on sensory perceptions of olive oil and its impact on consumers in terms of preferences. The experiment was designed to control for the effects of the possible spontaneous maturation of consumers due to a pre-seminar measurement of their preferences for the same olive oils evaluated (i.e., 'post' vs 'pre' measurement). The experimental design and the types of assessed olive oils are described in detail in Sect. 2. The seminar, lasting about one hour, was held after the first tasting session and aimed to inform participants about the commercial classification of olive oils, their olfactory and flavour characteristics and problems, their aftertaste, and how to identify qualified and/or fresh oil. Olfactory sensations referred to fruity olive as well as tomato, artichoke, apple, eucalyptus, and other flavours. Tasting properties referred to a significant but not extreme sour and spicy sensation. The participants were involved in the topical issue of the seminar as if they were expert judges able to sensorily evaluate the chemical and physical qualities of oils. The participants showed interest in the seminar issues, asking questions and generally interacting with the experts giving the seminar. Four categories of olive oils were included in the experiment: extra virgin olive oil, virgin olive oil, refined olive oil, and rancid olive oil. Of these, the extra virgin one is considered the best. The aim of this study was to determine whether the seminar on sensory perceptions of olive oil was able to change the structure of individuals' preferences. Specifically, if at the second tasting session, consumer ratings were higher for extra virgin olive oils and lower for defective oils.

Ranking data provides a more nuanced understanding of respondent opinions, as they are required to provide a ranking of preferences and priorities. The rank transformation allowed the application of Marden's [13] approach for comparing. To take into account the coupled nature of the data, the test was modified by integrating it with the approach proposed by Salmaso and Pesarin [16]. We can consider this approach as the nonparametric equivalent of the paired t-test. It makes no assumption about the distributions of the original population themselves.

In this paper, we present in detail our data analysis strategy in order to answer our research question (Sect. 3), a synthesis of the results of the statistical analysis performed (Sect. 4), and the conclusions drawn thereupon (Sect. 5).

2 Data

A pre-post sensory evaluation experiment was conducted on seven olive oils. The oils differed in quality, price, sensory profile, and peculiar flavor. Olive oil flavor depends primarily on the concentration and nature of volatile and phenolic compounds present in olive oil, which can change dramatically depending on agronomical and technological factors [8]. In particular, the experiment included the following oils: (i) an extra virgin olive oil (A), which is strongly fruity, with very intense bitter and spicy characteristics and a slight hint of almond; (ii) an extra virgin olive oil (B) which has a light olive fruitiness and is characterised by a sweet, spicy and slightly bitter flavour; (iii) a refined olive oil (C) characterised by the absence of a fruity aroma; (iv) virgin olive oil (D) characterised mainly by fusty and, to a lesser extent, t by rancid and Winery-vinegar defects; (v) a rancid olive oil (E); (vi) an extra virgin olive oil (F) which is intensely fruity, in whose aromatic profile it is possible to distinguish the aromas of tomato and fresh leaf; (vii) a virgin olive oil (G) characterised mainly by fusty and to a lesser extent by rancid and mold defects. In the first tasting session, 106 assessors tasted the seven oils randomly. For each oil tasted, the assessors were asked to express a preferential judgment by stating a value on a nine-point scale, where one means awful and nine means excellent. Furthermore, participants were asked to express judgments on three intrinsic attributes of the tasted oils: appearance, nose and palate; data on the background characteristics of the assessors were also collected. In this work, we confine the analysis to the overall judgement of each olive oil. The olive oil samples tasted, were served at room temperature in plastic cups. Each cup contained 3 ml of oil and had a transparent plastic lid. White bread and water were provided as a sample carrier. A commercial brand was selected in order to prevent taste or quality variability in the bread [18]. All participants tasted the full set of seven oils and the administration order was randomised so as not to condition the judgements on the tasting sequence. At the end of the first tasting session, all participant attended a seminar. The seminar, lasting about one hour, aimed to inform participants about the commercial classification of olive oils, their olfactory and flavour characteristics and problems, their aftertaste, and how to identify qualified and/or fresh oil (e.g., EVOO). Olfactory sensations referred to fruity olive as well as tomato, artichoke, apple, eucalyptus, and other flavours. Tasting properties referred to a significant but not extreme bitter and spicy sensation. After one week, the second tasting session involved 98 assessors who had participated in the first tasting session and the seminar. In the second tasting session, all judges evaluated the same olive oils that they had tasted in the first tasting session (in the same tasting order), even though they had been told that they were evaluating different olive oils. All tasting experiments were conducted under a double-blind control procedure, and the data analysis included only people who participated in both tasting sessions.

For each judge, two evaluations were collected: one before and one after the seminar, both in the form of ratings. The Wilcoxon signed-rank test, although commonly used for such data, was not sensitive enough to detect significant differences in consumer preferences before and after the seminar. To apply our strategy, we first converted the ratings into weak rankings, allowing for ties. This conversion was necessary to apply the test proposed by Marden [13].

3 Methodology

To highlight a possible change in the structure of preferences of individuals, in particular to verify if there was a change in the structure of preferences for each olive oil tasted, the Wilcoxon signed-rank test was performed to compare two related samples. However, in our data this test was unable to detect a significant change 'post' versus 'pre' for all olive oils tasted. In our case, the pre-post ratings provided by the judges may not provide enough differentiation to capture the nuances of a preference accurately, just as they may not accurately rank the olive oils rated. In our opinion, a more sensitive test is needed to fully exploit the information contained in the data and to be able to detect changes in the judges' preferences. For these reasons, in this work, we adopt a data analysis strategy based on the ranking assigned by the judge rather than on the ratings provided. The proposed strategy is specifically designed to detect differences in a population's preferences before and after an event. The test allows for comparisons on paired samples, enabling the evaluation of whether an event has induced a change in the aggregate preference structure of the judges.

This test is an adaptation of the method presented by Marden [13], based on the work of Feigin [6], by building it inside the framework proposed by Pesarin and Salmaso [16].

As briefly mentioned in Sect. 2, the data must be converted into ranks. The rank conversion will not break the ties; they will be broken only when the consensus rank is calculated. This is to make a clear statement of the ranking structure and on the possible alteration of it.

The general consensus is extracted from the data using the branch and bound algorithm, as described in [1].

3.1 Consensus Ranking: Branch and Bound Algorithm (BB-ALGORITHM)

Investigating the differences between the two groups requires extrapolating a ranking that best represents the common tendencies within each group. So, to evaluate solely the influence of the grouping. Originally, the test uses the modal ranking. In our proposal, we adopt an alternative ideal group ranking: the consensus ranking, which is easier to compute. The consensus ranking, also referred to as the median ranking, provides a representative ranking for the population of judges. The median ranking is defined as the ranking that minimizes the sum of the distances between itself and the data rankings.

Kemeny and Snell [10] proposed a distance metric based on an axiomatic approach, in which they defined the axioms and found the unique metric that satisfied them. Edmond and Mason [7] defined the correlation coefficient equivalent to the Kemeny distance, the τ_x. Their approach builds upon Kendall's τ. They modified the score matrix representation by assigning 1 not only when one object is ranked ahead of another but also when the objects are tied.

$$a_{ij} = \begin{cases} 1, & \text{if object } i \text{ is ranked ahead of object or if or tied with object } j, \\ -1, & \text{if object } i \text{ is ranked behind object } j, \\ 0, & \text{if } i = j. \end{cases} \tag{1}$$

The τ_x is then obtained by the equation:

$$\tau_x = \frac{\sum_{i=1}^{m}\sum_{j=1}^{m} a_{ij}b_{ij}}{m(m-1)} \tag{2}$$

The τ_x coefficient, as noted in Eq. (2), is equivalent to Kendall's τ when ties are not permitted. It evaluates the degree of similarity between two rankings, providing a measure of concordance. When $\tau_x = 1$, there is complete agreement between the rankings, which means that they are identical. Conversely, when $\tau_x = -1$, the rankings are in total disagreement, indicating that one ranking is the exact inverse of the other.

To calculate the consensus ranking is used the algorithm proposed by Emond and Mason which allows to avoid to investigate the entirety of the space defined by all the possible rank permutation. Precisely we use the version of the Emond and Mason BB-algorithm for full ranking present in the r package `ConsRank` [5].

The algorithm takes advantage of the τ_x correlation coefficient which allows to collect all the information of each of $m \times m$ score matrices into a single $m \times m$ matrix defined as the combined input matrix(CI), which contains the information of all the rankings in the data. Mathematically the entries of CI matrix are obtained as follows:

$$c_{ij} = \sum_{k=1}^{N} a_{ij}^{(k)} \tag{3}$$

where k is a generic ranking.

Thus the consensus ranking problem simply becomes a maximization:

$$max = \sum_{k=1}^{N} c_{ij}s_{ij} \tag{4}$$

where s_{ij} represents the score matrix elements of the consensus ranking candidates; the choice of the initial candidate influences the computation time. In this paper, this decision is based on the sign matrix of the CI matrix and is implemented by the `ConsRank` package [5].

The numerator is simply the sum of the term-by-term product of a score matrix for the CI matrix. Therefore, the maximum is obtained when the ranking score matrix matches the CI matrix in terms of sign, and thus, this particular ranking is the ideal solution. As a result of the structure of the score matrix, where the diagonal elements are zero and the off-diagonal elements are either 1 or -1, the maximum possible value is simply the sum of the absolute values of the elements in the CI matrix:

$$V = \sum_{i,j=1}^{n} |c_{ij}| \tag{5}$$

Considered a ranking σ with a score matrix elements s_{ij}, this vector is evaluated in the Eq. (6):

$$p = \sum_{i,j=1}^{n} c_{ij} s_{ij} \tag{6}$$

To solve the problem, it is necessary to define an initial penalty term, which is obtained by subtracting p from the maximum value V:

$$P = V - p \tag{7}$$

In the first stage of the algorithm, the σ ranking is considered as a candidate consensus ranking. The favorite object is fixed, and the second-ranked object is considered. This object can be ranked either ahead or behind the first one, creating two branches based on their relative positions. For each branch, an incremental penalty is computed using the combined input matrix. The penalty P is calculated based on the following conditions:

- If object i is preferred to object j: - If $c_{ij} > 0$ and $c_{ji} < 0$, then $\delta P = 0$ - If $c_{ij} < 0$ and $c_{ji} > 0$, then $\delta P = c_{ji} - c_{ij}$

- If object j is preferred to object i: - If $c_{ij} > 0$ and $c_{ji} < 0$, then $\delta P = c_{ij} - c_{ji}$ - If $c_{ij} < 0$ and $c_{ji} > 0$, then $\delta P = 0$.

Where δP is a term that indicate the incremental penalty. If the incremental penalty for a branch exceeds the initial penalty, it is discarded. If the incremental penalty is smaller or equal, the algorithm considers the next object and creates new branches by placing the object in all possible positions. The penalties are recalculated and summed. The algorithm continues, pruning branches and evaluating promising ones, until a solution is found.

Moreover, the algorithm provides a measure of the spread of rankings around their consensus ranking, known as the Averaged τ_x Rank Correlation Coefficient. A higher value indicates a stronger agreement between individual rankings and the representative ranking of the group. Formally the Averaged τ_x is defined by the equation:

$$\text{Averaged } \tau_x = \frac{\sum_{i=1}^{m} \sum_{j=1}^{m} c_{ij} s_{ij}}{mn(m-1)} \tag{8}$$

where m is the number of items.

For datasets with more than 15 items, the algorithm becomes computationally intensive. Therefore, it is recommended to use the `fast` or `quick` algorithms available in the `ConsRank` package [5].

3.2 Test Definition and Logic

To inquire whether or not there is a difference in the center of population relative to their preferences, [13] proposed a test based on [6]. This implies that the null hyphotesis is that the representative ranking for each group are not significantly different from the overall modal ranking. Formally, the null hypothesis is:

$$w^{(1)} = w^{(2)} = w^{(T)}$$

where $w^{(i)}$ indicate the consensus ranking relative to each group and $w^{(T)}$ is the global consensus ranking. Other alternative approaches have been discussed [19] and [12], which are rather based on the concordance and so to test if the population are uniform or there is "agreement" which Marden in [13] define as: "the expected distance between two random vectors from different populations is less than the expected distance under uniformity". Based instead on the first approach, this test evaluate whether or not there is a difference in center of populations. To obtain such result in [13] Marden propose to evaluate the sum of the distances between the observed rankings and the modal ranking of each group relative to the sum of the distance of each ranking and the population ranking. This exercise is used to evaluate the role of the within-group distances on the overall distance of the data. Our proposal merges the approach described in the [13] with the suggestion made by [16] to take in account that the two samples aren't indipendent.

Matematically the test statistic is:

$$R^2 = 1 - \frac{\sum_{l=1}^{L}\sum_{i=1}^{n} d(y^{(li)}, \hat{\omega}^{(l)})}{\sum_{i=1}^{L}\sum_{i=1}^{n} d(y^{(li)}, \hat{\omega}^{(T)})} \tag{9}$$

where $\hat{\omega}^{(T)}$ is the global consensus ranking and i is the single evaluation and l is the session to which correspond the evaluation, so before or after the seminary.

Higher the value of the test statistic R lower the role of the within distances in the explanation of the overall distance: if the group consensus ranking are equal then $R^2 = 0$, while if there is only one distinct ranking for each group but they are completely different from each other then $R^2 = 1$. As noted by Marden in [13] this statistic has not particularly useful asymptotic properties, in fact for $n \to \infty$ the probability that the consensus ranking of the distinct population are the same goes to 1. To address this issue he propose the randomization p-value [6].

The randomization p-value proposed by Feigin and Alvo [6] is based on the assumption that each rank in the full ranking space of $m!$ permutations has a certain probability of being chosen. As a result, each group originates from a population defined by a probability vector of size $m!$, meaning that each ranking has a specific probability of being selected for each group.

This implies a null hypothesis in which all these probability vectors are equal. However, the way the test is computed presents a significant issue when applied to paired sample, making it unsuitable for the dataset used in this study. In fact, by defining new matrices through the random selection of the rows of the overall matrix (as defined by [13]), the issue is potentially finding the same judge and both his evaluations in the newly defined group.

For this reason, the test will be built on a different framework [16], which allows addressing this issue. Their approach is designed for repeated observations, making it easily adaptable to the previously defined test for our strategy.

They consider that each judge's response is exchangeable within their own evaluations, implying that, for their randomized p-value, the null hypothesis assumes the judges' responses follow a uniform distribution.

Previously, the randomization was performed by assigning judges to groups. However, since each judge in this case has two different responses that are interchangeable, we will instead swap their evaluations between groups. This ensures that both responses from the same judge are never placed within the same group.

To perform the test, it is necessary to calculate the consensus ranking for each group and the overall dataset, which will then be used to evaluate Eq. (9).

The obtained R^2 statistic is not sufficient to draw any conclusions; therefore, a randomization p-value will be computed. To calculate the R^2 statistic for each iteration of the permutation, the two groups will be reconfigured using a binomial distribution with $p = 0.5$, which determines the individuals to be swapped between groups.

In this way, the assumption of uniformity of the response rank vector is satisfied. Since there are only two time periods, a binomial distribution ensures that the rank vectors are interchangeable within evaluations.

At each iteration, the R^2 statistic is computed for the newly formed groups. Finally, the randomized p-value for R^2 is obtained as:

$$\hat{p} = mean(r^2 > R^2) \tag{10}$$

If such value is above a certain imposed threshold then the consensus ranking of the two sample are equal, so the occasion in which the judges made their evaluation has no effect.

4 Results

The test rejects the hypothesis that the median ranking of the judges before and after the evaluation is not significantly different. In fact in our case, none of the r^2 statistics computed on the permuted datasets exceeded the initial statistic (see Fig. 1).

This confirms that the two general rankings are significantly different, meaning that the initial assumption of uniformity in responses is violated. Consequently, we can conclude that there is a time effect, most likely due to the informative seminar conducted before the second evaluation. The median ranking before the seminar was (E, D, B, F, A, G, C).

In this ranking, the first two oils, E and D, present rancid defects. The third place is occupied by oil B, which is an extra virgin olive oil (EVOO) with a light fruity flavor but slightly bitter. Oil F is also an EVOO, characterized by a strong fruity taste, and is ranked fourth. A is an EVOO with an intense fruity

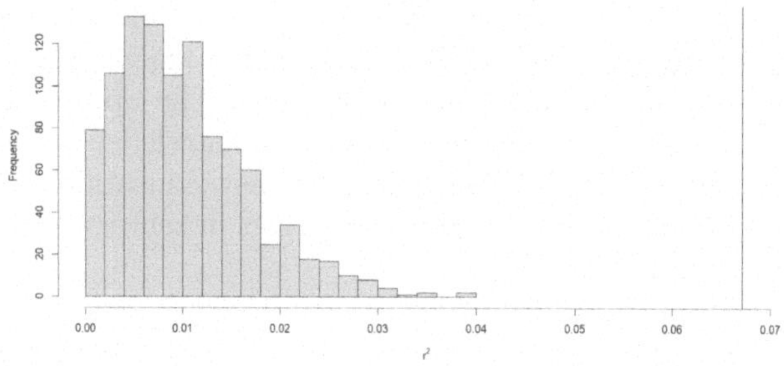

Fig. 1. Histogram of the distribution of r^2 statistic against the R^2, value of the statistic computed on the original dataset indicated by the vertical line

flavor, primarily characterized by bitterness and spiciness, placed fifth. Oil G is a virgin olive oil with a fusty flavor, ranked sixth. The last, oil C, is a refined olive oil, which lacks a fruity aroma. After the seminar the new median ranking is A, B, E, F, D, G, C.

The most significant ranking reversals occurred for oil A, an EVO characterized by a strong fruity flavor, and oils E and D, which exhibited rancid defects. Oil A, initially deemed not particularly good by the judges before the seminar, became the most appreciated oil afterward. In contrast, oils E and D, which were ranked first and second in the initial evaluations despite their rancid defects, were ranked lower after the seminar.

In particular, the evaluation of the quality of the oils of uninformed consumers appears to be influenced by the reduction of information asymmetry. Another interesting insight comes from the averaged τ_x coefficient for the consensus ranking within each group, which is a measure of the degree of consensus around the median ranking [4]. At time 1, the group exhibited an averaged τ_x of 0.2317. However, after the seminar, the same judges demonstrated greater cohesion in their preference structure, as indicated by a higher averaged τ_x value of 0.4572.

This finding not only confirms that the seminar effectively educated consumers on the qualities associated with higher values oil, but also suggests that this understanding was received homogeneously across participants.

5 Conclusions

This study presents a data analysis strategy to highlight differences in consumer judgments before and after exposure to seminar teachings. The proposed strategy, based on consensus rankings and randomized p-value testing, successfully captures shifts in preference structures between two evaluation periods. The results demonstrate that the informational intervention, which aimed to educate consumers on the qualities of olive oils, significantly altered their judgments. Specifically, oils that were initially less favored, such as oil A, saw an increase in preference after the seminar, while oils with defects, like oils E and D, were ranked lower in the second evaluation. This shift suggests that the new information reduced the information asymmetry among participants and allowed for more informed and cohesive judgments. The approach provides a robust framework for understanding how additional knowledge can reshape consumer preferences and offers valuable insights into how education or information dissemination can impact decision-making processes in sensory evaluations and beyond.

References

1. Amodio, S., D'Ambrosio, A., Siciliano, R.: Accurate algorithms for identifying the median ranking when dealing with weak and partial rankings under the Kemeny axiomatic approach. Eur. J. Oper. Res. **249**(2), 667–676 (2016). https://doi.org/10.1016/j.ejor.2015.08.048
2. Brentari, E., Zuccolotto, P.: The impact of chemical and sensory characteristics on the market price of Italian red wines. Electron. J. Appl. Stat. Anal. **4**, 265–276 (2011). https://doi.org/10.1285/i20705948v4n2p265
3. Charters, S., Pettigrew, S.: The dimensions of wine quality. Food Qual. Prefer. **18**(7), 997–1007 (2007). https://doi.org/10.1016/j.foodqual.2007.04.003
4. D'Ambrosio, A., Amodio, S., Iorio, C.: Two algorithms for finding optimal solutions of the Kemeny rank aggregation problem for full rankings. Electron. Notes Disc. Math. **8**(2), 198–213 (2015). https://doi.org/10.1285/i20705948v8n2p198
5. D'Ambrosio, A.: ConsRank: Compute the Median Ranking(s) According to the Kemeny's Axiomatic Approach. R package version 2.1.4 (2024). https://CRAN.R-project.org/package=ConsRank
6. Feigin, P.D., Alvo, M.: Intergroup diversity and concordance for ranking data: an approach via metrics for permutations. Ann. Stat. **14**(2), 691–707 (1986). https://doi.org/10.1214/aos/1176349947
7. Emond, E., Mason, D.: A new rank correlation coefficient with application to the consensus ranking problem. J. Multi-Criteria Decis. Anal. **11**, 17–28 (2002). https://doi.org/10.1002/mcda.313
8. Genovese, A., Caporaso, N., Sacchi, R.: Flavor chemistry of virgin olive oil: an overview. Appl. Sci. **11**(4), 1639 (2021). https://www.mdpi.com/2076-3417/11/4/1639
9. Grunert, K.G.: What's in a steak? A cross-cultural study on the quality perception of beef. Food Qual. Prefer. **8**(3), 157–174 (1997). https://doi.org/10.1016/S0950-3293(96)00038-9
10. Kemeny, J.G., Snell, L.J.: Preference ranking: an axiomatic approach. In: Mathematical Models in the Social Sciences, pp. 9–23 (1962)
11. Krystallis, A., Ness, M.: Consumer preferences for quality foods from a South European perspective: a conjoint analysis implementation on Greek olive oil. Int. Food Agribus. Manag. Rev. **8**, 62–91 (2005)
12. Li, L., Schucany, W.R.: Some properties of a test for concordance of two groups of rankings. Biometrika **62**(2), 417–423 (1975). https://doi.org/10.1093/biomet/62.2.417
13. Marden, J. I. (1996). *Analyzing and Modeling Rank Data* (pp. 101-102). Chapman & Hall/CRC. https://books.google.it/books?id=Dp24H21QHNoC

14. Mtimet, N., Zaibet, L., Zairi, C., Hzami, H.: Marketing olive oil products in the Tunisian local market: the importance of quality attributes and consumers' behavior. J. Int. Food Agribus. Mark. **25**, 134–145 (2013). https://doi.org/10.1080/08974438.2013.736044
15. Olson, J.C., Jacoby, J.: Cue utilization in the quality perception process. In: Proceedings of the Third Annual Conference of the Association for Consumer Research (1972). https://api.semanticscholar.org/CorpusID:148296156
16. Pesarin, F., Salmaso, L.: Permutation Tests for Complex Data: Theory, Applications and Software, pp. 59–62. Wiley (2010). https://books.google.it/books?id=9PWVTOanxPUC
17. Pomarici, E., Piscitelli, A., Fabbris, L., Sacchi, R.: A pre-post sensory experiment on the effect of a seminar on olive oil preferences of Italian consumers. In: Carpita, M., Fabbris, L. (eds.) ASA2019 - Scientific Conference on Statistics for Health and Well-being, pp. 181–184. CLEUP, Padova (2019)
18. Porretta, S.: Analisi sensoriale & consumer science. Chiriotti (2000)
19. Schucany, W.R., Frawley, W.H.: A rank test for two group concordance. Psychometrika **38**(2), 249–258 (1973). https://doi.org/10.1007/BF02291117
20. Steenkamp, J.E.M.: Dynamics in consumer behavior with respect to agricultural and food products. In: Wierenga, B., van Tilburg, A., Grunert, K., Steenkamp, J.-B.E.M., Wedel, M. (eds.) Agricultural Marketing and Consumer Behaviour in a Changing World, pp. 143–188. Kluwer Academic Publishers (1997)
21. Szybillo, G., Jacoby, J.: Intrinsic versus extrinsic cues as determinants of perceived product quality. J. Appl. Psychol. **59**(1), 74–78 (1974). https://doi.org/10.1037/h0035796
22. Veale, R., Quester, P.: Do consumer expectations match experience? Predicting the influence of price and country of origin on perceptions of product quality. Int. Bus. Rev. **18**(2), 134–144 (2009)
23. Van der Lans, I.A., Van Ittersum, K., De Cicco, A., Loseby, M.: The role of the region of origin and EU certificates of origin in consumer evaluation of food products. Eur. Rev. Agric. Econ. **28**(4), 451–477 (2001)
24. Verdu, A., Llorens Montes, F., Fuentes, M.: Measuring perceptions of quality in food products: the case of red wine. Food Qual. Prefer. **15**, 453–469 (2004). https://doi.org/10.1016/j.foodqual.2003.08.002
25. Zeithaml, V.: Consumer perceptions of price, quality and value: a means-end model and synthesis of evidence. J. Mark. **52**(3), 2–22 (1988)

Weighted Logistic Oblique Tree for Regression

Andrea Carta(✉) and Luca Frigau

Department of Economics and Business Sciences, University of Cagliari, Viale Sant'Ignazio 17, Cagliari, Italy
{andrea.carta88,frigau}@unica.it

Abstract. In this article, we propose a Weighted Oblique Logistic Tree for Regression (WOLTReg), a novel method to build regression oblique trees using a weighted logistic regression classifier to determine the optimal oblique splitting hyperplanes; this novel approach at each node integrates a weighting mechanism to reduce impurity and a feature selection process that avoids data overfitting. We also evaluate WOLTReg on multiple benchmark datasets and find out that WOLTReg works better when only a small subset of predictors is used to create the separating hyperplane. Furthermore, we compared its performance in a test set across all datasets with other oblique tree methods and a standard decision tree. The results indicate that WOLTReg has better performance with respect to the other methods in terms of root mean square error. In conclusion, this study highlights the potential of WOLTReg and suggests future applications in various domains.

Keywords: GLM · Decision Tree · Classification · Regression · Oblique Tree

1 Introduction

Among all the nonparametric models used both for regression and classification tasks, Decision Trees (DTs) are one of the most popular. Their main advantage is the straightforward interpretation of results and the ability to be visualized and understood even by the general public. To build a DT, the predictor space is progressively divided into non-overlapping sub-regions, known as nodes. The main goal of this partitioning is to maximize the change in a specific impurity function at each split. Usually, in a classification problem, standard measures are the Gini index or Entropy. On the other hand, for regression tasks, the goal of partitioning is to decrease the deviance of the tree with respect to a continuous target variable value. Nodes can then be classified into two major categories: terminal nodes, if they can not be split further, and these are also called leaves, and non-terminal nodes, if they can still be partitioned. The splitting process is applied recursively until a stopping criterion is met, such as reaching a maximum depth of the tree, a minimum node size, or creating a totality of pure terminal

A. D'Ambrosio et al. (Eds.): CLADAG-VOC 2025, SCDAKO, pp. 70–82, 2025.
https://doi.org/10.1007/978-3-032-03042-9_7

nodes. Finally, when the tree is complete, to each terminal node is assigned a class label for classification DTs or a predicted value for regression DTs. Traditional DTs, such as Classification and Regression Trees (CART) [7], split the feature space at each node with an axes-parallel hyperplane; that is, they identify a feature and a threshold that maximizes the reduction in node impurity and split the data accordingly [7].

Issues with traditional DTs arise when the data structure becomes more complex. Traditional DTs are often built with great depth to achieve better accuracy. However, this comes at the cost of reduced interpretability, one of their key advantages. Moreover, they overlook the correlation among variables by selecting only one variable for each node. To address these limitations, many researchers proposed the idea of splitting data using a hyperplane made by a linear combination of multiple predictors; these hyperplanes will then be oblique regarding the axes, and the generated DTs are called Oblique Trees (OTs) [7,21]. OTs are usually smaller and achieve higher accuracy with respect to the traditional axes-parallel DTs [15]. However, finding the hyperplane that maximizes the reduction of node impurity is an NP-complete problem [12], so the optimal oblique splits are usually found through a heuristic search. On this topic, we propose a novel methodology to build OTs for regression tasks. This method, called Weighted Oblique Logistic Tree for Regression (WOLTReg), separates the predictor space at each node using hyperplanes derived from the estimated coefficients of a weighted logistic regression classifier.

The main goal of this study is to introduce a novel method to build a DT in which the predictor space is partitioned recursively using oblique hyperplanes and to analyze its strengths and limitations by comparing its performance in terms of predictive power with respect to other nonparametric regression algorithms. Furthermore, as most OTs mainly focus on classification tasks [9], the novelty of our method lies in utilizing a weighted logistic regression classifier within a DT framework for regression problems. Our methodology could then be used as a valid method for predicting continuous variables in the case of a complex data structure. The remainder of the paper is structured as follows. In Sect. 2, we illustrate the literature on weighted logistic regression and oblique DTs. Section 3 describes the methodological framework of WOLTReg. In Sect. 4, we evaluate the characteristics of WOLTReg and compare it with other DT methods. The article terminates with concluding remarks in Sect. 5.

2 Background

In this section, we will present an overview of the weighted logistic regression, the classifier used in WOLTReg to separate the data at each node. Moreover, we give an introduction to the various types of OTs.

2.1 Weighted Logistic Regression

Logistic regression is one of the most popular supervised statistical models for classification tasks. It can predict the probability of a binary outcome based on one or more explanatory variables. Numerous scientific fields, including epidemiology, medicine, and economics, widely use its applications [28,30]. Given n training data points

$$\{(\mathbf{x}_i, y_i)\}_{i=1}^{n}, \quad \mathbf{x}_i \in \mathbb{R}^m, \quad y_i \in \{0, 1\}$$

the model calculates the probability of an event occurring as

$$p_i = \mathbb{P}(y_i = 1|\mathbf{x}_i) = \frac{1}{1 + \exp(-\mathbf{x}_i^\top \boldsymbol{\beta})},$$

where $\boldsymbol{\beta}$ is the vector of coefficients and $\mathbf{x}_i$ is the vector of predictors for observation i. Furthermore, logistic regression is based on the logit link function. This function represents the log odds as a linear combination of the predictors

$$\log\left(\frac{p_i}{1 - p_i}\right) = \mathbf{x}_i^\top \boldsymbol{\beta}.$$

The log-likelihood function ℓ of the observed data is maximized in order to estimate the coefficients β. Thus, the log-likelihood function, assuming that each observation is independent, is expressed as

$$\ell(\beta) = \sum_{i=1}^{n} \left[y_i \log(p_i) + (1 - y_i)\log(1 - p_i)\right].$$

This maximization can be achieved through numerical optimization techniques, such as Newton-Raphson methods [6].

Weighted logistic regression expands the standard logistic regression model by assigning a weight w_i to each observation i. This approach is beneficial when dealing with imbalanced datasets, stratified sampling, or observations requiring different levels of influence in the model estimation [18]. The main difference between the two models can be found in the log-likelihood function to be maximized, in the case of weighted logistic regression, this function ℓ_W can be expressed as:

$$\ell_W(\beta) = \sum_{i=1}^{n} w_i \left[y_i \log(p_i) + (1 - y_i)\log(1 - p_i)\right],$$

By adding weights specific to each observation, weighted logistic regression allows the model to consider data heterogeneity. This approach is widely applied in fields such as epidemiology, economics, and machine learning to improve model robustness in the presence of unbalanced or stratified data [1,18,19].

2.2 Oblique Decision Trees

Our algorithm uses a weighted logistic regression classifier to separate the data at each node. In fact, for OTs, an exhaustive search for the optimal hyperplane is rarely possible. Hence, good oblique splits are typically obtained heuristically. Depending on the algorithm used to find the optimal oblique split, there are three main categories of OTs: direct search, indirect search, and projection.

Direct Search OTs. The direct search approach aims to find the optimal oblique split by implementing optimization techniques. The first and most used OTs can be classified under this group. For instance, Breiman, already in the eighties, proposed the CART-Linear combination (CART-LC) [7]. CART-LC uses a deterministic hill-climbing algorithm to find the optimal oblique split at each non-terminal node. However, this leads to a problem of local minima, making the search not optimal. Another method that uses a hill-climbing algorithm is the Oblique Classifier 1 (OC1) [21]. OC1 perturbs the coefficients that define the oblique hyperplane at each node. These coefficients correspond to the weights in the linear combination of input features used to separate the data. The algorithm iteratively adjusts them to minimize impurity, combining deterministic optimization with random perturbations to escape local minima. The Evolutionary algorithm is another class used to find the optimal oblique split. For instance, [8] uses a technique based on natural selection and genetics to find the optimal split coefficients. On the other hand, [23] introduces the Continuous Optimization of Oblique Split (CO2) method, which optimizes the hyperplane coefficients using a latent Support Vector Machine (SVM) formulation.

Indirect Search OTs. The main intuition in the indirect search approach is that oblique hyperplanes can be identified as linear decision boundaries resulting from a linear classifier. Our novel method can be categorized in this group. The main advantage of this type of tree is that it is more computationally feasible, as there is no randomization or a hill climbing algorithm as with the OTs that use the direct search approach. Among this category of OTs, we can cite the Fast Algorithm for Classification Trees (FACT) [17] and Quick Unbiased Efficient Statistical Tree (QUEST) [16], which use linear discriminant analysis to find the best oblique split at each node. Moreover, reference [4] uses as a classifier an SVM, with the particularity that using a different kernel other than the linear in the SVM optimization process can also grow non-linear trees. Similarly, [20] uses SVM to find the best split handling and a multiclass problem. [33], on the other hand, uses logistic regression with a variable selection process to find the separating hyperplanes. In contrast to these last methods that apply the standard top-down approach of the DTs, [3] implement a Bottom-Up Oblique Decision Tree Induction Algorithm (BUTIA), in which they implement a clustering algorithm to generate the initial leaves and use an SVM to estimate the separating hyperplanes in a bottom-up approach.

Projection OTs. In this approach, the intuitive idea is to rotate the feature space and then apply the traditional axes-parallel split on this new rotated space. This splitting hyperplane will then be oblique in the original space. This type of OT is mainly used as a base learner for ensemble methods. The advantage of this method is that there is an increase in variability among the trees, assuring better results [5]. The difference in these models is how the feature space is rotated. For instance, [27] rotates the space using Principal Component Analysis (PCA) and applies a base learner tree. References [26,36] use a Householder matrix to find the new coordinates of the rotated space. References [14,37], instead, use the projection pursuit optimization to rotate the space at each node. Moreover, the rotation can also be random. In fact, [5,32], construct an ensemble of trees in which, for each tree, the space is randomly transformed.

3 Method

WOLTReg works as standard DTs: it implements a top-down approach, and at each node, it separates the data given a certain hyperplane that is a linear combination of a set of predictors. The main difference with respect to traditional DTs is how this hyperplane is computed. However, as the traditional DTs, WOLTReg can be explained through three key components: a splitting rule applicable to any node split, a stopping rule determining when to halt the recursive process, and a method for assigning values to each terminal node. The split criterion of WOLTReg is illustrated in Algorithm 1.

Let $\mathbf{X} \in \mathbb{R}^{n_t \times m}$ be the matrix of predictor variables at a given node t, where n_t is the number of observations and m is the total number of features. Let $\mathbf{y} \in \mathbb{R}^{n_t}$ be the corresponding vector of continuous response values. As the first step, we compute the absolute value of the Pearson correlation of $\mathbf{X}$ with the response variable $\mathbf{y}$. As for the second step, we apply the variable selection process; that is, we select the γ features most Pearson correlated in an absolute value sense, with the response variable $\mathbf{y}$. This creates the subspace $\mathbf{X}^{\star}$. Thus, γ is a tuning parameter for the model and is fixed and does not depend on the depth of the tree. This variable selection step aims to reduce dimensionality and retain the most relevant features in a computationally efficient manner. As shown in Sect. 4, it also helps prevent the OT from overfitting the data. While the use of linear correlation may overlook non-linear relationships with the target variable, it is intended as a preliminary filter to discard features with little or no predictive power.

Step 3 describes how to compute the weights for the classifier. The weights ω are computed as the absolute value of the scaled elements of $\mathbf{y}$. Steps from 4 to 10 describe the computation of the hyperplane and the dichotomization process. In fact, we need to use a dichotomous variable to find the estimated coefficients of a separating hyperplane using the classifier. So we begin this cycle by transforming the response variable $\mathbf{y}$ into a dichotomous variable $\mathbf{y}_k$ using a threshold given by a quantile value $q_k \in Q$ of $\mathbf{y}$, where $Q = \{q_1, \ldots, q_K\}$ is the set of selected quantile thresholds. For instance, if $|Q|$ is equal to one, the algorithm would apply the dichotomization using only as a threshold the median of $\mathbf{y}$; if $|Q|$ is equal to 9, on the other hand, the algorithm will dichotomize $\mathbf{y}$ for each decile of its distribution. After the dichotomization process, for each different $\mathbf{y}^k$ we apply a weighted logistic classifier $\Psi(\mathbf{X}^{\star}, \mathbf{y}^k, \omega)$ (Step 8), the resulting estimated coefficients will represent the separating hyperplane HP^k and a function $f^{dev}(\mathbf{y}, \mathrm{HP}^k)$ will compute the resulting split deviance δ_k. Upon completing this for loop, we will have computed $|Q|$ hyperplanes and the corresponding resulting deviance. In step 11, the algorithm selects the separating hyperplane $\mathrm{HP}^{\star}$ that results in the lowest node deviance and partitions the data accordingly into two children nodes, just as in a standard axes-parallel DT.

A key characteristic of this algorithm is the incorporation of weights in the logistic classifier to split the data. In fact, by categorizing a continuous variable, there is a significant loss of information [2,22]. To mitigate this loss, WOLTReg uses as a splitting method a weighted logistic classifier; thus, to each observation a weight ω_i proportional to the absolute value of the elements of y_i is assigned. This ensures that elements of $\mathbf{y}$ at the extreme tails of the distribution have a higher weight than those close to the average. Therefore, the algorithm identifies which observations have a more significant influence on the impurity of the node.

Furthermore, to highlight the importance of the weights, we generated a toy example dataset (Fig. 1) in which two predictor variables were generated by sampling from two distinct bivariate normal distributions with different mean and covariance structures and the target variable was then independently drawn from normal distributions with varying central tendencies and dispersions across the two groups. Figure 1 shows the optimal split using three different methods: on panel a, the split is done by CART; in panel b, we used the algorithm described in Algorithm 1 but using a standard logistic classifier without weights and $|Q|$ equal to nine; finally in panel c, we apply WOLTReg algorithm with again $|Q|$ equals to nine. The results show that there is a decrease in impurity with respect to the standard CART for both oblique splits (panels b, c). However, using the weights in the logistic classifier, the decrease in impurity is much more significant with respect to that of the CART, as it is around 15% less for the WOLTReg and 9% for the split using the unweighted logistic classifier.

Apart from the way in which the separating hyperplane is computed, WOLTReg is very similar to standard CART in terms of stopping criteria. WOLTReg stops growing the tree if the node has fewer instances than a certain threshold; similarly, if a split creates a leaf with fewer observations than a given threshold, that split does not occur, and the node becomes a leaf. Moreover, WOLTReg includes a tuning parameter to prevent overfitting. This parameter is called *deviance decrease*, and is indicated as D. It is identified as the decrease in deviance of the resulting split in percentage with respect to the root node deviance. In this manner, if a split does not bring enough decrease in impurity, then the tree stops its recursive partitioning process and creates a terminal node. In this context, it is worth clarifying that deviance is not computed from a standard linear regression model, but rather from the fit of the separating hyperplane used to perform the split. Therefore, the *deviance decrease* measures the gain in purity achieved by the hyperplane partitioning, not by a global regression. The algorithm of WOLTReg, by default, through cross-validation, selects the best D in terms of root mean square error. This procedure is conceptually similar to the cost-complexity pruning used in standard CART. Finally, WOLTReg assigns each leaf a value corresponding to the average of the dependent variable for the observations contained within that final node.

Algorithm 1. WOLTReg method for best oblique split

Require: $\mathbf{X} \in \mathbb{R}^{n_t \times m}$, $\mathbf{y} \in \mathbb{R}^{n_t}$, $Q = \{q_1, \ldots, q_K\} \subset \mathbb{R}$ with $K \in \mathbb{N}$, $\gamma \in \mathbb{N}$

$\text{cor}_X \leftarrow |\text{cor}(\mathbf{X}, \mathbf{y})|$

$X^\star \leftarrow$ selected γ features with highest cor_X ▹ Variable Selection Process

$\omega = \dfrac{|\mathbf{y} - \bar{\mathbf{y}}|}{sd(\mathbf{y})}$

for $q_k \in Q$ **do** ▹ Dicothomization Process

  if $y \leq q_k$ **then**

    $\mathbf{y}^k \leftarrow$ Low **else** $\mathbf{y}^k \leftarrow$ High

  end if

  $\text{HP}^k \leftarrow \Psi(\mathbf{X}^\star, \mathbf{y}^k, \omega)$ ▹ Weighted Logistic Classifier

  $\delta_k = f^{dev}(\mathbf{y}, \text{HP}^k)$

end for

$\text{HP}^\star = \text{HP}^k : \{\arg\min_{k=1,\ldots,|Q|} \delta_k\}$

Ensure: $\text{HP}^\star \leftarrow \sum_{j=1}^{m} a_j X_j + a_{m+1} = 0$

4 Experiment

To assess the strengths and weaknesses of WOLTReg, and to evaluate its performance, we tested WOLTReg on several benchmark datasets downloaded from the Penn repository [24]. The experiment has been carried out using the software R (version 4.4.3) [25].

4.1 Model Evaluation

We considered 60 datasets and divided each into a training set of 70% of observations and a test set. We applied WOLTReg to each training dataset through 5-fold cross-validation to evaluate, in terms of root mean square error (RMSE), how they perform with respect to Q, the set of quantiles to be used in the dichotomization process (in particular, we used the median, the quartiles, and the deciles), and γ, i.e., the number of features to be used at each node to find the separating hyperplane. Moreover, a grid has been created such that, for each possible value of γ and Q, the algorithm selects the optimal pruned tree in terms of RMSE, using the decrease in deviance parameter D. Finally, to enable comparison across datasets, RMSE values are scaled by dividing them by the standard deviation of the dependent variable in each dataset, transforming them into the Normalized Root Mean Squared Error (NRMSE) [29].

The results of this experiment are shown in Fig. 2. In panel a, the plot indicates that using different quantiles in the dichotomization process does not bring different results in terms of predictive power. However, using the median simplifies the computational efficiency of the tree, as it just needs to fit only one logistic model at each node, with respect to nine logistic regressions when the deciles are used. Panel b, on the other hand, shows that there is a tendency to perform worse when the number γ of selected features increases. This is probably due to the overfitting effect that can happen when we use a larger number of

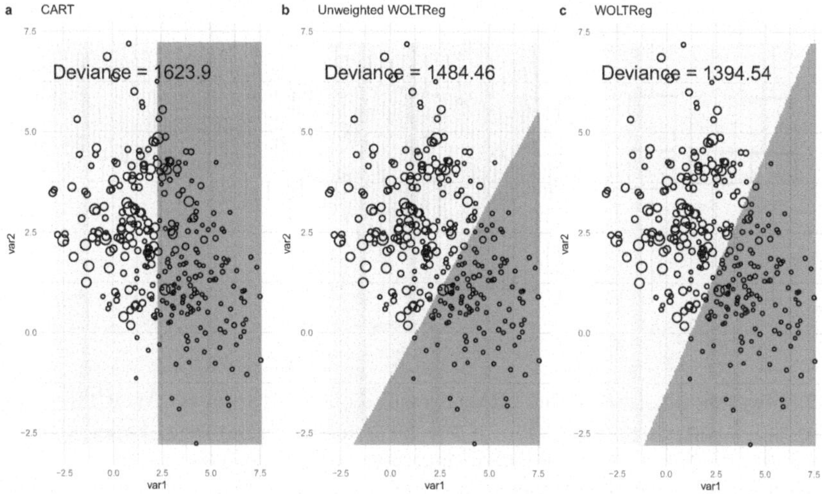

Fig. 1. Comparison of regression tree models. Panel **a** shows the optimal split by CART and the resulting node deviance. Panel **b** shows the optimal oblique split by the Unweighted WOLTReg. Panel **c** shows the optimal oblique split by WOLTReg. The size of circles is proportional to the value of the target variable.

features in the node to compute the separating hyperplane. Moreover, this info can help us when fine-tuning the model in a cross-validation approach, as we can identify the optimal γ among the lower values, even if the dataset has many features. It is worth noting that we eliminated the results for values of γ greater than 50, since only two datasets in our sample have more than 50 features. This limited number of cases was causing a sudden drop in the NRMSE and in its variability, making the resulting plot less interpretable and potentially misleading. Nonetheless, we also observed that increasing γ leads to poorer performance even for these two datasets.

The main take from this analysis is, however, the fact that a simple model that uses the median in the dichotomization process and 2 or 3 features in the variable selection process is the best in terms of the ratio between predictive power and model complexity.

4.2 Model Comparison

In this section, we compare the performance of WOLTReg with CART and other OT methods. Specifically, we consider CART [7,31], the BUTIA tree [3], the CO2 algorithm [23], and HHCART [36]. The latter three algorithms were implemented in Python (version 3.12.0) [35], and integrated into the R environment using the `reticulate` package [34]. The source codes were obtained from the `GitHub` repository available at https://github.com/maoqiangqiang/DecisionTreeBaseline, where the original algorithms were adapted to perform regression tasks.

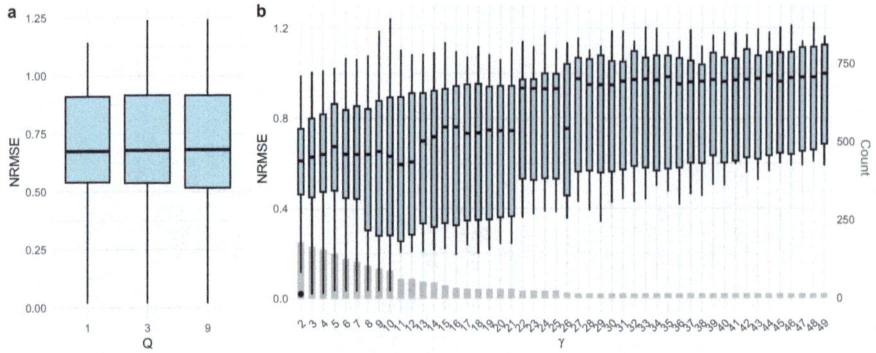

Fig. 2. Boxplots illustrating the NRMSE results obtained from the WOLTReg model during the 5-fold Cross-Validation of the 60 benchmark datasets. Panel **a** displays NRMSE values based on the median, quartiles, and deciles used to dichotomize the data within the nodes. The boxplots highlight the variability of the results. Panel **b** shows the NRMSE in relation to the number of features γ used for data splitting. The barplot at the bottom of panel **b** indicates the number of evaluated cases for each γ, accounting for all dataset and Q combinations.

We again used the benchmark datasets described in Sect. 4.1 for this experiment. Based on the results, we trained WOLTReg on the training set of each dataset, using the median as the quantile for the dichotomization process and setting γ equal to two. This constraint ensures the algorithm selects only two variables to define the separating hyperplane. The model's performance was then evaluated on the corresponding test set in terms of RMSE. Similarly, we trained, using the default parameters, the other models on the same training sets and assessed their performance on the test sets.

Figure 3 summarizes the results of the model comparison experiment. Specifically, panel *a* illustrates the frequency of each rank for the four models; in our case, the mode for the first rank is WOLTReg, in particular for 40% of datasets, WOLTReg is the best model, followed by HHCART. For panel *b* and panel *c*, we categorized the datasets into three classes using the terciles for the number of features and instances. Then, we analyzed the relative performance of WOLTReg with respect to these classes by checking the frequency of the model to rank first in terms of RMSE in the test set. From the panels emerges that WOLTReg performs relatively better than the other models for datasets with fewer features. On the other hand, WOLTReg performs relatively better for datasets with a medium level of instances with respect to the other tree methods analyzed here. Panel *d* shows the mean rank value and the variation of one standard deviation with respect to the mean rank. It is evident that across all datasets, WOLTReg has a lower value and, thus, the best performance in terms of ranking across all datasets. On the other hand, Panel *e* shows the mean NRMSE and the variation of one standard deviation with respect to the NRMSE across all datasets. It shows that WOLTReg has comparable performance to that of HHCART in terms of this metric, although it tends to perform slightly better on average. Finally, we can state that overall, WOLTReg has comparable performance in terms of ranking, if not better, with respect to the other OT methods. However, in terms of NRMSE, this difference is not extremely large.

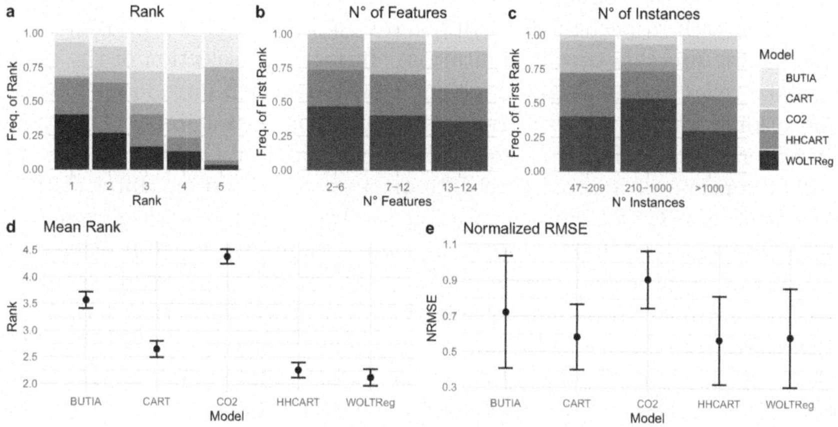

Fig. 3. Panel **a** reports the frequency of each model for being best, second, third, and worst for all the 60 datasets in the test set. Panel **b** illustrates the frequency of being the best model with respect to the number of features. Panel **c** shows the frequency of being the best model with respect to the number of instances. Panel **d** shows the error bar plot with the mean rank comparison of models across the benchmark datasets. Panel **e** shows the error bar plot with the Normalized Root Mean Square Error (NRMSE) comparison of models across the benchmark datasets.

5 Concluding Remarks

In this study, we proposed a novel method, WOLTReg, which is used to build oblique DTs for regression tasks. This OT applies an indirect search approach for finding each node's best separating hyperplane. In particular, it uses a weighted logistic classifier to estimate the hyperplane equation, and we showed that using weights leads to better splits in terms of a decrease in impurity. Moreover, it has the key aspect of implementing a variable selection process at each node that can prevent data overfitting.

Through a model evaluation on a set of benchmarking datasets, we show that WOLTReg brings the best performance regarding the predictive power and model complexity ratio, when the median is used as the quantile for the dichotomizing process and when the set of features selected is small. We also assessed the predictive performance of WOLTReg by comparing its RMSE with those obtained from other OT methods, namely HHCART, CO2, and BUTIA, as well as with a standard DT model like CART. The results show that WOLTReg, across the benchmark datasets, has the best performance in terms of ranking; in particular, there is a notable difference in the mean rank with respect to the other models across all datasets. In addition, WOLTReg has the highest percentage to be ranked first among all the models. This suggests that our novel model performs well in all combinations of numerosity of instances and features. On the other hand, when compared in terms of NRMSE, WOLTReg, has comparable performance with respect to HHCART and CART.

WOLTReg presents certain limitations. In fact, the selection of the oblique hyperplane relies on the weighted logistic classifier, which introduces additional computational complexity with respect to traditional axes-parallel DTs. Indeed, in our experiments in Sect. 4.2, cross-validating a single WOLTReg tree required a median runtime of approximately 2 s on a standard laptop (Intel i7, 16 GB RAM), while the CART implementation via the `rpart` R package required a median time of about 0.33 s under the same conditions. On top of that, the choice of the quantile and the number of selected variables could introduce a degree of sensitivity to parameter selection, especially with respect to CART. Furthermore, the variable selection process relies solely on linear correlation. While the goal is to retain the most relevant features and reduce the risk of overfitting, this approach may overlook nonlinear relationships. To address this limitation, alternative measures such as mutual information [10] could be employed. Another constraint lies in the handling of categorical variables: WOLTReg does not process them directly. They must be transformed into numerical form using techniques like one-hot encoding or frequency encoding. However, these transformations can increase the dataset's dimensionality or fail to capture certain relationships with the target variable [11, 13].

Extensions of WOLTReg may be investigated in future studies, such as incorporating it in ensemble methods such as random forests, improving feature selection approaches, or applying different classifiers to determine the oblique split. Additional understanding of its potential can also be gained through researching its applicability for specific fields, such as environmental modeling, healthcare, or finance. Overall, WOLTReg represents a promising development in nonparametric methodologies, offering a powerful tool for regression modeling in complex datasets.

Acknowledgments. This study was funded by the European Union - NextGenerationEU, Mission 4, Component 2, in the framework of the GRINS -Growing Resilient, INclusive and Sustainable project (GRINS PE00000018 CUP: F53C22000760007). The work of Luca Frigau was partially funded by the e.INS -Ecosystem of Innovation for Next Generation Sardinia (CUP: F53C22000430001). The views and opinions expressed are solely those of the authors and do not necessarily reflect those of the European Union, nor can the European Union be held responsible for them.

References

1. Agresti, A.: Categorical Data Analysis, vol. 792. John Wiley & Sons, Hoboken (2012)
2. Altman, D.G., Royston, P.: The cost of dichotomising continuous variables. BMJ **332**(7549), 1080 (2006)
3. Barros, R.C., Jaskowiak, P.A., Cerri, R., de Carvalho, A.C.: A framework for bottom-up induction of oblique decision trees. Neurocomputing **135**, 3–12 (2014)
4. Bennett, K.P., Blue, J.: A support vector machine approach to decision trees. In: 1998 IEEE International Joint Conference on Neural Networks Proceedings. IEEE World Congress on Computational Intelligence (Cat. No. 98CH36227), vol. 3, pp. 2396–2401. IEEE (1998)
5. Blaser, R., Fryzlewicz, P.: Random rotation ensembles. J. Mach. Learn. Res. **17**(1), 126–151 (2016)
6. Böhning, D.: Multinomial logistic regression algorithm. Ann. Inst. Stat. Math. **44**(1), 197–200 (1992)
7. Breiman, L., Friedman, J.H., Olshen, R.A., Stone, C.J.: Classification and Regression Trees. Chapman & Hall, New York (1984)
8. Cantú-Paz, E., Kamath, C.: Using evolutionary algorithms to induce oblique decision trees. In: Proceedings of the 2nd Annual Conference on Genetic and Evolutionary Computation, pp. 1053–1060 (2000)
9. Chaturvedi, S., Patil, S.: Oblique decision tree learning approaches-a critical review. Int. J. Comput. Appl. **82**(13) (2013)
10. Cover, T., Hart, P.: Nearest neighbor pattern classification. IEEE Trans. Inf. Theory **13**(1), 21–27 (1967)
11. Hancock, J.T., Khoshgoftaar, T.M.: Survey on categorical data for neural networks. J. Big Data **7**(1), 1–41 (2020). https://doi.org/10.1186/s40537-020-00305-w
12. Hyafil, L., Rivest, R.L.: Constructing optimal binary decision trees is NP-complete. Inf. Process. Lett. **5**(1), 15–17 (1976). https://doi.org/10.1016/0020-0190(76)90095-8
13. Kuhn, M., Johnson, K.: Feature Engineering and Selection: A Practical Approach for Predictive Models. Chapman and Hall/CRC, Boca Raton (2019)
14. Lee, Y.D., Cook, D., Park, J.W., Lee, E.K.: Pptree projection pursuit classification tree. Electron. J. Stat. **7**, 1369–1386 (2013)
15. Li, X.B., Sweigart, J.R., Teng, J.T., Donohue, J.M., Thombs, L.A., Wang, S.M.: Multivariate decision trees using linear discriminants and tabu search. IEEE Trans. Syst. Man Cybern. A Syst. Hum. **33**(2), 194–205 (2003)
16. Loh, W.Y., Shih, Y.S.: Split selection methods for classification trees. Stat. Sin. 815–840 (1997)
17. Loh, W.Y., Vanichsetakul, N.: Tree-structured classification via generalized discriminant analysis. J. Am. Stat. Assoc. **83**(403), 715–725 (1988)
18. Maalouf, M., Siddiqi, M.: Weighted logistic regression for large-scale imbalanced and rare events data. Knowl.-Based Syst. **59**, 142–148 (2014)
19. McCullagh, P.: Generalized Linear Models. Routledge, Abingdon (2019)
20. Montañana, R., Gámez, J.A., Puerta, J.M.: STree: a single multi-class oblique decision tree based on support vector machines. In: Alba, E., et al. (eds.) CAEPIA 2021. LNCS (LNAI), vol. 12882, pp. 54–64. Springer, Cham (2021). https://doi.org/10.1007/978-3-030-85713-4_6
21. Murthy, S.K., Kasif, S., Salzberg, S.: A system for induction of oblique decision trees. J. Artif. Intell. Res. **2**, 1–32 (1994)

22. Naggara, O., Raymond, J., Guilbert, F., Roy, D., Weill, A., Altman, D.G.: Analysis by categorizing or dichotomizing continuous variables is inadvisable: an example from the natural history of unruptured aneurysms. Am. J. Neuroradiol. **32**(3), 437–440 (2011)
23. Norouzi, M., Collins, M.D., Fleet, D.J., Kohli, P.: Co2 forest: Improved random forest by continuous optimization of oblique splits (2015). https://arxiv.org/abs/1506.06155
24. Olson, R.S., La Cava, W., Orzechowski, P., Urbanowicz, R.J., Moore, J.H.: Pmlb: a large benchmark suite for machine learning evaluation and comparison. BioData Min. **10**, 1–13 (2017)
25. R Core Team: R: A Language and Environment for Statistical Computing. R Foundation for Statistical Computing, Vienna, Austria (2013). https://www.R-project.org/
26. Robertson, B.L., Price, C.J., Reale, M.: Cartopt: a random search method for nonsmooth unconstrained optimization. Comput. Optim. Appl. **56**(2), 291–315 (2013)
27. Rodriguez, J.J., Kuncheva, L.I., Alonso, C.J.: Rotation forest: a new classifier ensemble method. IEEE Trans. Pattern Anal. Mach. Intell. **28**(10), 1619–1630 (2006)
28. Schober, P., Vetter, T.R.: Logistic regression in medical research. Anesth. Analg. **132**(2), 365–366 (2021)
29. Shcherbakov, M.V., Brebels, A., Shcherbakova, N.L., Tyukov, A.P., Janovsky, T.A., Kamaev, V.A., et al.: A survey of forecast error measures. World Appl. Sci. J. **24**(24), 171–176 (2013)
30. Strzelecka, A., Kurdyś-Kujawska, A., Zawadzka, D.: Application of logistic regression models to assess household financial decisions regarding debt. Procedia Comput. Sci. **176**, 3418–3427 (2020)
31. Therneau, T., Atkinson, B.: rpart: Recursive Partitioning and Regression Trees (2023). https://CRAN.R-project.org/package=rpart. R package version 4.1.23
32. Tomita, T.M., et al.: Sparse projection oblique randomer forests. J. Mach. Learn. Res. **21**(1), 4193–4231 (2020)
33. Truong, A.: Fast growing and interpretable oblique trees via logistic regression models. Ph.D. thesis, Oxford University, UK (2009)
34. Ushey, K., Allaire, J., Tang, Y.: reticulate: Interface to 'Python' (2023). https://CRAN.R-project.org/package=reticulate. R package version 1.34.0
35. Van Rossum, G., Drake, F.L.: Python 3 Reference Manual. CreateSpace, Scotts Valley, CA (2009)
36. Wickramarachchi, D.C., Robertson, B.L., Reale, M., Price, C.J., Brown, J.: Hhcart: an oblique decision tree. Comput. Stat. Data Anal. **96**, 12–23 (2016)
37. Zhan, H., Liu, Y., Xia, Y.: Consistency of oblique decision tree and its boosting and random forest (2024). https://arxiv.org/abs/2211.12653

ISP Index: A Parsimonious Method to Predict Defaults

Roberto Casarin(✉), Fausto Corradin, and Antonio Peruzzi

Ca' Foscari University of Venice, Venice, Italy
{r.casarin,fausto.corradin,antonio.peruzzi}@unive.it

Abstract. The necessity of determining the probability of default often conflicts with the requirement for employing parsimonious methodologies. The inclusion of a large number of regressors in Logit models or Machine Learning approaches can lead to overfitting, thereby introducing biases that distort the results. This study aims to examine the extent to which an indicator that synthesizes information related to balance sheet metrics can achieve a performance comparable to that obtained through a comprehensive set of indicators. In this paper, we introduce the Synthetic Performance Indicator (ISP), which is derived from specific balance sheet indicators. We demonstrate its effectiveness as a synthetic measure of financial stability in localized settings. Furthermore, we assess its potential to serve as a viable alternative to the broader panel of indicators from which it is constructed. Finally, we provide further evidence of how Machine Learning approaches, despite being effective in-sample, perform poorly out-of-sample.

Keywords: Default · Probability of Default · Logit · Machine Learning · Overfitting

1 Introduction

Credit risk analysis is a critical task for financial institutions, as it enables the assessment of the likelihood of borrower default and informs risk management decisions. The foundational contributions in this field include the work of Altman (1968), who employed linear discriminant analysis on financial indices within a multivariate framework to develop default prediction models. Subsequent studies explored alternative statistical methodologies, including cluster analysis (Gupta and Huefner, 1972), and Logit regression models (Westgaard and Van der Wijst,

This study was funded by the European Union - NextGenerationEU, in the framework of the GRINS - Growing Resilient, INclusive and Sustainable project (GRINS PE00000018 - CUP H73C22000930001), National Recovery and Resilience Plan (NRRP) - PE9 - Mission 4, C2, Intervention 1.3. The views and opinions expressed are solely those of the authors and do not necessarily reflect those of the European Union, nor can the European Union be held responsible for them.

A. D'Ambrosio et al. (Eds.): CLADAG-VOC 2025, SCDAKO, pp. 83–94, 2025.
https://doi.org/10.1007/978-3-032-03042-9_8

2001). Other prominent methodologies include Probit models (Casey et al., 1968) and survival analysis (Luoma and Laitinen, 1991).

In addition to traditional econometric approaches, researchers have increasingly compared these models with benchmark linear regression techniques such as Lasso (Tibshirani, 1996), Ridge regressions (Hoerl and Kennard, 1970), and ElasticNet (Zou and Hastie, 2005). These methods have proven effective in handling high-dimensional datasets, making them suitable for exploring the relationship between financial indicators and default probabilities.

Over the past two decades, the rise of artificial intelligence (AI) has significantly influenced credit risk modeling (Montevechi et al., 2024). Machine Learning (ML) techniques have gained widespread popularity, particularly in light of advancements in generative AI and the increasing emphasis on data-driven decision-making (Van Thiel and Van Raaij, 2019). (Azzone et al., 2022) investigated the application of ML techniques, such as Random Forest, for lapse prediction in life insurance, demonstrating their potential for financial risk modeling.

Despite these advances, three critical factors must be carefully considered in credit risk modeling: i) the selection of explanatory variables, ii) the influence of the sample on model outcomes, and iii) the choice of the cutoff threshold. Economics and financial variables, particularly financial ratios, are commonly employed in such studies. However, it is generally inadvisable to mix absolute and relative measures.

Given the wide array of available variables, factor analysis is often utilized to reduce dimensionality, preserve degrees of freedom, and mitigate multicollinearity issues while ensuring that essential financial dimensions are properly represented. Nevertheless, factor analysis is inherently sample-dependent; variations in the sample composition can lead to different variable selections and, consequently, different model specifications.

Balancing parsimony and informativeness in model specification has been a long-standing challenge in credit risk analysis. More complex models increase the risk of overfitting, necessitating careful selection of explanatory variables. This issue is particularly relevant for Logit regression models, where reducing the number of regressors helps mitigate overfitting, as well as for ML methodologies, which are equally susceptible to the curse of dimensionality. To address this issue, we introduce the Synthetic Performance Indicator (ISP), which aggregates information from eight financial indicators over a five-year period to provide a comprehensive measure of a firm's health and financial stability, relying on expert surveys. It is worth noting that expert surveys are well-established in economics and finance (see Bassetti et al., 2023 and Chen et al., 2022) and offer valuable insights, especially in cases where data alone may not capture the full complexity of domain-specific knowledge. This study evaluates the ISP as a predictor of default, comparing its performance against the forty individual indicators from which it is derived. Furthermore, we assess the ISP's ability to function as a synthetic measure that captures the informational content of multiple financial indicators. Additionally, we compare various ML methodologies, including

Decision Trees, Random Forests, and LSBoost (Breiman, 2001), against benchmark models such as Logit, ElasticNet, and the Two-Stage Adaptive Lasso.

The remainder of this paper is structured as follows. Section 2 introduces the Synthetic Performance Indicator (ISP) and outlines its construction methodology. Section 3 describes in detail the Logit model and the selected ML approaches. Section 4 presents the empirical findings, and Sect. 5 concludes the work.

2 The Synthetic Performance Indicator (ISP)

We propose the Synthetic Performance Indicator (ISP) as a composite measure designed to assess the overall performance of companies. ISP is obtained by evaluating firms' equity, financial stability, and earnings conditions over the past five fiscal years. The primary objective of ISP is to identify top-performing firms based on a standardized set of criteria.

The analysis relies on Italian financial statement data extracted from the AIDA - Bureau Van Dijk database (Bureau van Dijk, 2023) as of October 17, 2024. The dataset is comprehensive of over 180,000 firms and includes approximately 80% (by value) of enterprises within each province of the Triveneto macro-region, where *province* refers to the Italian NUTS3 administrative unit. The study focuses on active companies, i.e. firms currently in operation and not subject to insolvency or default procedures. These companies must have their registered offices in one of the following districts in the Triveneto macro-region which are: Belluno, Padova, Rovigo, Treviso, Venice, Verona, and Vicenza for the Veneto region; Bolzano and Trento for the Trentino-Alto Adige region; and Gorizia, Pordenone, Trieste, and Udine for the Friuli-Venezia Giulia region. To be included in the analysis, firms must have reported annual revenues exceeding one million euros and possess complete balance sheet data for the last five financial years.

To quantify a firm's performance using a single numerical value, we employ balance sheet indicators deemed relevant by members of Triveneto ODCEC, i.e. the Regional Professional Association of Chartered Accountants. Each fiscal year indicator is normalized so that the company with the highest value for a given indicator is assigned a score of 1,000, while the firm with the lowest value is assigned a score of 0. Once the indicators are standardized, it becomes necessary to determine the appropriate weights for each indicator. The weighting scheme was established based on responses from more than 900 chartered accountants across the 13 provinces of the Triveneto macro-region. The relative importance of the balance sheet indicators was quantified in percentage terms using a statistical methodology employed by researchers at the European Commission's Joint Research Center for constructing synthetic indicators (JRC, 2008).

The ISP comprises two main components: the Economic Performance Indicator (ISP-E) and the Financial Stability Indicator (ISP-F). The ISP-E component aggregates four financial indicators, collectively assessing a company's economic performance: i.e. ROE, EBITDA to Revenue, ROI, and Revenue Growth. The weights assigned to these indicators sum to one, and their weighted sum constitutes ISP-E. The ISP-F component includes four indicators related to the enterprise's financial stability: Leverage, Debt-to-Equity, Asset Growth and Net Financial Position to EBTDA. Also, in this case, the weights assigned to these indicators sum to one. These eight indices are well established and represent the workhorse of the balance sheet analysis (see Ross et al., 2018). The final ISP score is then obtained as a weighted sum of ISP-E and ISP-F, incorporating their respective importance determined by the surveyed chartered accountants.

Table 1. Weights assigned to ISP-E and ISP-F indicators.

Indicator	Weight	Sector Weight
Economic Performance Indicators (ISP-E)		0.49
ROE (Return on Equity)	0.2727	
EBITDA/Revenue	0.3560	
ROI (Return on Investment)	0.2386	
Revenue Growth (ΔRevenue, y_t/y_{t-1})	0.1327	
Financial Stability Indicators (ISP-F)		0.51
Leverage	0.3162	
D/E (Debt to Equity)	0.2703	
Asset Growth (ΔAsset, y_t/y_{t-1})	0.1583	
NFP/EBITDA (Net Financial Position to EBITDA)	0.2552	

Financial statement data for the last 5 fiscal years was considered to reward the company's performance over time rather than in a single year. The financial statements considered are statutory financial statements, so the rankings may include companies in the same group. Indicators for the years closest to the analysis were also weighted more heavily. The weights assigned to the various years are 0.33 for 2023, 0.27 for 2022, 0.20 for 2021, 0.13 for 2020, and 0.07 for 2019. After removing companies with missing data over the five years, we retain a resulting dataset of 14,768 companies. The set of companies can be categorised into 3 distinct groups identified by size (small, medium, and large) according to revenue, balance sheet assets and the number of employees following the European Union parameters (EU Commission Directive 2023/2775, EUComission, 2023).

To examine how the ISP behaves as a standard indicator, we consider the 14,768 enterprises with a Legal Status (the exact term in Italian is *Stato Giuridico*) of Active enterprise, i.e. that are in full working order. Then, we assess how their legal status changed in the fiscal year 2023.

To be able to carry out a more detailed analysis, we consider different Legal Statuses as default indicators. We define a dependent variable Y_i, which assumes value 1 for companies in a severely deteriorated status in 2023 and 0 otherwise (see Table 2). This way, we add the information on whether the company defaulted or not in 2023, and we link this information with their ISP value in 2019, which will be the main dependent variable of our analysis. Moreover, we include the 8 indicators used to create the ISP index for the preceding 5 years (40 regressors in total). This allows us to analyze the transition of firms' legal status over time.

Table 2. Classification of Legal Status in 2023 and corresponding dependent variable definition.

Legal Status in 2023	Dependent Variable (Y_i)
Active	0
Active (with insolvency status)	1
Active (with insolvency proceedings)	1
Discontinued	0
Liquidated (finalized bankruptcy)	1
Cancelled (merger)	0
Cancelled (in liquidation)	1
Cancelled (demerger)	0
Bankrupt	1
In liquidation	1

As a preliminary analysis, we examine the relationship between ISP values in 2019 and the subsequent firm status in 2023. A simple descriptive comparison, presented in Table 3, indicates that, on average, firms that remained Active in 2023 had a higher ISP value in 2019, compared to those that transitioned into the default status. To assess more effectively whether the ISP can explain a firm's financial stability, we employ three Logit models and three ML models, which are briefly described in Sect. 3. The dataset consists of 14,768 observations, among which only 230 correspond to instances of default, representing 1.56% of the total sample.

Table 3. Comparison of ISP values in 2019 with Legal Status in 2023.

Legal Status in 2023	Mean ISP (2019)	Number of Firms
Large Active	504.6	648
Large Default	374.4	5
Medium Active	507.4	2,818
Medium Default	449.3	24
Small Active	496.6	11,072
Small Default	465.8	201

3 Logit and Machine Learning Models

3.1 Logit, ElasticNet and 2sAdaptive Lasso

The Logit model is, without a doubt, the workhorse model for default prediction. Denote the vector of explanatory variables $\boldsymbol{x}_i$ for company i and of dimension $1 \times k$, where the first element is equal to one, and denote with Y_i the binary dependent variable, taking value 1 if firm i is in default and 0 otherwise. The probability of default is modelled using the logistic function:

$$P\left[Y_i = 1|\boldsymbol{x}_i\right] = \frac{e^{\boldsymbol{x}'_i\boldsymbol{\beta}}}{1 + e^{\boldsymbol{x}'_i\boldsymbol{\beta}}}, \tag{1}$$

where $\boldsymbol{\beta} = (\beta_1, \ldots, \beta_k)'$ is a $k \times 1$ vector of parameters. To estimate the model parameters, we maximize the log-likelihood function:

$$\log L(\boldsymbol{\beta}) = \sum_{i=1}^{N} \left[y_i \log\left(\frac{e^{\boldsymbol{x}'_i\boldsymbol{\beta}}}{1 + e^{\boldsymbol{x}'_i\boldsymbol{\beta}}}\right) + (1 - y_i)\log\left(1 - \frac{e^{\boldsymbol{x}'_i\boldsymbol{\beta}}}{1 + e^{\boldsymbol{x}'_i\boldsymbol{\beta}}}\right)\right]. \tag{2}$$

The estimated parameter vector $\hat{\boldsymbol{\beta}}$ is then used for classification, while the predicted value $\hat{Y}_i$ assumes value 1 if the estimated probability exceeds a predefined threshold, known as the *cut-off*, and assumes value 0 otherwise.

To enhance predictive performance and address multicollinearity, we employ the ElasticNet regularization, which extends the log-likelihood function by incorporating a penalty term:

$$ElasticNet(\boldsymbol{\beta}) = \log L(\boldsymbol{\beta}) - \lambda(1 - \alpha)\sum_{j=2}^{k} \beta_j^2 - \alpha \sum_{j=2}^{k} |\beta_j|. \tag{3}$$

For $\lambda = 0$, the model reduces to Lasso regression, which applies an ℓ_1 penalty, while for $\alpha = 0$, it reduces to a Ridge regression with an ℓ_2 penalty.

Ridge regression retains all predictors in the model but shrinks their coefficients, whereas Lasso performs both continuous shrinkage and automatic variable selection by forcing some coefficients to zero. ElasticNet combines these properties, effectively selecting groups of correlated variables.

We also consider the two-stage adaptive Lasso (2sAdaptiveLasso) approach to improve variable selection consistency. The objective function is given by:

$$2sAdaptiveLasso(\boldsymbol{\beta}) = \log L(\boldsymbol{\beta}) - \lambda \sum_{j=2}^{k} \left| \frac{\beta_j}{\hat{\beta}_{j,Lasso}} \right|, \tag{4}$$

where λ is determined from ElasticNet optimization and $\hat{\boldsymbol{\beta}}_{Lasso}$ represents the estimated coefficients obtained from Lasso regression.

3.2 Decision Trees, Random Forests and LSBoost

Non-parametric learning algorithms, such as decision trees, random forests, and LSBoost, have proven effective in classification tasks as they can capture non-linearities in the data structure. For this reason, we decided to adopt them as the first alternative to the logistic regression techniques. A decision tree is a tree-like structure in which each internal node represents a decision rule based on a feature, each branch represents the decision's outcome, and each leaf node represents the final prediction.

In our specification, we construct 200 decision trees, where the first tree has a maximum of one split, the second has two splits, the third has three splits, and so forth. Again, the dependent variable is Y_i, which indicates whether the firm is in default or not in 2023, and the regressor is either the ISP index for 2019 or the set of 40 balance sheet indices on which the ISP is based.

As illustrated in Fig. 1, each tree is evaluated using in-sample tests, where key performance metrics such as AUC-ROC, sensitivity, specificity, false positive rate (FPR), and false negative rate (FNR) are computed.

The optimal tree is selected based on the balance between predictive accuracy and model complexity, ensuring that the number of splits is minimized to prevent overfitting. As a result, the maximum number of splits is set to 111 for both the ISP and the 40-regressor cases.

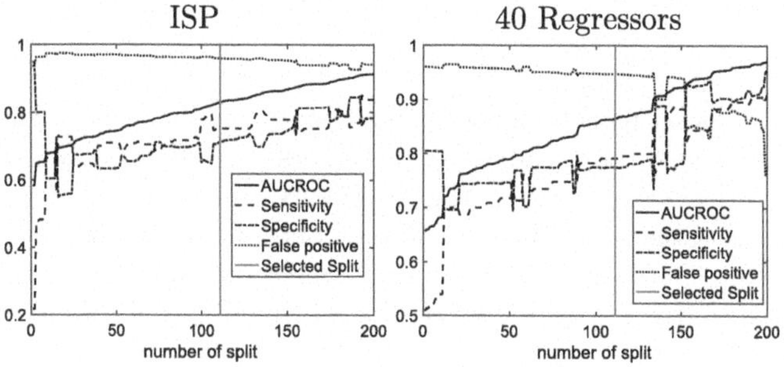

Fig. 1. Performance measures against number of splits for the ISP and the set of 40 regressors.

A Random Forest consists of multiple decision trees, where each tree depends on a randomly sampled vector of features. The final prediction is obtained by aggregating the outputs of all trees. As shown by Breiman (2001), Random Forest exhibit converging generalization error with increasing number of trees. The ensemble model's performance depends on the strength of individual trees and the correlation between them.

In addition, we employ LSBoost (Breiman, 2001), a boosting algorithm that iteratively improves predictive accuracy by minimizing the mean squared error. At each iteration, the algorithm fits a new weak learner, generally a decision tree, to the residuals representing the difference between the observed response and the aggregate prediction from previously derived learners. Unlike traditional boosting algorithms that optimize for classification loss functions, LSBoost explicitly targets regression-based loss minimization. This approach enables the model to better capture the relationship between financial indicators and firm stability while mitigating overfitting through iterative refinement.

4 Results

We report in Table 4 the coefficients obtained regressing the binary default variable on the ISP index for the Logit model, ElasticNet and 2sAdaptiveLasso. As expected, the coefficients are quite similar when only the ISP index is included as a regressor. Moreover, the slope coefficient is negative, and this is also expected: the lower the ISP index, the worse the economic and financial conditions of the firm and the higher the probability of default. For the Logit model, the coefficient is significant at 0.1%.

Table 4. Estimated coefficients for the ISP case across different models.

Coefficient	Logit	ElasticNet	2sAdLasso
Constant	0.49654	0.49639	0.49641
ISP	-0.00963^{***}	-0.00963	-0.00963

We employ all six methodologies discussed above, using both the ISP index and the 40 balance sheet variables as regressors. In each case, we compute the cutoff as the intersection between the Specificity and Sensitivity curve.

We report in Table 5 the cutoff for the six methodologies with the two distinct sets of regressors. The results indicate that the cutoff values are consistently low for both cases and across all methodologies, highlighting the extent of data imbalance.

The optimal threshold is context-dependent, as a low threshold may result in an excess of false positives, whereas a high threshold could lead to missed opportunities for retention. This trade-off emphasizes the need for a dynamic and context-aware approach to model evaluation and threshold setting. The model can be fine-tuned by carefully adjusting these thresholds to perform default detection.

4.1 In-Sample Forecast

We report the results for the ISP case and for the 40 regressors case in Table 6. We can notice that using ISP as a regressor allows us to obtain results comparable to the case in which we use the original 40 balance sheet regressors. Some discrepancies in terms of AUCROC can be found using the two sets of regressors with the Logit model, ElasticNet and 2sAdaptiveLasso. At the same time,

Table 5. Cutoff values for ISP and 40-regressor models across different methodologies.

Regressors	Logit	ElasticNet	2sAdLasso	Random Forest	LSBoost	Decision Tree
ISP	0.0140	0.0140	0.0140	0.0010	0.0120	0.0130
40 Regressors	0.0140	0.0140	0.0140	0.0010	0.0691	0.0050

Decision Tree, Random Forest and LSBoost display a lower discrepancy. Further results with the introduction of polynomial terms have been conducted without relevant performance improvements and are available upon request.

Table 6. Performance comparison of different models for the ISP and 40-regressor cases.

Case	Parameters	Logit	ElasticNet	2sAdLasso	Random Forest	LSBoost	Decision Tree
ISP	Sensitivity	0.64348	0.64348	0.64348	1.00000	0.83043	0.75217
	Specificity	0.61824	0.61824	0.61824	1.00000	0.82618	0.71385
	FPR	0.97403	0.97403	0.97403	0.00000	0.92973	0.96007
	FNR	0.00904	0.00904	0.00904	0.00000	0.00324	0.00546
	RCC	0.61863	0.61863	0.61863	1.00000	0.82625	0.71445
	AUC-ROC	0.66297	0.66297	0.66297	1.00000	0.92961	0.83151
40 Regressors	Sensitivity	0.69565	0.66522	0.66522	1.00000	0.84348	0.79565
	Specificity	0.69253	0.69542	0.69164	1.00000	0.94339	0.76083
	FPR	0.96544	0.96660	0.96700	0.00000	0.79135	0.95000
	FNR	0.00690	0.00756	0.00760	0.00000	0.00095	0.00423
	RCC	0.69258	0.69495	0.69122	1.00000	0.94339	0.76138
	AUC-ROC	0.74058	0.73713	0.73915	1.00000	0.99153	0.86255

4.2 Resampling Excercise

As we acknowledge that our dataset may be unbalanced, we try to mitigate its impact on estimation. To do so, we implement a resampling strategy on the pool of observations, as suggested in Salas-Eljatib et al. (2018).

Since we observe only 230 default events, we set a sample size of 690 observations (3 times the number of defaults), and we sample 460 non-default observations while retaining the 230 default events. This way, we maintain a fixed proportion of default and no-default events. We run the Logit regression within the subsample, and we repeat this procedure for $s = 10,000$ simulation. Finally, we compute $\hat{\boldsymbol{\beta}}_{re} = s^{-1} \sum_{h=1}^{s} \hat{\boldsymbol{\beta}}_{re,h}$ where $\hat{\boldsymbol{\beta}}_{re,h}$ is the vector of coefficient values for the h-th iteration and we use this value to assess the Logit performance. We report in Table 7 the performance assessment of the Logit using this resampling strategy. We can see how this procedure provides results similar to the ones obtained by the standard Logit.

Table 7. Resampling Performance for ISP and 40 regressors.

Parameters	Resampling ISP	Resampling 40
Sensitivity	0.63043	0.67826
Specificity	0.63365	0.67726
FPR	0.97350	0.96782
FNR	0.00914	0.00750
RCC	0.63360	0.67728
AUCROC	0.66300	0.72630

4.3 Out-Sample Forecast

It is essential to assess the performance of the proposed models using appropriate validation techniques. One commonly used method is k-fold cross-validation, which is used to estimate the performance of a model by training it multiple times on different subsets of the data (training set) and testing it on the excluded portions (learning set). This technique provides a more reliable estimate of how

Table 8. Out-of-sample performance comparison for ISP and 40-regressor models across different methodologies. The top panel reports confusion matrix counts, while the bottom panel reports performance metrics using k-fold validation.

Top Panel: Confusion Matrix Counts							
Case	**Parameters**	**Logit**		**Decision Tree**		**Random Forest**	
		$Y=1$	$Y=0$	$Y=1$	$Y=0$	$Y=1$	$Y=0$
ISP	$\hat{Y}=1$	143	5380	84	3392	7	208
	$\hat{Y}=0$	87	9158	146	11146	223	14330
	Total	230	14538	230	14538	230	14538
40 Regressors	$\hat{Y}=1$	142	4414	66	2335	0	0
	$\hat{Y}=0$	88	10124	164	12203	230	14538
	Total	230	14538	230	14538	230	14538

Bottom Panel: k-fold Validation Metrics				
Case	**Parameters**	**Logit**	**Decision Tree**	**Random Forest**
ISP	Sensitivity	0.62174	0.36522	0.03043
	Specificity	0.62994	0.76668	0.98569
	FPR	0.97411	0.97583	0.96744
	FNR	0.00941	0.01293	0.01532
	RCC	0.62981	0.76043	0.97082
40 Regressors	Sensitivity	0.61739	0.28686	0.00000
	Specificity	0.69638	0.83939	1.00000
	FPR	0.96883	0.97251	-
	FNR	0.00862	0.01326	0.01557
	RCC	0.69515	0.83078	0.98443

the model will perform on unseen data. The basic idea of k-$fold$ is to divide the dataset into k equal-sized folds (subsets), and for each iteration, we trained the model on $k-1$ folds and tested on the remaining fold; in our case, $k=5$. In Table 8, we report the k-fold performance assessment. We noticed that the ML procedure does not perform as well as in the in-sample exercise because of overfitting issues; we confirm that the performance of ISP is similar to that of the 40 regressors.

5 Discussion and Conclusion

In this work, we introduced ISP as an expert-based indicator of economic and financial stability that can be applied to different localized settings. Our ISP indicator represents a parsimonious alternative to the use of many regressors in predicting defaults, also in rebalanced scenarios. We have proven the effectiveness of our approach in the localized context of the Triveneto macro-region in Italy. Nonetheless, we think that our approach can be easily exported to other contexts at the regional or national level whenever experts' understanding of the environment plays a crucial role.

While our methodology leverages expert input to reflect local knowledge, we recognize that this introduces a degree of subjectivity. This risk may be justified when domain-specific insights cannot be fully inferred from available data. Moreover, the ISP relies on a fixed-weight structure derived from expert judgment. We recognize that this may not fully capture shifts in financial conditions or broader economic dynamics over time. However, this choice reflects a practical trade-off: conducting new expert surveys annually would be resource-intensive, while the relative importance of the core indicators tends to evolve gradually. To address this, we incorporated a dynamic adjustment component that places greater weight on more recent observations, allowing the index to partially adapt to changing conditions. Of course, other indices can always be added, either directly (as a control variable) or indirectly (via experts' assessment), considering the trade-off between effectiveness and parsimony.

The ISP framework could be expanded in several directions. First, integrating data-driven or ML approaches to update indicator weights could enhance the index's responsiveness to changing economic conditions while preserving its interpretability. Second, future implementations may benefit from collecting and incorporating demographic information about the experts involved in the weighting process. Third, one could introduce non-linearities, such as multiplicative combinations, and threshold effects between financial ratios. Finally, access to a larger set of firms would allow for a more granular sectoral analysis, enabling the development of sector-specific ISPs.

References

Altman, E.I.: Financial ratios, discriminant analysis and the prediction of corporate bankruptcy. J. Financ. **23**(4), 589–609 (1968)

Azzone, M., Barucci, E., Moncayo, G.G., Marazzina, D.: A machine learning model for lapse prediction in life insurance contracts. Expert Syst. Appl. **191**, 116261 (2022)

Bassetti, F., Casarin, R., and Del Negro, M.: Inference on probabilistic surveys in macroeconomics with an application to the evolution of uncertainty in the survey of professional forecasters during the covid pandemic. In: Handbook of Economic Expectations, pp. 443–476. Elsevier (2023)

Breiman, L.: Random Forests. Mach. Learn. **45**, 5–32 (2001)

van Dijk, B.: AIDA Database (2023). Accessed 17 Oct 2025

Casey, C.J., McGee, V.E., and Stickney, C.P.: Discriminating between Reorganized and Liquidated Firms in Bankruptcy. Account. Rev., 249–262 (1986)

Chen, R., et al.: Composite Index Construction with Expert Opinion. J. Bus. Econ. Statist. **41**(1), 67–79 (2022)

EU-Comission,: Commission Delegated Directive (EU) 2023/2775 of 17 October 2023 amending Directive 2013/34/EU of the European Parliament and of the Council As Regards the Adjustments of the Size Criteria for Micro. Small, Medium-sized and Large Undertakings or Groups (2023)

Gupta, M.C., Huefner, R.J.: A cluster analysis study of financial ratios and industry characteristics. J. Accounting Res. 77–95 (1972)

Hoerl, A.E., Kennard, R.W.: Ridge regression: applications to nonorthogonal problems. Technometrics **12**(1), 69–82 (1970)

JRC. Handbook on Constructing Composite Indicators: Methodology and User Guide. OECD publishing (2008)

Luoma, M., Laitinen, E.K.: Survival analysis as a tool for company failure prediction. Omega **19**(6), 673–678 (1991)

Montevechi, A.A., de Carvalho Miranda, R., Medeiros, A.L., Montevechi, J.A.B.: Advancing credit risk modelling with machine learning: a comprehensive review of the State-of-the-art. Eng. Appl. Artif. Intell. **137**, 109082 (2024)

Ross, S.A., Westerfield, R.W., Jordan, B.: Corporate Finance, 2010. Edition Hardcover (2018)

Salas-Eljatib, C., Fuentes-Ramirez, A., Gregoire, T.G., Altamirano, A., Yaitul, V.: A study on the effects of unbalanced data when fitting logistic regression models in ecology. Ecol. Ind. **85**, 502–508 (2018)

Tibshirani, R.: Regression shrinkage and selection via the lasso. J. R. Stat. Soc. Ser. B Stat Methodol. **58**(1), 267–288 (1996)

Van Thiel, D., Van Raaij, W.F.F.: artificial intelligence credit risk prediction: an empirical study of analytical artificial intelligence tools for credit risk prediction in a digital era. J. Risk Manag. Financial Institut. **12**(3), 268–286 (2019)

Westgaard, S., Van der Wijst, N.: Default probabilities in a corporate bank portfolio: a logistic model approach. Eur. J. Oper. Res. **135**(2), 338–349 (2001)

Zou, H., Hastie, T.: Regularization and variable selection via the elastic net. J. R. Stat. Soc. Ser. B Stat Methodol. **67**(2), 301–320 (2005)

Stability Post-processing for Items Importance in Preference Learning via the Bayesian Mallows Model

Luca Coraggio[1(✉)] and Valeria Vitelli[2]

[1] Department of Economics and Statistics, University of Naples Federico II, Naples, Italy
luca.coraggio@unina.it
[2] Oslo Centre for Biostatistics and Epidemiology, Department of Biostatistics, Institute of Basic Medical Sciences, University of Oslo, Oslo, Norway
https://luca-coraggio.com

Abstract. Rank and preference data are becoming increasingly ubiquitous, stimulating continuous advances in preference learning methodologies and their wider adoption across different domains. The Lower-dimensional Bayesian Mallows Models with Mixtures (LowBM3) is a recent extension of the Bayesian Mallows Models, which was originally developed as a unifying Bayesian framework to estimate the Mallows model. LowBM3 extends the Bayesian Mallows Model to ultra-high-dimensional settings, allowing to estimate a clustering of the assessors, and the within-cluster sets of relevant items and their consensus rankings. In this paper, we propose a novel post-processing strategy for LowBM3, named stability post-processing. We validate our methodology through experimental analysis and demonstrate its superior performance in specific settings characterized by significant variability in absolute rankings across assessors.

Keywords: Bayesian inference · clustering · variable selection · rankings · high-dimensional data

1 Introduction

Preference data arise in applications where items are organized by preference or importance by assessors, and are common in social sciences, marketing research, recommendation systems, and political science. The growing availability of preference data from online platforms has renewed interest in the field of preference learning. Key tasks in preference learning include estimating consensus rankings, grouping assessors into classes, and classifying new assessors based on preferences. Preference learning methods based on probabilistic models typically rely on the Placket-Luce [6,10] or Mallows model [7], even if settings based on, e.g., the Bradley-Terry-Luce model [9] are being proposed.

A. D'Ambrosio et al. (Eds.): CLADAG-VOC 2025, SCDAKO, pp. 95–106, 2025.
https://doi.org/10.1007/978-3-032-03042-9_9

The increasing popularity of ranking data, along with advances in preference learning, has stimulated the adoption of these techniques across various domains. A notable example is the analysis of genomics data [1–3,11], where transforming gene expression measurements in rankings handles non-normal distributions and preserves information about relative expression levels. The ultra-high-dimensional nature of genomics data poses a substantial challenge for preference learning. The Lower-dimensional Bayesian Mallows Model with Mixtures (LowBM3) [4] was developed to analyze clustered ranked data in such settings, building on the Bayesian Mallows Model [12,13] and Lower-dimensional Bayes Mallows Model [5].

LowBM3 employs a Metropolis-Hastings scheme to estimate parameters such as assessor clusters and within-cluster sets of relevant items, with their corresponding consensus rankings. Posterior summaries based on a *selection post-processing* criterion were proposed for LowBM3 [4], and proved to be effective in giving parameter estimates in settings where relevant items are top-ranked, but achieved poorer performances in other analyzed scenarios. This paper proposes a novel *stability post-processing* approach to posterior summary, devised to address these limitations. Replicating the empirical analysis in Eliseussen et al. [4], we show that this new proposal achieves superior performance results in settings with high noise levels in absolute item rankings.

The rest of the paper is organized as follows. Section 2 reviews LowBM3 and introduces the stability post-processing. Section 3 describes simulation designs and compares both post-processing approaches. Finally, Sect. 4 concludes the manuscript.

2 Methodology

Consider a finite set of n items, denoted as $\mathcal{A} = \{A^1, A^2, \ldots, A^n\}$. Assume that all of the n items are ranked by a set of N assessors, giving rise to complete rankings $\boldsymbol{R}_j = \{R_{1j}, R_{2j}, \ldots, R_{nj}\}$, for $j \in \{1, \ldots, N\}$. In ultra-high-dimensional settings, n is extremely large, and it is reasonable to assume that only relatively few items, $n^* \ll n$, are being ranked in the mind of assessors, while the other $n - n^*$ items are not. Assume that there is a set $\mathcal{A}^*$ of *relevant items* ($\mathcal{A}^* \subset \mathcal{A}$ and $|\mathcal{A}^*| = n^* \ll n$), and a set of irrelevant items $\mathcal{A} \setminus \mathcal{A}^*$ ($\mathcal{A} \setminus \mathcal{A}^* \subset \mathcal{A}$ and $|\mathcal{A} \setminus \mathcal{A}^*| = n - n^*$). In the setting of Eliseussen et al. [4], $\mathcal{A}^*$ contains items that the assessors consider worth ranking, and their rank is modeled with the Mallows model, around an unknown consensus ranking, $\boldsymbol{\rho}$. In contrast, $\mathcal{A} \setminus \mathcal{A}^*$ contains items that assessors do not wish to rank, but rather receive a noise ranking, modeled with a uniform distribution over the relevant permutation set. As both the set of relevant items ($\mathcal{A}^*$) and the consensus ranking ($\boldsymbol{\rho}$) are unknown, the aim is to infer them from the observed ranking ($\boldsymbol{R}$) from all the N assessors.

LowBM3 extends the framework above, allowing for the possibility of assessors being clustered together into C different groups. Each group of assessors has a cluster-specific set of n^* relevant items ($\mathcal{A}^*_c$) and consensus ranking ($\boldsymbol{\rho}_c$) both to be estimated for each group of assessors ($c \in \{1, \ldots, C\}$). Under these conditions, the likelihood of observing $\mathbb{D}_N = \{\boldsymbol{R}_1, \ldots, \boldsymbol{R}_N\}$ ranking is:

$$P\left(\mathbb{D}_N|(\boldsymbol{\rho}_c, \mathcal{A}_c^*, \tau_c)_c, (z_j)_j\right) = \prod_{j=1}^{N} \frac{\exp\left(-\frac{\alpha_{z_j}}{n^*} d_{\mathcal{A}_{z_j}^*}(\boldsymbol{R}_j, \boldsymbol{\rho}_{z_j})\right)}{Z_{n^*}(\alpha_{z_j})} U_{\mathcal{P}_{n-n^*}}(\boldsymbol{R}_{j|\mathcal{A}\setminus\mathcal{A}^*}) \mathbb{I}_{\{\boldsymbol{R}_j \in \mathcal{P}_n\}}. \tag{1}$$

In Eq. (1), the subscript z_j indicates the group to which assessor j belongs to: $z_j \in \{1, \ldots, C\}$; all quantities indexed by it are cluster-specific. τ_c is the unconditional probability for an assessor to belong to the c-th cluster. $\mathbb{I}_{\{.\}}$ is the indicator function and $\mathcal{P}_n$ is the set of all permutations of n integers; the last term in (1) requires $\boldsymbol{R}_j$ to be a proper complete ranking of the objects in $\mathcal{A}$, even if extension to partial rankings would be trivial. The notation $\boldsymbol{R}_{j|\mathcal{X}}$ indicates the projection of the ranks $\boldsymbol{R}_j$ to the lower-dimensional space $\mathcal{X} \subset \mathcal{A}$. That is, the complete ranking in $\boldsymbol{R}_j$ induces a ranking on the items in $\mathcal{X}$, by means of their partial ordering in the complete ranking.[1] Thus, while $\boldsymbol{R}_j \in \mathcal{P}_n$, if $|\mathcal{X}| = n' \leq n$, it follows that $\boldsymbol{R}_{j|\mathcal{X}} \in \mathcal{P}_{n'}$. The distribution $U_{\mathcal{P}_{n-n^*}}$ represents the uniform distribution over the space $\mathcal{P}_{n-n^*}$, reflecting the assumption that irrelevant items receive a noise ranking. The fractional term in (1) corresponds to the Mallows model applied to relevant items. The term $d_{\mathcal{A}_{z_j}^*}$ denotes a right-invariant distance [8]; the footrule distance is used in Eliseussen et al. [4]) in the lower-dimensional space, $\mathcal{A}_{z_j}^*$. This distance measures the deviation between $\boldsymbol{R}_{j|\mathcal{A}_{z_j}^*}$, the projected rank, and $\boldsymbol{\rho}_{z_j}$, the consensus rank. Finally, α_{z_j} and $Z_{n^*}(\alpha_{z_j})$ are the Mallows model precision parameter and partition function, respectively.

We assume uniform priors for $\boldsymbol{\rho}_c$ and $\mathcal{A}_c^*$:

$$P(\boldsymbol{\rho}_c|\mathcal{A}^*) = \frac{1}{n^*!}\mathbb{I}_{\{\rho \in \mathcal{P}_{n^*}\}}; \qquad P(\mathcal{A}^*) = \frac{1}{\mathcal{C}}\mathbb{I}_{\{\mathcal{A}^* \in \mathcal{C}\}},$$

where $\mathcal{C}$ is the collection of all $\binom{n}{n^*}$ possible sets of n^* items chosen from n objects. The prior related to the mixture parameters and cluster assignment are:

$$P(z_1, \ldots, z_N|\tau_1, \ldots, \tau_C) = \prod_{j=1}^{N} \tau_{z_j}; \qquad P(\tau_1, \ldots, \tau_C) = \Gamma(\psi C)\Gamma(\psi^{-C}).$$

Under the assumption that data are full rankings, and removing terms not depending on model parameters, the posterior takes the form [4]:

$$P((\boldsymbol{\rho}_c, \mathcal{A}_c^*, \tau_c)_c, (z_j)_j|\mathbb{D}_N) \propto \prod_{j=1}^{N} \exp\left(-\frac{\alpha_{z_j}}{n^*} d_{\mathcal{A}_{z_j}^*}(\boldsymbol{R}_j, \boldsymbol{\rho}_{z_j})\right) \prod_{c=1}^{C} \mathbb{I}_{\{\boldsymbol{\rho}_c \in \mathcal{P}_{n^*}\}} \mathbb{I}_{\{\mathcal{A}^* \in \mathcal{C}\}}. \tag{2}$$

[1] For example, say $\mathcal{A} = (A, B, C, D, E)$, and $\boldsymbol{R}_j = (1, 4, 2, 3, 5)$. If $\mathcal{X} = (B, C, D)$, then $\boldsymbol{R}_{j|\mathcal{X}} = (3, 1, 2)$.

The LowBM3 algorithm, detailed in Algorithm 1, is employed for posterior inference on the parameters. It is a Markov-Chain Monte Carlo (MCMC) algorithm, where Gibbs sampling steps are employed to update τ_c and z_j, while Metropolis-Hastings steps update $\boldsymbol{\rho}_c$ and $\mathcal{A}_c^*$. New consensus ranking values are proposed via the so-called *leap-and-shift* distribution, introduced in Vitelli et al. [13]; new $\mathcal{A}_c^*$'s are instead proposed by swapping uniformly at random few items in $\mathcal{A}_{c,m}^*$ with items in $(\mathcal{A} \setminus \mathcal{A}^*)_{c,m}$, where m denotes the m-th iteration in the MCMC chain [4].

In what follows, we describe a novel post-processing strategy, named *stability*, that we use to produce posterior summaries (or equivalently, estimates of the posterior parameters) of interest from the MCMC chain.

2.1 Stability Post-processing

Algorithm 1 produces a series of M samples from the posterior distribution of parameters, which we denote with an m subscript: $(\boldsymbol{\rho}_{c,m}, \mathcal{A}_{c,m}^*)_{m=1}^M$. For any given cluster $c \in \{1, \ldots, C\}$ and iteration $m \in \{1, \ldots, M\}$, we have $\boldsymbol{\rho}_{c,m} = (\rho_{c,m}^{(i_1)}, \ldots, \rho_{c,m}^{(i_{n^*})})$ and $\mathcal{A}_{c,m}^* = (A_{c,m}^{(i_1)} \ldots, A_{c,m}^{(i_{n^*})})$, where the paramter $\rho_{c,m}^{(i_1)}$ is the rank assigned to item $A_{c,m}^{(i_1)}$ in the consensus ranking of the c-th cluster at the m-th iteration. At each MCMC iteration, we also estimate $z_{j,m}$ for all assessors and $\tau_{c,m}$ for all clusters.

Analogously to the *selection* post-processing [4,5], the *stability* post-processing method produces posterior summaries $\hat{\boldsymbol{\rho}}_c$ and $\hat{\mathcal{A}}_c^*$ from the MCMC chain. It is essentially composed of four steps. To facilitate the description of the quantities of interest, let us define the matrix $W^c \in [0,1]^{M \times n}$ to be such that the (m, i) entry is an indicator of item A^i being selected and ranked in the m-th posterior sample and in the c-th cluster: $W_{m,i}^c = \mathbb{I}_{\{A^i \in \mathcal{A}_{c,m}^*\}}$. Additionally, let $\text{rank}(x_1, \ldots, x_n)$, for $x_i \in \mathbb{R}$, be the rank vector of vector $(x_1, \ldots, x_n)$, defined as

$$\text{rank}(x_1, \ldots, x_n) = (r_1, \ldots, r_n), \text{ with } r_i = \sum_{j=1}^{n} \mathbb{I}_{\{x_i - x_j \geq 0\}}.$$

Step 1. First, identify the Highest Probability Set (HPS) $\mathcal{A}_c'$ as the set of the items that appear most frequently in the posterior sample sets $\mathcal{A}_{c,m}^*$. Formally, for some fixed $k \in \{n^*, \ldots, n\}$,

$$\mathcal{A}_c' = \left\{ A^i \in \mathcal{A},\, i \in \{1, \ldots, n\} : \text{rank}(-\bar{\boldsymbol{w}}^c)_i \leq k \right\}, \tag{5}$$

where $\bar{\boldsymbol{w}}^c = (\bar{w}_1^c, \ldots, \bar{w}_n^c)$, and $\bar{w}_i^c = \frac{1}{M} \sum_{m=1}^{M} W_{m,i}^c$. The HPS set can be equivalently defined using a threshold $p \in [0,1]$ on vector $\bar{\boldsymbol{w}}^c$, filtering out those items whose frequency of inclusion is lower than p. For any k, it is always possible to find a value p that includes up to k items in $\mathcal{A}_c'$, and viceversa.

Algorithm 1: LowBM3

Input: $\mathbb{D}_N$; N; n^*; C; $d(\cdot,\cdot)$; l; L; M; ψ; $\alpha_1,\ldots,\alpha_C$; $Z_{n^*}(\alpha_1),\ldots,Z_{n^*}(\alpha_C)$.

Initialization: randomly generate $\boldsymbol{\rho}_{1,0},\ldots,\boldsymbol{\rho}_{C,0}$; $\mathcal{A}^*_{1,0},\ldots,\mathcal{A}^*_{C,0}$;

$\tau_{1,0},\ldots,\tau_{C,0}$; $z_{1,0},\ldots,z_{N,0}$.

for $m \leftarrow 1$ ***to*** M **do**

- **Gibbs step: update** $\tau_1,\ldots,\tau_C$
- compute: $n_c = \sum_{j=1}^{N} \mathbb{I}_{\{z_{j,m-1}=c\}}$, for $c \in \{1,..,C\}$
- sample: $\tau_{1,m},\ldots,\tau_{C,m} \sim \mathcal{D}(\psi+n_1,\ldots,\psi+n_C)$
- **for** $c \leftarrow 1$ ***to*** C **do**
  - **M-H step: update** $\boldsymbol{\rho}_c$
  - sample: $\boldsymbol{\rho}'_c \sim$ *leap-and-shift*$(\boldsymbol{\rho}_{c,m-1}, l)$ restricted on $\mathcal{A}^*_{c,m-1}$; $u \sim \mathcal{U}(0,1)$
  - compute: *ratio* $\leftarrow$ Equation (3)
  - (with $\boldsymbol{\rho}_c \leftarrow \boldsymbol{\rho}_{c,m-1}$, $\mathcal{A}^*_c \leftarrow \mathcal{A}^*_{c,m-1}$, $z_j \leftarrow z_{j,m-1}$)
  - **if** $u <$ *ratio* **then**
    - $\boldsymbol{\rho}_{c,m} \leftarrow \boldsymbol{\rho}'_c$
  - **else**
    - $\boldsymbol{\rho}_{c,m} \leftarrow \boldsymbol{\rho}_{c,m-1}$
  - **M-H step: update** $\mathcal{A}^*_c$
  - sample: $L_c \sim \mathcal{U}(0,L)$, $\mathcal{A}^*_{prop} \sim q(\mathcal{A}^*_{prop}|\mathcal{A}^*_{c,m-1}, L_c)$; $u \sim \mathcal{U}(0,1)$
  - compute: *ratio* $\leftarrow$ Equ. (4)
  - (with $\boldsymbol{\rho}_c \leftarrow \boldsymbol{\rho}_{c,m}$, $\mathcal{A}^*_c \leftarrow \mathcal{A}^*_{c,m-1}$, $z_j \leftarrow z_{j,m-1}$)
  - **if** $u <$ *ratio* **then**
    - $\mathcal{A}^*_{c,m} \leftarrow \mathcal{A}^*_{prop}$
  - **else**
    - $\mathcal{A}^*_{c,m} \leftarrow \mathcal{A}^*_{c,m-1}$
- **Gibbs step: update** $z_1,\ldots,z_N$
- **for** $j \leftarrow 1$ ***to*** N **do**
  - compute: $p_{cj} = \frac{\tau_{c,m}}{Z_{n^*}(\alpha_c)} \exp\left[-\frac{\alpha_c}{n^*} d_{\mathcal{A}^*_c}(\mathbf{R}_j, \boldsymbol{\rho}_{c,m})\right]$ for $c = 1,\ldots,C$
  - sample: $z_{j,m} \sim \mathcal{M}(p_{1j},\ldots,p_{Cj})$

Output: posterior distribution of $\boldsymbol{\rho}_c$, $\mathcal{A}^*_c$ and τ_c, for $c \in \{1,\ldots C\}$, and z_j for $j \in \{1,\ldots N\}$

Reference Eqs.[4]:

$$\min\left\{1, \frac{P_l\left(\boldsymbol{\rho}_c \mid \boldsymbol{\rho}'_c\right)}{P_l\left(\boldsymbol{\rho}'_c \mid \boldsymbol{\rho}_c\right)} \exp\left[-\frac{\alpha_c}{n^*}\left(\sum_{j:z_j=c} d_{\mathcal{A}^*_c}\left(\boldsymbol{R}_j, \boldsymbol{\rho}'_c\right) - \sum_{j:z_j=c} d_{\mathcal{A}^*_c}\left(\boldsymbol{R}_j, \boldsymbol{\rho}_c\right)\right)\right]\right\}, \tag{3}$$

(P_l refers to the probability mass function associated with moving from one ranking vector to another ranking vector in an n^*-dimensional space, according to the leap-and-shift proposal).

$$\min\left\{1, \exp\left[-\frac{\alpha_c}{n^*}\left(\sum_{j:z_j=c} d_{\mathcal{A}^*_{c,\text{prop}}}\left(\boldsymbol{R}_j, \boldsymbol{\rho}_c\right) - \sum_{j:z_j=c} d_{\mathcal{A}^*_c}\left(\boldsymbol{R}_j, \boldsymbol{\rho}_c\right)\right)\right]\right\}, \tag{4}$$

($\mathcal{A}^*_{c,\text{prop}}$ is a proposal obtained swapping uniformly at random L items in $\mathcal{A}^*_{c,m-1}$ with items in $(\mathcal{A} \setminus \mathcal{A}^*)_{c,m-1}$).

Step 2. Given $\mathcal{A}'_c$, the estimate of the set of n^* relevant items ($\hat{\mathcal{A}}^*_c$) is based on the item's posterior probabilities of being assigned neighboring ranks. More specifically, fix a cluster c, and denote with $\rho_{c,m}(A^i)$ the rank assigned to object A^i in the m-th MC sample. That is $\rho_{c,m}(A^i) = \rho_{c,m}^{(i_{A^i})}$ if $A^i \in \mathcal{A}_{c,m}$, and is otherwise equal to 0. Then, for each object $A^i \in \mathcal{A}'_c$, compute the probability of object A^i of being assigned rank $j \in \{1, \dots, n^*\}$ as

$$p^c_{ji} = \frac{\sum_{m=1}^{M} \mathbb{I}_{\{\rho_{c,m}(A^i)=j\}} W^c_{m,i}}{\sum_{m=1}^{M} W^c_{m,i}}, \qquad i \in \{1, \dots, k\}, \; k = |\mathcal{A}'_c|$$

(multiplying by elements of W^c allows considering only those samples where the object A^i is actually assigned a rank). For all c, fix an integer $\xi \in \{0, 1, \dots, n^* - 1\}$ and compute the matrix Γ_c, of dimensions $(n^* - \xi) \times k$ as follows:

$$\Gamma_c = (\gamma^c_{ji}) = \sum_{l=j}^{j+\xi} p^c_{ji}, \qquad j \in \{1, \dots, n^* - \xi\}, i \in \{1, \dots, k\}, \tag{6}$$

so that element γ^c_{ji} is the sample probability that object $A^i \in \mathcal{A}'_c$ is being assigned a rank in $[j, j + \xi]$. Finally, for each object A^i, consider only those ranks for which this probability is maximal, that is, compute the vector

$$\bar{\gamma}_c = (\bar{\gamma}_c^{(1)}, \dots, \bar{\gamma}_c^{(k)}), \qquad \bar{\gamma}_c^{(i)} = \max_{j \in \{1, \dots, n^* - \xi\}} \gamma^c_{ji}.$$

Step 3. Then, the estimate $\hat{\mathcal{A}}^*_c$ is obtained considering only the n^* objects achieving the highest value of $\bar{\gamma}_c^{(i)}$:

$$\hat{\mathcal{A}}^*_c = \left\{ A^i \in \mathcal{A}'_c : \text{rank}(-\bar{\gamma}_c)_i \leq n^* \right\}.$$

Remark 1. Note that the parameter ξ is an actionable parameter of the stability approach, with increasing ξ leading to greater stability (as more neighboring rank values are considered), and with $\xi = 0$ corresponding to no stability considered.

Step 4. Once the set $\hat{\mathcal{A}}^*_c$ is defined, the consensus rank ($\hat{\boldsymbol{\rho}}_c$) is estimated by ranking the mean ranks of objects $A^i \in \hat{\mathcal{A}}^*_c$:

$$\hat{\boldsymbol{\rho}}_c = \text{rank}(\rho_c(A^1), \dots, \rho_c(A^{n^*})), \qquad \rho_c(A^i) = \frac{\sum_{m=1}^{M} \rho_{c,m}(A^i) W^c_{m,i}}{\sum_{m=1}^{M} W^c_{m,i}}.$$

The stability approach prioritizes items whose ranks remain relatively unchanged along the MCMC chain. This contrasts with the selection approach [4], which identifies as relevant the items with the lowest average rank throughout the MCMC chain. The two approaches are appropriate for different scenarios, as will be clear in Sect. 3.

Remark 2. While the selection approach is most effective in datasets where relevant items are ranked first, the stability approach is more suitable for scenarios in which the relevant items are not necessarily those with the lowest ranks but rather those consistently ranked across assessors. This second scenario may arise, for instance, in situations where certain items clearly dominate others, yet the primary focus is on mid-preferred items. In such cases, the stability post-processing approach accounts for the fact that LowBM3 estimates the set of relevant items ($\hat{\mathcal{A}}_c^*$) with greater accuracy compared to the consensus ranks ($\hat{\boldsymbol{\rho}}_c$) [4], as estimated ranks may deviate slightly from their true values; this effect is more pronounced when the relevant items are not among the top-ranked ones.

3 Experimental Analysis

This section presents an experimental analysis comparing stability and selection post-processing methodologies. The comparison employs various simulated designs specifically constructed to elucidate the relative strengths and limitations of both approaches.

3.1 Simulated Designs

In our experiments, we devise two main Data Generating Processes (DGPs). They are named *toprank* and *consistency*, and they are designed to investigate the intuition exposed in Remark 2. The toprank DGP has relevant items ranked in top positions. In the consistency DGP, the relevant items are a set of objects that have adjacent ranks, irrespective of their absolute ranking. This mimics the idea that while assessors may agree on the relative ranking of relevant items, they may disagree on their absolute ranking relative to the full set of relevant and irrelevant items.

Both DGPs randomly divide N assessors into C clusters, each containing N_c assessors, with $z_j = c$ for $c \in \{1, \ldots, C\}$ and $j \in \{1, \ldots, N\}$. For each cluster, a n^*-dimensional consensus ranking $\boldsymbol{\rho}_c$ is generated uniformly at random, $\boldsymbol{\rho}_c \sim \mathcal{U}(\mathcal{P}_{n^*})$, along with a true set of relevant items $\mathcal{A}_c^*$, chosen randomly from n items. The ranking $\boldsymbol{\rho}_c$ determines the order of items in $\mathcal{A}_c^*$.

In the toprank DGP, N_c i.i.d. rank vectors are generated for each cluster using a Mallows model, $\boldsymbol{R}_{j|\mathcal{A}_c^*} \sim \text{Mallows}(\boldsymbol{\rho}_c, \alpha_c)$. The $n - n^*$ irrelevant items in $\mathcal{A} \setminus \mathcal{A}_c^*$ are assigned higher ranks uniformly at random: $\boldsymbol{R}_{j|\mathcal{A}\setminus\mathcal{A}_c^*} \sim \mathcal{U}(\mathcal{P}_{n-n^*}) + n^*$. Each rank vector $\boldsymbol{R}_j$ consists of n items, with relevant items (those in $\mathcal{A}_c^*$) ordered according to an n^*-dimensional Mallows model, while irrelevant items receive uniformly random higher ranks. All clusters share this structure, but top-ranked items and their rankings are cluster-specific.

The consistency DGP differs by allowing relevant items in $\mathcal{A}_c^*$ to appear in non-top positions. Here, $\boldsymbol{R}_{j|\mathcal{A}_c^*} \sim \text{Mallows}(\boldsymbol{\rho}_c, \alpha_c)$ is sampled and shifted by $s_j \sim \mathcal{U}(n - n^*)$, giving $\tilde{\boldsymbol{R}}_{j|\mathcal{A}_c^*} = \boldsymbol{R}_{j|\mathcal{A}_c^*} + s_j$. Irrelevant items are assigned available ranks from $\{1, \ldots, n\} \setminus \tilde{\boldsymbol{R}}_{j|\mathcal{A}_c^*}$ uniformly at random. The shift s_j is independently sampled for each assessor. Assessors in the c-th cluster thus share the same relevant items $\mathcal{A}_c^*$, ranked contiguously in a restricted space and following a Mallows model with consensus $\boldsymbol{\rho}_c$. However, their overall ranks, relative to irrelevant items, are assessor-specific. Figure 1 provides a visual representation of both DGPs.

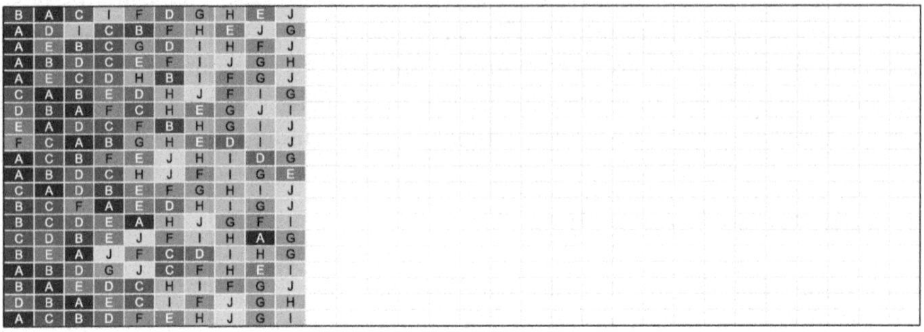

(a) Toprank DGP: all relevant items are ranked in top positions, obtaining lower ranks and being shifted to the left of the graph.

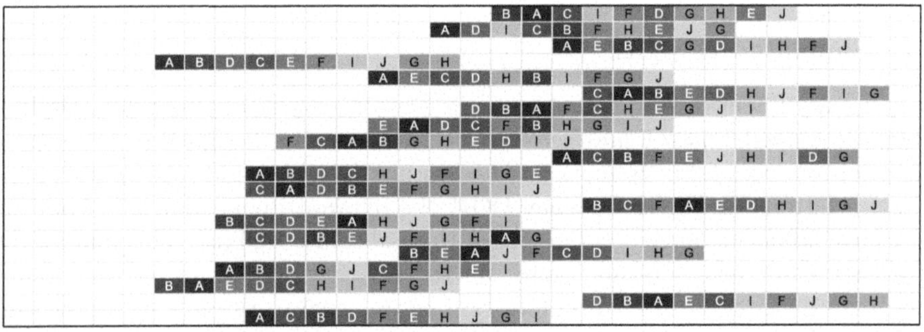

(b) Consistency DGP: relevant items have contiguous ranks, but the whole sequence is shifted by a random assessor-specific amount.

Fig. 1. The figure illustrates data generated using two distinct data-generating processes (DGPs): toprank (top panel) and consistency (bottom panel). Each panel displays a representative cluster comprising 20 assessors (rows). The columns present item rankings in ascending order from 1 to 30. Among these, only the 10 items belonging to $\mathcal{A}_c^*$ are highlighted, each denoted by a unique shade of gray and identified by a capital letter.

As LowBM3 is devised to handle high-dimensional settings, we test its performance in a smaller and simpler data setup, as well as in a larger one. This leads us to consider four setups. Setup 1 and Setup 2 use toprank and consistency DGP respectively, with: $n = 30$ items, $N = 60$ assessors, $n^* = 10$ relevant items, $C = 3$ balanced clusters, and $\alpha_c = 3$ for all c. Setup 3 and Setup 4 are larger setups, respectively generated under toprank and consistency DGPs, with: $n = 300$, $N = 150$, $n^* = 30$, $C = 5$ balanced clusters, and $\alpha_c = 3$ for all c. We add a fifth setup, referred to as Setup 5, where the DGP represents a combination of toprank and consistency DGPs. Specifically, for each cluster, half of the items in $\mathcal{A}_c^*$ are top-ranked, while the other half have their ranks shifted by an assessor-specific value $s \sim \mathcal{U}(n - \lfloor n^*/2 \rfloor)$. Within the restricted space $\mathcal{A}_c^*$, the ranks of the relevant items adhere to a Mallows model with a consensus rank $\boldsymbol{\rho}_c$. In addition, this setup is affected by noise: 10% of the items in $\mathcal{A}_c^*$ are randomly selected to have their ranks swapped with items from $\mathcal{A} \setminus \mathcal{A}_c^*$, independently for each assessor. Parameters used for Setup 5 are the same as for Setup 3 and Setup 4.

3.2 Discussion of the Results

For each of the five setups introduced in Sect. 3.1, we run LowBM3 to estimate $\hat{\boldsymbol{\rho}}_c$, $\hat{\mathcal{A}}^*{}_c$, $\hat{\tau}_c$, and $\hat{z}_j$, for all clusters. For Setup 1 and Setup 2, we set the total number of iterations for Algorithm 1 to $M = 25000$; the MCMC chain is first thinned taking every other 25th sample, and then the initial 30% of the thinned samples are discarded as burn-in. In Setup 3, Setup 4, and Setup 5, we set $M = 100000$, consider every other 100th observation and discard the initial 30% as burn-in. As for the other parameters in input to LowBM3, we set [4,5,13]: $L = 1$; $l = n^*/5$; $\psi = N/C$; n^*, C, and α are set equal to the true ones.

To compare the results of the two post-processing strategies, we consider the following metrics:

- the proportion of assessors correctly assigned to their originating cluster: $\hat{p}_z = \frac{1}{N}\sum_{j=1}^{N} \mathbb{I}_{\{\hat{z}_j = z_j\}}$;
- the mean over assessors of the posterior probability of assigning each assessor to the correct cluster: $Z_{\text{post}} = \frac{1}{N}\sum_{j=1}^{N} \hat{P}(\hat{z}_j = z_j)$;
- the average fraction across clusters of correctly identified relevant items: $\hat{p}_{\mathcal{A}^*} = \frac{1}{C}\sum_{c=1}^{C} \frac{|\hat{\mathcal{A}}_c^* \cap \mathcal{A}_c^*|}{n_c^*}$;
- the cluster-average normalized footrule distance between the true consensus ranking and the estimated one, normalized by the maximum footrule distance between n^*-dimensional rankings: $d(\boldsymbol{\rho}, \hat{\boldsymbol{\rho}}) = \frac{1}{C}\sum_{c=1}^{C} d(\boldsymbol{\rho}_c, \hat{\boldsymbol{\rho}}_c)/\bar{d}_{n^*}$.

Table 1 and Table 2 respectively show the results for LowBM3 post-processing with stability and selection approaches. As the two post-processing methods only affect the estimation of the set of relevant items and the consensus ranks, the quantities $\hat{p}_z$ and Z_{post} do not vary between them.

Table 1 shows results on Setup 1 to Setup 5 for the stability post-processing, at different configurations of the hyperparameters $p \in \{0, 0.1, 0.2\}$ (see discussion of Eq. (5)), and $\xi \in \{1, 2, 3\}$ (see discussion of Eq. (6)). From the table, the overall best results are obtained setting $(p = 0.2, \xi = 1)$ and $(p = 0.2, \xi = 2)$, with the latter configuration yielding marginally superior outcomes. It is clear from the results that the stability post-processing requires filtering-out infrequently selected items across MCMC samples before estimating the $\mathcal{A}_c^*$ sets; $\xi = 1$ and $\xi = 2$ are used to select the items which have the highest probability of receiving ranks within 2 or 3 neighboring values. A moderate value of ξ seems to achieve

Table 1. Stability post-processing approach. The table shows the values of $\hat{p}_{\mathcal{A}^*}$ and $d(\boldsymbol{\rho}, \hat{\boldsymbol{\rho}})$ for setups 1 to 5 (rows), at varying hyperparameters $\xi \in \{1,2,3\}$ and $p \in \{0, 0.1, 0.2\}$ (columns). Optimal values are hilighted in bold. Estimated $\hat{p}_z$ and Z_{post} do not vary with the hyperparameters.

	$\xi=1$ $p=0$	$\xi=1$ $p=0.1$	$\xi=1$ $p=0.2$	$\xi=2$ $p=0$	$\xi=2$ $p=0.1$	$\xi=2$ $p=0.2$	$\xi=3$ $p=0$	$\xi=3$ $p=0.1$	$\xi=3$ $p=0.2$
Setup 1 ($\hat{p}_z = 0.93$, $Z_{\text{post}} = 0.78$)									
$\hat{p}_{\mathcal{A}^*}$	0.67	0.67	0.70	0.63	0.70	**0.73**	0.50	0.50	0.50
$d(\boldsymbol{\rho}, \hat{\boldsymbol{\rho}})$	0.49	0.48	0.37	0.45	**0.29**	0.35	0.59	0.69	0.49
Setup 2 ($\hat{p}_z = 0.7$, $Z_{\text{post}} = 0.7$)									
$\hat{p}_{\mathcal{A}^*}$	0.37	**0.70**	**0.70**	0.40	**0.70**	**0.70**	0.40	0.67	**0.70**
$d(\boldsymbol{\rho}, \hat{\boldsymbol{\rho}})$	0.59	0.60	0.60	0.60	0.61	0.61	0.75	0.60	**0.56**
Setup 3 ($\hat{p}_z = 1$, $Z_{\text{post}} = 1$)									
$\hat{p}_{\mathcal{A}^*}$	0.45	0.66	**0.73**	0.48	0.63	**0.73**	0.52	0.62	**0.73**
$d(\boldsymbol{\rho}, \hat{\boldsymbol{\rho}})$	0.69	0.59	0.60	**0.58**	0.65	0.60	0.62	0.63	0.60
Setup 4 ($\hat{p}_z = 0.48$, $Z_{\text{post}} = 0.48$)									
$\hat{p}_{\mathcal{A}^*}$	0.17	0.53	**0.58**	0.21	0.53	**0.58**	0.19	0.55	0.57
$d(\boldsymbol{\rho}, \hat{\boldsymbol{\rho}})$	0.66	0.62	0.62	0.65	0.65	**0.61**	0.63	0.66	0.65
Setup 5 ($\hat{p}_z = 0.95$, $Z_{\text{post}} = 0.93$)									
$\hat{p}_{\mathcal{A}^*}$	0.21	0.43	**0.45**	0.21	0.44	**0.45**	0.19	0.43	0.44
$d(\boldsymbol{\rho}, \hat{\boldsymbol{\rho}})$	0.69	0.66	0.60	0.75	0.67	0.57	0.73	0.70	**0.54**

optimal results, especially for the larger consistency design, while excessively high values ($\xi = 3$) may harm the performance.

Table 2 presents analogous results for the selection post-processing approach, with p varying within the set $\{0, 0.1, 0.2\}$ (note that the hyperparameter ξ is specific to the stability post-processing approach). For $p = 0$, the results are generally superior to those obtained by the stability approach. Interestingly, the selection approach achieves better performance when infrequent items are not filtered-out in the formation of the $\mathcal{A}'_c$ sets. The results demonstrate that the selection approach performs effectively on toprank designs (setups 1 and 3) and on the smaller consistency design (Setup 2). However, its performance deteriorates in the large-scale consistency design (Setup 4), which is definitely the most challenging setup for this post-processing approach. In Setup 5, the selection approach exhibits moderate performance, where the improvement relative to Setup 4 is mainly due to the recovery of the parameters relative to the top-ranked items, which account for 50% of the relevant items.

Comparing Table 1 and Table 2, we conclude that the selection post-processing approach should be preferred in those scenarios where relevant items are all expected to be ranked in first positions. If applied in different scenarios, one can still expect good results as long as the departure from the toprank

Table 2. Selection post-processing approach. The table shows the values of $\hat{p}_{\mathcal{A}^*}$ and $d(\boldsymbol{\rho}, \hat{\boldsymbol{\rho}})$ for setups 1 to 5 (rows), at varying hyperparameter $p \in \{0, 0.1, 0.2\}$ (columns). Optimal values are hilighted in bold. Estimated $\hat{p}_z$ and Z_{post} do not vary with the hyperparameters.

	$p = 0$	$p = 0.1$	$p = 0.2$
Setup 1 ($\hat{p}_z = 0.93$, $Z_{\text{post}} = 0.78$)			
$\hat{p}_{\mathcal{A}^*}$	**1.00**	0.90	0.83
$d(\boldsymbol{\rho}, \hat{\boldsymbol{\rho}})$	**0.20**	0.33	0.35
Setup 2 ($\hat{p}_z = 0.7$, $Z_{\text{post}} = 0.7$)			
$\hat{p}_{\mathcal{A}^*}$	0.80	**0.90**	0.87
$d(\boldsymbol{\rho}, \hat{\boldsymbol{\rho}})$	0.37	**0.24**	0.47
Setup 3 ($\hat{p}_z = 1$, $Z_{\text{post}} = 1$)			
$\hat{p}_{\mathcal{A}^*}$	**0.91**	0.79	0.75
$d(\boldsymbol{\rho}, \hat{\boldsymbol{\rho}})$	**0.48**	0.51	0.57
Setup 4 ($\hat{p}_z = 0.48$, $Z_{\text{post}} = 0.48$)			
$\hat{p}_{\mathcal{A}^*}$	0.19	0.33	**0.46**
$d(\boldsymbol{\rho}, \hat{\boldsymbol{\rho}})$	**0.63**	0.68	0.68
Setup 5 ($\hat{p}_z = 0.95$, $Z_{\text{post}} = 0.93$)			
$\hat{p}_{\mathcal{A}^*}$	**0.59**	0.52	0.45
$d(\boldsymbol{\rho}, \hat{\boldsymbol{\rho}})$	0.63	0.66	**0.56**

assumption is not too strong or the design itself is not excessively complex. However, the selection post-processing approach does not work well on other complex settings that significantly depart from the toprank assumption, as it is the case for Setup 4 and, to a lesser extent, for Setup 5. In such cases, while it is possible to improve selection performance by filtering-out more items in the construction of $\mathcal{A}'$, another post-processing approach should be used. Indeed, on Setup 4, the performance obtained by the stability post-processing approach is unmatched and, in Setup 5, it provides a more precise estimation of the ranks of relevant items, $\hat{\boldsymbol{\rho}}_c$. This new post-processing method is designed to retrieve items that consistently receive a rank within a few neighboring values. This greatly helps in cases where (i) relevant items may not be ranked in top positions, and (ii) there is more noise in the observed ranks. Moreover, the stability post-processing approach appears to have more robust performance to the specification of its hyperparameters in these cases.

4 Conclusion

In this paper, we introduced the *stability approach*, a new MCMC post-processing method for obtaining posterior summaries of the set of relevant items and the related consensus ranking from the output of LowBM3. The stability approach processes the MCMC chain by selecting items that consistently receive an estimated ranking within a few neighboring values. We validated our proposal with an experimental analysis on different scenarios, based on two designs: (i) one setting where relevant items are expected to be ranked in top positions (toprank);

(ii) another setting where relevant items may not be ranked in top positions but are consistently ranked relative to one another (consistency). We explored the performance of the stability approach at different values of the hyperparameters controlling the methodology and compared the results against the selection method, already introduced in Eliseussen et al. [4]. The stability post-processing appears more robust and achieves better performance than the selection approach under complex settings following the consistency design.

Evidently, each post-processing is designed to excel under particular conditions. A selection approach without filtering ($p = 0$) should be used whenever relevant items are expected to be ranked in top positions. More generally, we advocate defaulting to the selection approach, possibly with a small-to-moderate filtering ($p \approx 10\%$) in more complex cases that depart from the top-rank assumption. A stability approach should be preferred under scenarios where the data generating process is assumed to be closer to a consistency scheme, using a moderate filtering ($p \approx 20\%$) and a small-to-moderate number of neighboring ranks ($\xi \in \{1, 2\}$).

References

1. Afsari, B., Braga-Neto, U., Geman, D.: Rank discriminants for predicting phenotypes from RNA expression. Annals Appli. Statist. **8**(3), 1469–1491 (2014)
2. Badgeley, M.A., Sealfon, S.C., Chikina, M.D.: Hybrid Bayesian-rank integration approach improves the predictive power of genomic dataset aggregation. Bioinformatics **31**(2), 209–215 (2015)
3. Deng, K., Han, S., Li, K., Liu, J.: Bayesian aggregation of order-based rank data. J. Am. Stat. Assoc. **109**(507), 1023–1039 (2014)
4. Eliseussen, E., Coraggio, L., Muggerud, H., Scheel, I., Fleischer, T., Vitelli, V.: Bayesian genome-wide clustering and variable selection of transcriptomic data via rank-based mixtures. Technical report, arXive Working Paper (2025)
5. Eliseussen, E., Fleischer, T., Vitelli, V.: Rank-based bayesian variable selection for genome-wide transcriptomic analyses. Stat. Med. **41**(23), 4532–4553 (2022)
6. Luce, R.D.: Individual choice behavior: A Theoretical Analysis. John Wiley & Sons, Inc., New York, USA (1959)
7. Mallows, C.L.: Non-null ranking models. i. Biometrika **44**, 114–130 (1957)
8. Marden, J.I.: Analyzing and Modeling Rank Data. Chapman and Hall/CRC (January 2014)
9. Pearce, M., Erosheva, E.A.: Bayesian rank-clustering (2024)
10. Plackett, R.L.: The analysis of permutations. Appl. Stat. **24**(2), 193 (1975)
11. Shang, H., Liu, Z.-P.: Network-based prioritization of cancer genes by integrative ranks from multi-omics data. Comput. Biol. Med. **119**, 103692 (2020)
12. Sørensen, Ø., Crispino, M., Liu, Q., Vitelli, V.: Bayesmallows: an r package for the bayesian mallows model. R J. **12**(1), 324 (2020)
13. Vitelli, V., Sørensen, Ø., Crispino, M., Frigessi, A., Arjas, E.: Probabilistic preference learning with the mallows rank model. J. Mach. Learn. Res. **18**(158), 1–49 (2018)

Decoding Locomotor Intentions: Neural Networks and Probabilistic Machine Learning for Customizable Exoskeletons

Lorenza Cotugno[1(✉)], Stefano Pellegrino[1], and Roberta Siciliano[2]

[1] Department of Physics, University of Naples Federico II, Naples, Italy
lorenza.cotugno@unina.it

[2] Department of Electrical Engineering and Information Technology, University of Naples Federico II, Naples, Italy

Abstract. Lower limb assistive technologies can help transfemoral amputees restore locomotion and perform diverse daily activities. The ENcyclopedia of Able-bodied Bilateral Lower Limb Locomotor Signals (ENABL3S) serves as a benchmark dataset, capturing neuro-mechanical signals via wearable sensors during locomotor tasks such as sitting, standing, walking and ascending or descending stairs and ramps, where these tasks constitute an imbalanced target variable. This paper aims to offer a unified perspective on both predictive accuracy and interpretability by applying two complementary methodologies to the same complex dataset. Real time locomotor prediction is accomplished using Recurrent Neural Networks (RNN), including Long Short-Term Memory Networks (LSTM) and Gated Recurrent Units (GRU). In parallel, the Latent Budget Tree, a probabilistic machine learning model, is employed to optimize feature settings for all tasks, including underrepresented classes, providing valuable insights into locomotion dynamics. This dual perspective highlights the need for personalized exoskeletons and effective management of complex motor transitions.

Keywords: Recurrent Neural Network · Long Short-Term Memory network · Gated Recurrent Unit · Latent Budget Model

1 Introduction

Medical robotics is changing the face of healthcare, from diagnostics to surgery and rehabilitation. Exoskeletons, active (actuator-driven with real-time control) or passive (mechanically driven), are among the most promising rehabilitation solutions. In active systems, artificial intelligence can predict patients' motor trajectories in real-time, allowing the exoskeleton to dynamically adapt to each individual's motor intent.

Despite growing interest in intelligent rehabilitation, most existing approaches remain primarily focused on maximizing prediction accuracy. As a result, critical factors necessary for real-world clinical deployment are often

A. D'Ambrosio et al. (Eds.): CLADAG-VOC 2025, SCDAKO, pp. 107–118, 2025.
https://doi.org/10.1007/978-3-032-03042-9_10

overlooked, including subject-specific adaptability, model efficiency and interpretability.

This study addresses these challenges through a dual contribution: first, by developing accurate and lightweight recurrent neural network (RNN)-based models for subject-specific motion classification and second, by introducing a novel interpretable probabilistic model that quantifies the contribution of inertial measurement unit (IMU) features across a variety of locomotor tasks, thereby enabling a deeper understanding of the underlying movement dynamics.

To date, the majority of motion classification approaches rely on complex deep learning architectures [24,25]. In contrast, we demonstrate that simpler models, such as RNNs, LSTMs and GRUs, can achieve comparable and, in some cases, superior performance when adapted to a single individual, which is attributable to the meticulous pre-processing and in-depth study of the data.

To further enhance the interpretability of patient-specific locomotor patterns, we propose the Latent Budget Tree (LBT), a novel probabilistic machine learning model. LBT quantifies the relative importance of each IMU feature in the classification process and reveals which components are most relevant to each locomotion mode. This level of explainability is largely absent in the current literature, where traditional machine learning algorithms (e.g., SVM, Random Forest, k-NN, LDA, XGBoost [2,6,11,23]) are typically optimized solely for prediction accuracy, with little attention given to transparency or clinical interpretability. In contrast, our approach places interpretability and subject-specific customization at the center of the model design.

To explore these aspects in a realistic and constrained scenario, we base our evaluation on a single healthy subject from the ENABL3S dataset, a widely recognized benchmark in human locomotion analysis. This setting allows us to assess both classification performance and model interpretability in the context of personalized exoskeleton control. Even in this case, prior work [4,17] predominantly employs deep architectures with limited transparency and no emphasis on interpretability, underscoring the need for new, more explainable approaches.

The dataset further highlights one of LBT's core advantages: its ability to maintain strong performance even under class imbalance, particularly for critical but underrepresented tasks such as stair ascent, stair descent and ramp transitions, which are typically challenging to model using standard methods.

In summary, this pilot study seeks to inform the design of next-generation prosthetic and assistive systems that can intelligently adapt to individual motor behavior using lightweight neural architectures, while offering fine-grained, patient-specific interpretability at the feature level.

2 ENABL3S Dataset

This dataset, the ENcyclopedia of Able-bodied Bilateral Lower Limb Locomotor Signals (ENABL3S) [10], serves as a benchmark for lower limb neuromechanical signals from 10 able-bodied individuals during unassisted locomotion. Beyond sitting and standing, participants performed level walking, ramp and stair ascent/descent at self-selected speed. Bilateral instrumentation included surface EMG on seven lower-limb muscles, goniometers at knee and ankle, and IMUs on the shank and thigh, with an additional IMU at the waist (Fig. 1).

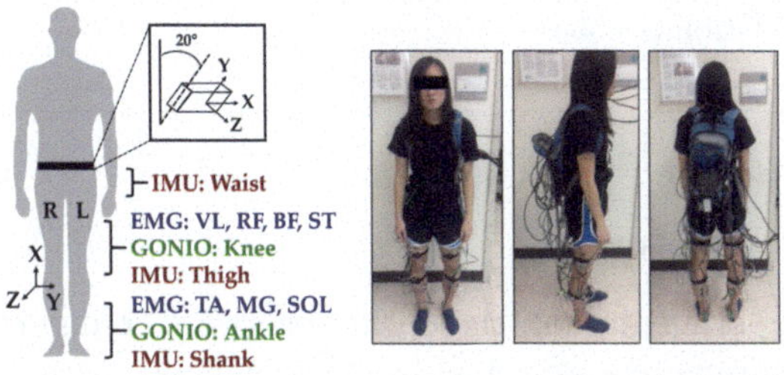

Fig. 1. Placement of the bilateral sensors in the instrumentation system. Coordinate axes are used to display the orientations of the inertial measurement units (IMUs) in the waist, thigh, and shank.

3 Locomotor Intention Prediction by Deep Learning

To address the problem of locomotion prediction, a series of deep learning methodologies are employed, specifically a Recurrent Neural Network (RNN), a Long Short-Term Memory (LSTM) and a Gated Recurrent Unit (GRU), as these models are well-suited for sensor-based data in which capturing complex, dynamic and often noisy patterns is essential.

3.1 Experimental Setup

The application dataset consists of raw data from a single patient and includes 1,139,130 observations across seven modes of locomotor activity such as sitting, standing, ascending or descending ramps and stairs.

Before applying the deep learning models, the dataset was pre-processed to increase the algorithm's performance effectiveness. This has included scaling, encoding and feature engineering. Moreover, as the original data consisted of signals recorded at different sampling rates (EMG signals sampled at 1000 Hz, while joint kinematics and IMU sensors at 500 Hz) and to ensure uniformity of all modalities, a binning procedure was applied to match the sampling frequency of the higher frequency signals to the lower one by averaging consecutive pairs of samples from the 1000 Hz signals. This resulted in the creation of 150-sample time sequences with a sampling rate of 500 Hz and 300 ms windows.

The dataset was then divided into training, validation and testing sets and the same approach was used for all models (Table 1).

Table 1. Dataset split into time windows. Each time window consists of 150 samples (300 ms).

Dataset	Time Windows	Total Samples	Percentage
Training Set	2,655	398,250	70%
Validation Set	569	85,350	15%
Test Set	569	85,350	15%

3.2 Model Architectures

After extensive experimentation, the optimal network configurations were identified, and hyperparameters were fine-tuned to enhance performance. The final architectures for the three neural networks are defined as follows:

- **RNN**: A single recurrent layer of 32 units (tanh activation function), with a dropout rate of 0.3. The output layer consists of 7 units with a softmax activation function to generate class probabilities. The model was trained for 100 epochs with a batch size of 64, using the Adam optimizer and categorical cross-entropy loss.
- **LSTM**: A single recurrent layer of 24 LSTM units (tanh activation function), with a dropout rate of 0.3. The output layer consists of 7 units with a softmax activation function. The training conditions were identical to those of the RNN model.
- **GRU**: A single recurrent layer of 32 GRU units (ReLU activation function), with a dropout rate of 0.4 and a Batch Normalization layer to stabilize training. The output layer consists of 7 units with a softmax activation function, producing class probabilities. The training setup was identical to that of the RNN and LSTM models, with 100 epochs, a batch size of 64, the Adam optimizer, and categorical cross-entropy loss.

3.3 Results and Analysis

The results obtained highlight the discovery of particularly effective GRU and LSTM configurations, capable of capturing the different locomotor activities, including the most complex ones, with remarkable precision.

Table 2 summarizes the performance of the evaluated neural network architectures in terms of weighted F1-score, which accounts for class imbalance by computing class-specific scores and weighting them according to their frequency. Among the models, the GRU architecture yielded the best result, achieving a weighted F1-score of 94.31% and outperforming the others. By contrast, the vanilla RNN showed high accuracy only on class 0 (Sitting), the least complex activity, with an F1-score of 92.86%, but failed to generalize to other classes. This is consistent with the known limitations of standard RNNs in retaining long-term dependencies due to the vanishing gradient. The LSTM partially addresses this issue through its gated memory cells and the GRU, a simplified variant of the LSTM, strikes a more efficient balance between model capacity and generalization in short-sequence scenarios, making it particularly well-suited for the real-time locomotion classification task addressed in this study.

Importantly, these results were supported by a carefully structured preprocessing pipeline, including the use of short, fixed-length input windows (300 *ms*) to efficiently capture relevant locomotor patterns and signal normalization to ensure scale consistency across samples.

Figure 2 shows the plots of accuracy on the validation and training sets and it is possible to observe an overall good trend in performance, with both training and validation accuracy improving consistently over epochs. When compared to existing approaches in the literature, such as those employing deeper or hybrid CNN-RNN architectures or longer input sequences [4,17], it is important to note that direct benchmarking remains limited due to differences in experimental setup. Nonetheless, the performance observed in this study is in line with results reported in more complex systems, thus demonstrating that lightweight recurrent architectures like GRU can serve as a strong, computationally efficient foundation for real-time locomotor intention recognition.

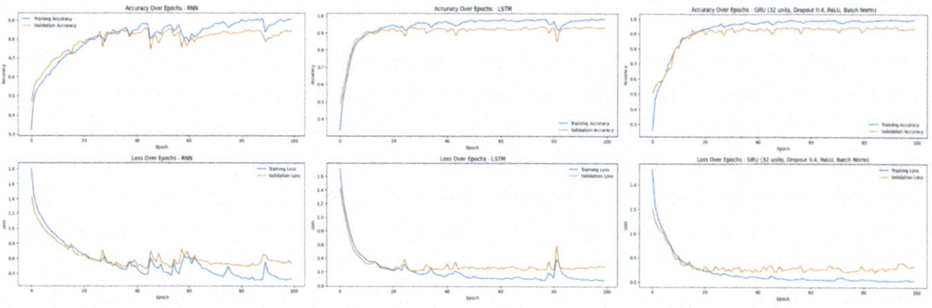

Fig. 2. Training and validation accuracy (top) and loss (bottom) over 100 epochs for RNN, LSTM, and GRU models.

Table 2. Comparison of F1-Scores for RNN, LSTM, and GRU Models

Mode	RNN (%)	LSTM (%)	GRU (%)
0 (Sitting)	92.86	94.74	94.82
1 (Level walking)	84.34	92.15	95.98
2 (Ramp ascent)	77.88	89.09	97.30
3 (Ramp descent)	76.47	94.12	94.66
4 (Stair ascent)	76.60	88.46	94.34
5 (Stair descent)	74.51	92.31	96.00
6 (Standing)	75.52	82.19	86.11
Weighted F1-score	82.68	91.16	94.31

4 Probabilistic Machine Learning for Imbalanced Multi-class Response

The development of intelligent exoskeletons and prostheses hinges on predictive accuracy and explainability. Although deep learning methods offer strong performance, they function as black-box models, limiting interpretability. As a complementary approach applied to the ENABL3S dataset, the Latent Budget Tree (LBT) provides a transparent and probabilistic methodology [20]. This method interprets complex, imbalanced response modes and reveals the underlying probabilities in the decision-making process, thus improving the explanation.

4.1 Enhancements in Explainability of Tree-Based Methods

Tree-based methods are interpretable machine learning models when they provide a single tree and perform well for prediction when they are averaged, as in ensemble methods such as random forest and boosting [12].

A single tree structure for classification consists of a recursive partitioning of a training sample into subgroups to reduce the heterogeneity of the response. Cross-validation can be used to select the decision tree that can be generalized to predict fresh data. As an alternative, visual pruning allows one to identify reliable tree paths [14].

In standard tree-growing procedures [3], splitting the predictor's modalities into two subgroups brings observations of the parent node $\{t\}$ into two child nodes $\{2t\}$ and $\{2t+1\}$. The best split $s(t)$ is found by maximizing the decrease in heterogeneity of the response variable.

In classification trees, it is possible to distinguish the global predictability of a predictor X and the partial predictability of any split s generated by it to explain the heterogeneity of the response Y [21]. Suitable statistical indices called the Global Predictability Index (GPI) and the Partial Predictability Index (PPI) can be defined on the basis of the predictability τ-index of Goodman and Kruskal. This is obtained from the decomposition of the Gini heterogeneity index of Y into the explained component due to the predictor X (or the slit s) and the residual one. The τ-index measures the percentage of improvement in predictability of the response Y provided by the predictor X or the split s. It can be shown that $\tau_{Y|X}(t) \geq \tau_{Y|s}(t)$ for any split s generated by X. The gap measures the loss in terms of predictability when compressing the predictor into a split. This adds some insight to the interpretation of the tree structure.

4.2 Latent Budget Tree

Latent Budget Tree (LBT) [20] can be understood as a probabilistic tree-growing procedure for multiclass response variable Y with an unbalanced distribution of observations between $J > 2$ classes. Its name is due to the goodness-of-split criterion based on the latent budget model for compositional or budget data [8].

LBT considers the child nodes as two latent members of a mixture model assigned to the parent node $\{t\}$. At the parent node $\{t\}$, the best predictor X is selected maximizing the predictability τ-index of Goodman and Kruskal for two-way cross-classifications, thus providing the Global Predictability Index (GPI) measure $\tau_{Y|X}$.

Let $\mathbf{p}_{Y|i}(t)$ be the i-th observed budget (row profile), which is a row vector of J conditional proportions $p_{j|i}(t)$ such that $\sum_j p_{j|i}(t) = 1$ where $i = 1, \ldots, I; j = 1, \ldots, J$.

The Latent Budget Model assumes that the i-th observed budget $\mathbf{p}_{Y|i}(t)$ (for $i = 1, \ldots, I$) can be approximated by the theoretical (row vector) budget $\pi_{Y|i}(t)$ defined as:

$$\pi_{Y|i}(t) = \alpha_{2t|i}\beta_{Y|2t} + \alpha_{2t+1|i}\beta_{Y|2t+1} \tag{1}$$

where:

- $\alpha_{2t|i}$ and $\alpha_{2t+1|i}$ are the mixing parameters describing the probability that the i-th row profile belongs to the left and to the right child node respectively, with $\alpha_{2t|i} + \alpha_{2t+1|i} = 1$:
- $\beta_{Y|2t}$ and $\beta_{Y|2t+1}$ are J-dimensional column vectors of latent budget parameters $\beta_{j|2t}$ and $\beta_{j|2t+1}$ respectively, with $\sum_j \beta_{j|2t} = 1$ and $\sum_j \beta_{j|2t+1} = 1$.

The mixing parameters quantify the different contributions of each latent budget to reconstruct the theoretical budget. The parameters are estimated by an alternating least-squares method.

The latent budget model can be understood as a supervised neural network model [22], specifically a double layer perceptron, with linear activation functions, where the weights are interpreted as conditional probabilities. LBT considers two latent members as child nodes (Fig. 3). In particular, the mixing parameters connect the input layer (predictor) to the hidden layer, while the latent budget parameters link the hidden layer to the output layer (response). Each predictor modality brings observations to the child where the mixing parameter is the highest. Each response class is better represented in the child where the latent budget parameter estimate is the highest.

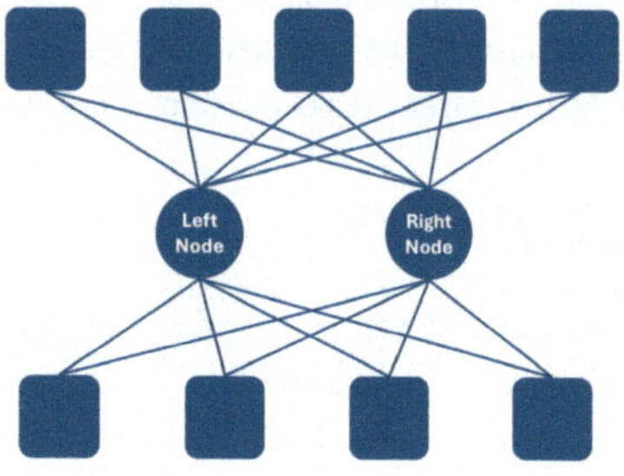

Fig. 3. Splitting rule in the Latent Budget Tree

The splitting rule is derived assigning each observation to the child node with the highest mixing parameter estimate. The τ-index can be evaluated for the splitting variable or split s, thus evaluating the Partial Predictability Index (PPI) measure $\tau_{Y|s}$.

A Bayesian interpretation of the splitting rule is provided. The $\hat{\beta}_{j|2t}$ can be understood as the estimate of the posterior probability to assign the j-th response class once the observation is assigned to the child node $\{2t\}$. Analogously is for $\hat{\beta}_{j|2t+1}$. These estimates can be compared with the prior probability estimate at the parent node $\{t\}$ to identify the response classes that are better described within the child node.

Starting from the root node $\{1\}$ of the tree with the prior probability estimates $p_j(1)$ for each response class, the posterior probability estimates are recursively updated and related through a chain of conditional probability estimates until the definition of the final posterior classification in the leaves of the final tree structure. In the final tree $\mathcal{T}$ with the set of leaves $h \in H$, the comparison of the probability estimates $p_j(h)$ for each response class $j \in J$ allows one to identify the overrepresented response classes in each leaf, helping to interpret the path that produces the minority classes.

4.3 Experimental Setup

To better understand leg movement, the focus is on the kinematics data captured by the IMU sensor, specifically the continuous variables, namely acceleration and gyroscopic rotation along the X, Y and Z axes of the left shank. These are transformed into categorical variables for the application of the LBT. To perform this transformation, a decision tree is developed to iteratively determine the optimal thresholds, identifying the values that best divide the data into groups in terms of homogeneity of the target variable ('Mode'). The observation is labeled with an alphabetic character based on the interval to which it belongs, delimited by the threshold. This approach identifies meaningful categories that reflect the variability in the data, without incurring categories that are too rare or difficult to interpret. The new categorical variables are then used to build the Latent Budget Tree, initialized with stopping criteria by setting a Gini impurity threshold of 0.01. Furthermore, the tree's growth is constrained with a minimum of 1000 observations for parent nodes and 500 for child nodes to prevent overfitting. This combination of parameters has proven to be the most effective configuration.

4.4 Results and Analysis

Table 3 summarizes the output of the LBT for each locomotor class in terms of class probabilities. Each row corresponds to a node, with parent nodes (such as n1) where the splitting rule occurs, and child nodes (such as n3) which are leaves.

The columns in the table are as follows:

- 'n' represents the number of observations for each node
- 'GPI' indicates the Global Predictability Index, which measures the overall predictability of the predictor within the node before the split.

- 'PPI' is the Partial Predictability Index, which quantifies the proportion of predictability explained by the split. Higher values mean that the split effectively separates observations from the parent node into two more predictable subgroups.
- 'P(j)' indicates the probability of the occurrence of each 'Mode' within the node.

Table 3. Summary of Latent Budget Tree Nodes with Class Probabilities by Category

Node	Type	Splitting Variable	n	GPI	PPI	P(0)	P(1)	P(2)	P(3)	P(4)	P(5)	P(6)
n1	Parent	Left_Thigh_Ax	911304	0.213	0.198	0.22	0.35	0.1	0.11	0.04	0.05	0.13
n3	Child	None	729403	–	–	0.04	0.43	0.13	0.13	0.05	0.06	0.16
n2	Parent	Left_Thigh_Az	181901	0.266	0.131	0.91	0.01	0.01	0.0	0.01	0.0	0.04
n4	Parent	Left_Shank_Ax	3620	0.161	0.141	0.01	0.4	0.14	0.15	0.17	0.12	0.02
n8	Child	None	1177	–	–	0.0	0.11	0.04	0.13	0.52	0.2	0.0
n9	Parent	Left_Thigh_Ay	2443	0.099	0.054	0.01	0.54	0.18	0.15	0.0	0.08	0.03
n18	Child	None	562	–	–	0.0	0.26	0.3	0.22	0.0	0.22	0.01
n19	Child	None	1881	–	–	0.02	0.62	0.15	0.13	0.0	0.04	0.04
n5	Parent	Left_Shank_Az	178281	0.192	0.138	0.93	0.01	0.01	0.0	0.01	0.0	0.04
n11	Child	None	175341	–	–	0.95	0.0	0.01	0.0	0.0	0.0	0.04
n10	Parent	Left_Shank_Gz	2940	0.344	0.279	0.06	0.31	0.22	0.0	0.41	0.0	0.0
n20	Child	None	760	–	–	0.01	0.09	0.83	0.0	0.06	0.0	0.0
n21	Parent	Left_Thigh_Gx	2180	0.288	0.226	0.08	0.39	0.0	0.0	0.53	0.0	0.0
n42	Child	None	553	–	–	0.08	0.82	0.01	0.0	0.09	0.0	0.0
n43	Child	None	1627	–	–	0.08	0.25	0.0	0.0	0.68	0.0	0.0

4.5 Identification of Minority Classes

One of the main challenges of this study is the representation of minority classes. The LBT identifies nodes where these classes are better represented compared to their parent node: as an example, 'Mode 4' ('Stair Ascent') rises from 4% of representations in the root node n1 to 52% in the leaf n8 (Fig. 4) and 'Mode 2' ('Ramp Ascent'), from 10% in the root to 83% in node n20. This highlights the LBT's ability to emphasize the importance of underrepresented classes, ultimately providing a more comprehensive understanding of movement patterns.

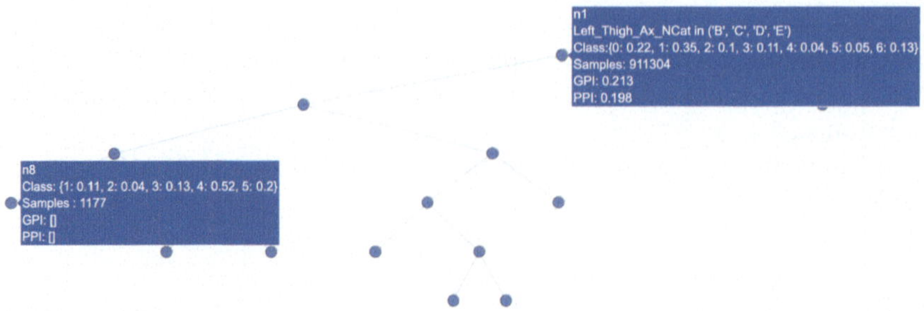

Fig. 4. Latent Budget Tree output

5 Concluding Remarks

This paper introduces two complementary approaches for decoding locomotor intentions from sensor data: deep learning models (RNN, GRU, LSTM) and the Latent Budget Tree (LBT). Pre-processing steps, such as signal alignment, cleaning, scaling, encoding, and categorization, were crucial to achieving optimal performance.

Among the deep learning models, the GRU achieved the highest weighted F1-score of 94.31%, demonstrating the extraordinary accuracy of deep learning models in predicting locomotor activity. Indeed, especially when dealing with imbalanced datasets, the F1-score is the harmonic mean of precision and recall, which balances the importance of both. Despite their excellent performance in overall prediction accuracy, deep learning methods are limited by their nature as black-box models, leading to a lack of explainability in the decision-making process. Although LSTM and GRU models can reach high accuracy in predicting also minority classes, there is no way to identify the key sensors and their best setting for a particular task. As a complementary contribution, the Latent Budget Tree for multi-class with imbalanced response offered an interpretable probabilistic framework to identify which parts of the body are involved and how sensor features combine to determine the likelihood of making a certain movement. Acceleration along the vertical axis of the thigh is a key factor since it discriminates between walking and sitting. For climbing a ladder, both the horizontal acceleration of the thigh and the shank play a significant role. LBT provides to focus on variables that are crucial in balance, such as lateral acceleration in the descent phase, which suggests that an individual going down a ramp or a staircase tends to keep the legs as aligned as possible. For the sitting position, high horizontal forces are required in both the lower leg and the upper leg to control and balance. These insights support the effective development of assistive technologies by highlighting critical forces involved in various movements, ensuring that resources are allocated strategically to essential components. Hence, these approaches enable the development of intelligent and adaptable exoskeleton systems that strike a compromise between accuracy and transparency. Although these results demonstrate the potential of our approach and provide a focused analysis of model behavior and sensor relevance, this also limits the generalizability of the findings. Future work will include evaluations across multiple subjects to validate the robustness of the approach. Furthermore, it will be necessary to apply this methodology to a clinical population such as individuals with gait impairments or neuromotor disorders to confirm its effectiveness in real-world rehabilitation scenarios. Ultimately, this study lays the foundations for future research that involves deep reinforcement learning in dynamic locomotion control of personalized exoskeletons.

Acknowledgments. The study is part of the national initiative Fit4MedRob Fit for Medical Robotics, supported by the Italian Ministry of Research under the subproject EXODAI4: Exoskeleton Optimization and Control by Data and Artificial Intelligence 4.0.

References

1. Bevilacqua, A., MacDonald, K., Rangarej, A., Widjaya, V., Caulfield, B., Kechadi, T.: Human activity recognition with convolutional neural networks. In: Brefeld, U., et al. (eds.) ECML PKDD 2018. LNCS (LNAI), vol. 11053, pp. 541–552. Springer, Cham (2019). https://doi.org/10.1007/978-3-030-10997-4_33
2. Bhakta, K., Camargo, J., Donovan, L., Herrin, K., Young, A.: Machine learning model comparisons of user independent & dependent intent recognition systems for powered prostheses. IEEE Rob. Autom. Lett. **5**(4), 5393–5400 (2020)
3. Breiman, L., Friedman, J., Stone, C.J., Olshen, R.A.: Classification and Regression Trees. Taylor & Francis (1984). https://books.google.it/books?id=JwQx-WOmSyQC
4. Bruinsma, J., Carloni, R.: IMU-based deep neural networks: prediction of locomotor and transition intentions of an osseointegrated transfemoral amputee. IEEE Trans. Neural Syst. Rehabil. Eng. **29**, 1079–1088 (2021)
5. Camargo, J., Flanagan, W., Csomay-Shanklin, N., Kanwar, B., Young, A.: A machine learning strategy for locomotion classification and parameter estimation using fusion of wearable sensors. IEEE Trans. Biomed. Eng. **68**(5), 1569–1578 (2021)
6. Chen, B., Zheng, E., Wang, Q.: A locomotion intent prediction system based on multi-sensor fusion. Sensors **14**(7), 12349–12369 (2014)
7. Cho, K., et al.: Learning phrase representations using RNN encoder-decoder for statistical machine translation. In: Proceedings of the 2014 Conference on Empirical Methods in Natural Language Processing (EMNLP), pp. 1724–1734. Association for Computational Linguistics (2014)
8. De Leeuw, J., Van der Heijden, P. G. M.: The Analysis of Time-Budgets with a Latent Time-Budget Model. Elsevier, North-Holland (1988). https://dspace.library.uu.nl/handle/1874/23188
9. Dehzangi, O., Taherisadr, M., ChangalVala, R.: IMU-based gait recognition using convolutional neural networks and multi-sensor fusion. Sensors **17**(12), 2735 (2017)
10. Hu, B., Rouse, E., Hargrove, L.: Benchmark Datasets for Bilateral Lower Limb Neuromechanical Signals from Wearable Sensors During Unassisted Locomotion in Able-Bodied Individuals (2018). https://figshare.com/articles/dataset/5362627. Accessed 24 Nov 2024

11. Figueiredo, J., Carvalho, S.P., Goncalve, D., Moreno, J.C., Santos, C.P.: Daily locomotion recognition and prediction: a kinematic data-based machine learning approach. IEEE Access **8**, 33250–33262 (2020)
12. Hastie, T., Tibshirani, R., Friedman, J.: The Elements of Statistical Learning: Data Mining, Inference, and Prediction, Second Edition. Springer Series in Statistics. Springer (2009)
13. Hochreiter, S., Schmidhuber, J.: Long short-term memory. Neural Comput. **9**(8), 1735–1780 (1997)
14. Iorio, C., Aria, M., D'Ambrosio, A., Siciliano, R.: Informative trees by visual pruning. Expert Syst. Appl. **127**, 228–240 (2019). https://doi.org/10.1016/j.eswa.2019.03.018
15. Li, H., Derrode, S., Pieczynski, W.: An adaptive and on-line IMU-based locomotion activity classification method using a triplet Markov model. Neurocomputing **362**, 94–105 (2019)
16. Lim, B., Arik, S.O., Loeff, N., Pfister, T.: Temporal fusion transformers for interpretable multi-horizon time series forecasting. Int. J. Forecast. (2021)
17. Mazon, D.M., Groefsema, M., Schomaker, L.R.B., Carloni, R.: IMU-based classification of locomotion modes, transitions, and gait phases with convolutional recurrent neural networks. Sensors **22**(22), 8871 (2022)
18. Papapicco, V., et al.: A classification approach based on directed acyclic graph to predict locomotion activities with one inertial sensor on the thigh. IEEE Trans. Med. Rob. Bionics **3**(2), 436–445 (2021)
19. Ravi, D., Wong, C., Lo, B., Yang, G.-Z.: Deep learning for human activity recognition: a resource efficient implementation on low-power devices. In: 2016 IEEE 13th International Conference on Wearable and Implantable Body Sensor Networks (BSN), pp. 71–76. IEEE (2016)
20. Siciliano, R.: Latent budget trees for multiple classification. In: Vichi, M., Opitz, O. (eds) Classification and Data Analysis. Studies in Classification, Data Analysis, and Knowledge Organization. Springer, Heidelberg (1999)
21. Siciliano, R., Mola, F.: Multivariate data analysis and modeling through classification and regression trees. Comput. Stat. Data Anal. **32**(3–4), 285–301 (2000)
22. Siciliano, R., Mooijaart, A.: Unconditional latent budget analysis: a neural network approach. In: Advances in Classification and Data Analysis, pp. 127–134. Springer, Heidelberg (2001). https://doi.org/10.1007/978-3-642-59471-7_16
23. Stolyarov, R., Burnett, G., Herr, H.: Translational motion tracking of leg joints for enhanced prediction of walking tasks. IEEE Trans. Biomed. Eng. **65**(4), 763–769 (2017)
24. Su, B.-Y., Wang, J., Liu, S.-Q., Sheng, M., Jiang, J., Xiang, K.: A CNN-based method for intent recognition using inertial measurement units and intelligent lower limb prosthesis. IEEE Trans. Neural Syst. Rehabil. Eng. **27**(5), 1032–1042 (2019)
25. Wang, J., Chen, Y., Hao, S., Peng, X., Hu, L.: Deep learning for sensor-based activity recognition: a survey. Pattern Recogn. Lett. **119**, 3–11 (2019)
26. Williams, R.J., Zipser, D.: A learning algorithm for continually running fully recurrent neural networks. Neural Comput. **1**(2), 270–280 (1989)
27. Young, A.J., Simon, A.M., Hargrove, L.J.: A training method for locomotion mode prediction using powered lower limb prostheses. IEEE Trans. Neural Syst. Rehabil. Eng. **22**(3), 671–677 (2013)
28. Zebin, T., Scully, P.J., Ozanyan, K.B.: Human activity recognition with inertial sensors using a deep learning approach. In: 2016 IEEE Sensors, pp. 1–3. IEEE (2016)

Clustering Using Multidimensional Indicators: An Approach without Feature Reduction

Alessia D'Ambrosio[1(✉)], Giuseppe Gismondi[1], Leonardo Alaimo[2], and Alfonso Piscitelli[1]

[1] Università degli Studi di Napoli Federico II, Naples, Italy
{alessia.dambrosio,alfonso.piscitelli}@unina.it
[2] Università di Roma La Sapienza, Rome, Italy
leonardo.alaimo@uniroma1.it

Abstract. The complexity of social phenomena can be decomposed into different dimensions, each measured by a set of basic indicators. These elementary measures capture distinct aspects of each dimension, summarizing them into a composite indicator. In this framework, one of the main goals of statistical analysis is to partition individuals using these dimensional indicators directly rather than all available sets of indicators, possibly leading to a loss of information. In this work, we propose a new methodological approach that allows the clustering of individuals using all the elementary measures, preserving information about the dimensional structure of the phenomenon under study when characterizing the clusters obtained from the partitioning algorithm. Our method relies on distance matrices between individuals belonging to different groups with elements that possess the desirable property of multivariate additivity. The results of applying this method to an example dataset are compared with those obtained using traditional methods.

Keywords: Latent Structure Contribution · Distance matrices · Clustering · Social Phenomena

1 Introduction

Almost all the phenomena studied in psychology, economics, sociology, and political science are not directly observable and have multidimensional characteristics. To measure a complex social science phenomenon, it is necessary to *(i)* find a set of observable characteristics that cover different aspects of the phenomenon, *(ii)* collect data for the different aspects of the phenomenon, and *(iii)* choose an aggregation method that transforms the gathered data into a unidimensional index [8–10,17,21]. When the objective is to study complex social phenomena and classify individuals into homogeneous groups, clustering may be performed directly on the original high-dimensional variables; however, if underlying latent

A. D'Ambrosio et al. (Eds.): CLADAG-VOC 2025, SCDAKO, pp. 119–131, 2025.
https://doi.org/10.1007/978-3-032-03042-9_11

dimensions are assumed, a common strategy is to reduce dimensionality (creating a composite indicator or via factor analysis) and then apply clustering to the reduced data.

The simplest additive aggregation method to construct a composite indicator, called the **Ordinal Approach** [3], entails calculating the ranking of each i-th individual according to each l-th elementar indicator and summing the resulting rankings:

$$CI_i = \sum_{l=1}^{p} \text{Rank}_{il} \quad \forall i = 1, 2, \ldots, n \tag{1}$$

where:

- n are the individuals,
- p are the elementary indicators.

Regarding factor analysis, the confirmative approach (i.e., when the information about the relationships between indicators and latent dimensions is known in advance) is the **Confirmative Factor Analysis (CFA)**. CFA is a statistical technique used to test whether a hypothesized factor structure fits the observed covariance matrix of a dataset. Factor analysis represents observed variables as linear combinations of latent variables, unobserved constructs, and residual terms, estimating parameters such as factor loadings, residual variances, and latent covariances. In psychological research, these strategies are used for various purposes, such as examining patterns of coping strategies [7], profiling consumers [12,13], or studying consumer behavior [6]. In economic research, it is used to measure the relative development and integration of capital markets [11], quantify trend differences and similarities between countries [14,16], or measure progress in environmental areas [15]. In sociological research, a combination of dimensionality reduction methods and clustering techniques has been used in health studies to derive a posteriori dietary patterns [2,20], in education [1,18,22], or for social work studies [19]. In political science, the combination of these methods has been used to classify political cultures [4] or for welfare research [5].

The traditional methods mentioned in this paragraph (in particular Ordinal Approach and CFA) will serve as benchmark to compare the analysis results obtained with the application of the proposed methodology.

2 Methodological Framework: An Overview

The methodology is based on grouping individuals using an approach that preserves the hierarchical dimensional structure of the data without resorting to dimensionality reduction. Consider a data matrix $\mathbf{X} \in \mathbb{R}^{n \times p}$, with n individuals in rows *(i=1...n)* and p variables in columns *(l=1...p)*. The p variables are in turn grouped into k blocks, each representing a latent dimension of the phenomenon being analyzed. Formally, $\mathbf{X}$ can be written as:

$$\mathbf{X} = [\mathbf{X}^{(1)}, \mathbf{X}^{(2)}, \ldots, \mathbf{X}^{(k)}] \quad (2)$$

where $X^{(b)} \in \mathbb{R}^{n \times p_b}$ indicates the matrix of the b-th block of variables $(b=1...k)$, with p_b being the number of variables in block b, such that $\sum_{b=1}^{k} p_b = p$. The objective is to group the individuals taking into account the dimensional hierarchical structure of the problem, using all available variables without resorting to any type of dimensionality reduction, while still being able to quantify the contribution of the latent blocks of variables to the differences among the units. The methodological process consists of two steps: first, the clustering of individuals and then the quantification of the contribution of the blocks of variables to the obtained partition. The clustering step is conducted using all available variables. To this end, any distance-based unsupervised clustering method can be applied. In particular, the developed methodology requires that the chosen distance respects the property of multivariate additivity. Formally, given a generic distance d defined between two individuals i and j in the dataset, this property implies that the overall distance between the two individuals $d(i, j)$ can be expressed as the sum of the distances calculated separately on the individual blocks of variables:

$$d(i, j) = \sum_{b=1}^{k} dist^{(b)}(i, j) \quad (3)$$

where $d^{(b)}(i, j)$ represents the distance between individuals i and j calculated exclusively on the variables belonging to block b. This property ensures that each block of variables contributes independently and additively to the overall measure of distance. Consequently, it is possible to analyze and quantify the relative importance of each block in defining the distances between individuals and, therefore, in the formation of groups, while preserving the hierarchical structure of the data. We denote by $C = \{c_1, c_2, \ldots, c_m, \ldots, c_G\}$ the set of group membership labels, where $m = 1, \ldots, G$ indicates the group to which the m-th individual belongs, and G is the total number of groups. Labeling individuals using all variables allows for the preservation of the original variability of the data, leveraging it for the clustering of the data without introducing distortions.

To quantify the contribution of variable blocks to the partition (i.e., to the differences between individuals belonging to different groups), we resort to the concept of inter and intra-cluster distance.

Calculation of Total Distance. After the clustering process, it is possible to calculate the global distance matrix $D^{(tot)} \in \mathbb{R}^{n \times n}$, that is, the matrix obtained by calculating the distances between individuals considering all available variables. This matrix has a particular structure in which observations belonging to the same cluster occupy adjacent positions along the rows and columns, such that:

- the diagonal sub-matrices constitute the distance matrices between individuals belonging to the same cluster. Lower distance values in these matrices indicate low variability and strong cluster cohesion

– the extra-diagonal sub-matrices instead are the distance matrices between individuals belonging to different clusters, that is, they represent the intra-cluster variability. Higher values suggest significant differences between clusters

As a distance matrix, $D^{(tot)}$ is symmetric and square, and in light of the described structure, it can be represented in compact form as:

$$D^{(tot)} = \begin{bmatrix} D_{1,1}^{(tot)} & D_{1,2}^{(tot)} & \cdots & D_{1,G}^{(tot)} \\ D_{2,1}^{(tot)} & D_{2,2}^{(tot)} & \cdots & D_{2,G}^{(tot)} \\ \vdots & \vdots & \ddots & \vdots \\ D_{G,1}^{(tot)} & D_{G,2}^{(tot)} & \cdots & D_{G,G}^{(tot)} \end{bmatrix}$$

In this matrix, each sub-matrix $D_{m,m'}^{(tot)}$ is formed by the distances between individuals of clusters m and m', and each element of the sub-matrices is specified by the formula:

$$(D_{m,m'}^{(tot)})_{ij} = d_{i,m;j,m'}^{(tot)} = dist^{(tot)}(i, m; j, m')$$

with m and m' ranging from 1 to G . If $m = m'$, the sub-matrices are positioned along the diagonal; otherwise, the sub-matrices are found in the extra-diagonal positions.

Calculation of the Distance Matrix for Blocks of Variables. Next, the distance matrices for each block of variables are calculated, whose elements are the distances between two individuals considering only the variables belonging to the relative block. The distance matrix associated with block X_b, with $b = 1, \ldots, k$, is defined as $D^{(b)} \in \mathbb{R}^{n \times n}$, which will have the same structure as the global distance matrix:

$$D^{(b)} = \begin{bmatrix} D_{1,1}^{(b)} & D_{1,2}^{(b)} & \cdots & D_{1,G}^{(b)} \\ D_{2,1}^{(b)} & D_{2,2}^{(b)} & \cdots & D_{2,G}^{(b)} \\ \vdots & \vdots & \ddots & \vdots \\ D_{G,1}^{(b)} & D_{G,2}^{(b)} & \cdots & D_{G,G}^{(b)} \end{bmatrix}$$

The elements of the matrix can be defined as:

$$(D_{m,m'}^{(b)})_{ij} = d_{i,m;j,m'}^{(b)} = dist^{(b)}(i, m; j, m')$$

which corresponds to the additive distance introduced in 3, now computed specifically between clusters m and m', with m and $m' \in \{1, \ldots, G\}$. In this way, k distance matrices $D^{(b)} \in \mathbb{R}^{n \times n}$ are obtained that can be compared with each other, and that, given the previously described property of multivariate additivity, when summed, yield the global distance matrix:

$$D^{(tot)} = \sum_{b=1}^{k} D^{(b)} \tag{4}$$

Average Matrices. To obtain a more compact and interpretable representation of the distances, the average of the distance sub-matrices is calculated, defined as:

$$\bar{D}_{mm'} = \frac{1}{|c_m| \cdot |c_{m'}|} \sum_{i \in c_m} \sum_{j \in c_{m'}} d_{i,m;j,m'} \tag{5}$$

where $|.|$ indicates the cardinality of the group. It is possible to perform this averaging operation on the distance matrices for k blocks of variables:

$$\bar{D}^{(b)} = \begin{bmatrix} \bar{D}^{(b)}_{1,1} & \bar{D}^{(b)}_{1,2} & \cdots & \bar{D}^{(b)}_{1,G} \\ \bar{D}^{(b)}_{2,1} & \bar{D}^{(b)}_{2,2} & \cdots & \bar{D}^{(b)}_{2,G} \\ \vdots & \vdots & \ddots & \vdots \\ \bar{D}^{(b)}_{G,1} & \bar{D}^{(b)}_{G,2} & \cdots & \bar{D}^{(b)}_{G,G} \end{bmatrix}$$

Thanks, again, to the multivariate additivity property, the sum of k average distance matrices for blocks of variables returns the total average distance matrix. Each element on the diagonal blocks of such matrices represents the average distance between individuals belonging to the same cluster concerning only the variables belonging to the reference block; each element on the extra-diagonal blocks of such matrices represents the distance between two individuals belonging to different clusters concerning and concerning only the variables belonging to the reference block.

Average Contributions Matrix. Finally, for each block of variables, it is possible to calculate the average contributions by relating the average distance matrices for each block to the overall average distance. We define the average contribution of the block as:

$$C^{(b)} = \frac{\bar{D}^{(b)}}{\bar{D}^{(tot)}} \tag{6}$$

it represents the relative contribution of block b, directly linked to the overall similarity structure among individuals. This allows for the analysis of the relative weight of each block of variables in the differentiation between groups.

By calculating the contribution of each block of variables, one can obtain the average contribution matrix $C \in \mathbb{R}^{G \times G}$:

$$C = \begin{bmatrix} C^{(b)}_{11} & \cdots & C^{(b)}_{1G} \\ \vdots & \ddots & \vdots \\ C^{(b)}_{G1} & \cdots & C^{(b)}_{GG} \end{bmatrix}$$

3 Showcase of the Methodology

The methodological framework described in the previous paragraph has been applied to an example dataset and the results obtained have been compared with those resulting from the application of Ordinal Approach and Confirmatory Factor Analysis.

3.1 The Dataset

The dataset used in the study was specifically constructed to determine whether the methodology would behave as expected, compared to traditional approaches, in a simple controlled environment. The data for each block of variables were generated using multinomial samples, adjusting the support according to the desired range of the ordinal variables, and also modifying the probability of each category to distinguish two clusters of observations. The ranges of the ordinal variables are as follows: from 1 to 6 for the variables in Block X, and from 1 to 4 for the variables in Blocks Y and Z. The probabilities were assigned in such a way that samples belonging to Cluster 1 were more likely to exhibit higher values for the variables in Blocks X and Y, whereas the opposite trend was observed for samples in Cluster 2. Conversely, the probabilities assigned to the categories of Block Z are uniform.:

$$prob\_X\_1 = (0.3, 0.3, 0.3, 0.033, 0.033, 0.033);$$

$$prob\_X\_2 = (0.033, 0.033, 0.033, 0.3, 0.3, 0.3);$$

$$prob\_Y\_1 = (0.4, 0.4, 0.1, 0.1);$$

$$prob\_Y\_2 = (0.1, 0.1, 0.4, 0.4);$$

$$prob\_Z = (0.25, 0.25, 0.25, 0.25).$$

3.2 Results: Traditional Approaches

For the purposes of the analysis, a CFA model was developed in which each variable, according to the model specification shown in Eq. 7, contributes to its a priori assigned block. The model's fit indices (Tucker Lewis Index and Comparative Fit Index equal to one) suggest that it is well-identified, indicating that the latent variables and their corresponding blocks of elementary indicators can be considered as equivalent in terms of their interpretive contribution to the clustering process. The same conclusion can be derived from the Table 1 of the factor loadings. Using the CFA's factors, a cluster analysis is carried out using the k-means algorithm with the Euclidean distance. The number of groups chosen is two, based on the higher value of silhouette returned during the phase of internal validation.

In order to describe the obtained clusters descriptive statistics of the factors are employed. By looking at the first two rows of the Tables 2 and 3, it is clear

Table 1. Estimated factor loadings for each indicator on three latent factors, including standard errors, z-values, and p-values.

	Indicator	Estimate	Std. Err	z-value	P-value
Factor 1	X1	1.000	–	–	–
	X2	0.958	0.048	20.154	0.000
	X3	0.946	0.047	20.180	0.000
	X4	0.990	0.047	21.036	0.000
	X5	0.990	0.046	21.363	0.000
Factor 2	Y1	1.000	–	–	–
	Y2	0.949	0.071	13.418	0.000
	Y3	1.065	0.073	14.558	0.000
	Y4	0.946	0.070	13.421	0.000
Factor 3	Z1	1.000	–	–	–
	Z2	–0.872	0.713	–1.223	0.221
	Z3	1.063	0.907	1.172	0.241

that Cluster 1 has higher mean values of Factor 1 and Factor 2 than Cluster 2, while the mean value of Factor 3 seems to be irrelevant in the interpretation of the differences between the clusters.

$$\begin{aligned} &\textit{Factor 1} \sim X_1 + X_2 + X_3 + X_4 + X_5 \\ &\textit{Factor 2} \sim Y_1 + Y_2 + Y_3 + Y_4 \\ &\textit{Factor 3} \sim Z_1 + Z_2 + Z_3 \end{aligned} \tag{7}$$

Table 2. Descriptive statistics for Cluster 1—CFA.

	mean	min	max	range	skew	kurtosis
Factor 1	0.213	–0.002	0.417	0.419	–0.093	–0.234
Factor 2	0.179	0.007	0.342	0.335	–0.123	–0.280
Factor 3	–0.003	-0.084	0.081	0.165	–0.131	–0.550

The same cluster analysis is also carried out on the factors derived from the Ordinal Approach. The results are quite consistent with the partition derived from the CFA, Table 4 and Table 5, in the sense that the first two factors characterize the groups the most, while the third one seems to be less determinant.

Table 3. Descriptive statistics for Cluster 2—CFA

	mean	min	max	range	skew	kurtosis
Factor 1	-0.216	-0.423	0.417	0.422	0.050	-0.351
Factor 2	-0.182	-0.348	0.342	0.348	0.078	-0.439
Factor 3	0.003	0.077	0.081	0.155	0.065	-0.651

Table 4. Descriptive Statistics Cluster 1—Ordinal Approach

	mean	min	max	range	skew	kurtosis
Block X	3500.17	1905.5	4594.5	2689.0	-0.28	-0.02
Block Y	2619.92	983.5	3507.5	2524.0	-0.29	-0.13
Block Z	1491.29	391.0	2642.5	2251.5	0.01	-0.49

Table 5. Descriptive Statistics Cluster 2—Ordinal Approach

	mean	min	max	range	skew	kurtosis
Block X	1480.60	570.0	3089.5	2519.5	0.27	-0.22
Block Y	1369.07	490.0	2993.0	2503.0	0.46	0.08
Block Z	1511.95	391.0	2642.5	2251.5	-0.13	-0.57

3.3 Results: Proposed Methodology

The proposed methodology starts with the choice of a distance measure, which has to satisfy the property in Eq. 3. In this case is used the Manhattan Distance both for the clustering process and the calculation of the contribution. The adopted clustering algorithm is Partitioning Around Medoids (PAM), due to the ordinal nature of the data: the optimal number of clusters according to the silhouette value, also in this case, is two. This ensures the possibility of a comparison between the results of the traditional approaches, without the need to artificially restrict the analysis to a specific number of clusters. In this case, as hinted before, the focus of description relies on all the elementary indicators, that reflect the behavior of the latent variables. This means that the variables X1 to X5, namely the *Block X*, can be associated to Factor 1 (either of the CFA or the ordinal approach), and the same is true for *Block Y* and *Block Z*. The first two blocks reflect the previous configurations. As shown in Table 6, Block X and Block Y have higher median values over Cluster 1 than over Cluster 2. Block Z seems to remain non-impactful apart from the median of Z1 on Cluster 2.

A more comprehensive look at the distributions (Figs. 1, 2 and 3) of the variables over the clusters, helps to highlight what has been already stated.

Finally, the core of this method is the Average Contribution Matrix in Table 7 which helps to quantify the contribution of the blocks to the intra and inter-clusters variances. In this analysis, the intra clusters contribution is around 30%,

Table 6. Descriptive statistics for variables of Cluster 1 and Cluster 2

	Median (C1)	Median (C2)	Min	Max
X1	0.800	0.200	0.000	1.000
X2	0.800	0.200	0.000	1.000
X3	0.800	0.200	0.000	1.000
X4	0.800	0.200	0.000	1.000
X5	0.800	0.200	0.000	1.000
Y1	0.667	0.333	0.000	1.000
Y2	0.667	0.333	0.000	1.000
Y3	0.667	0.333	0.000	1.000
Y4	0.667	0.333	0.000	1.000
Z1	0.333	0.667	0.000	1.000
Z2	0.333	0.333	0.000	1.000
Z3	0.333	0.333	0.000	1.000

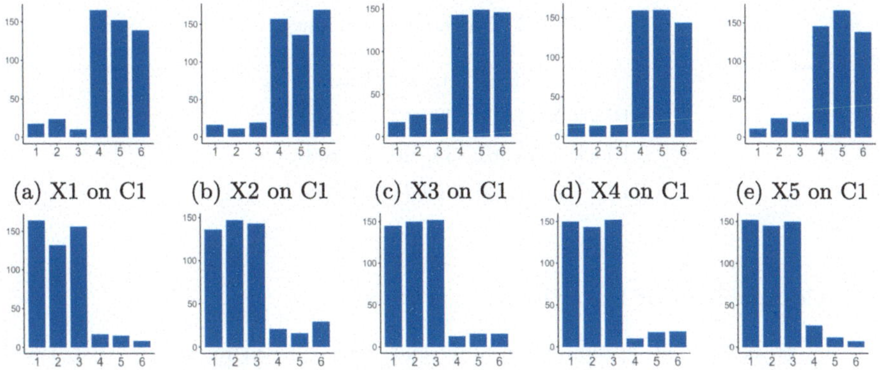

(a) X1 on C1 (b) X2 on C1 (c) X3 on C1 (d) X4 on C1 (e) X5 on C1

(f) X1 on C2 (g) X2 on C2 (h) X3 on C2 (i) X4 on C2 (j) X5 on C2

Fig. 1. Block X: frequency plots for each variable and cluster

suggesting that no block is predominant in the characterization. In fact, a lower contribution of some block would have hinted that individuals belonging to the cluster were more similar according to the variables of that specific block, thus allowing to characterize the cluster using these variables. The inter cluster contribution, instead, leads to confirm that the principal responsible for the differences between Cluster 1 and 2 is Block X, followed by Block Y and Block Z. In conclusion, with the proposed method it is possible to achieve equivalent results in terms of clustering and groups interpretation with respect to the traditional methods, without any dimensionality reduction, but still maintaining information relative to the latent variables.

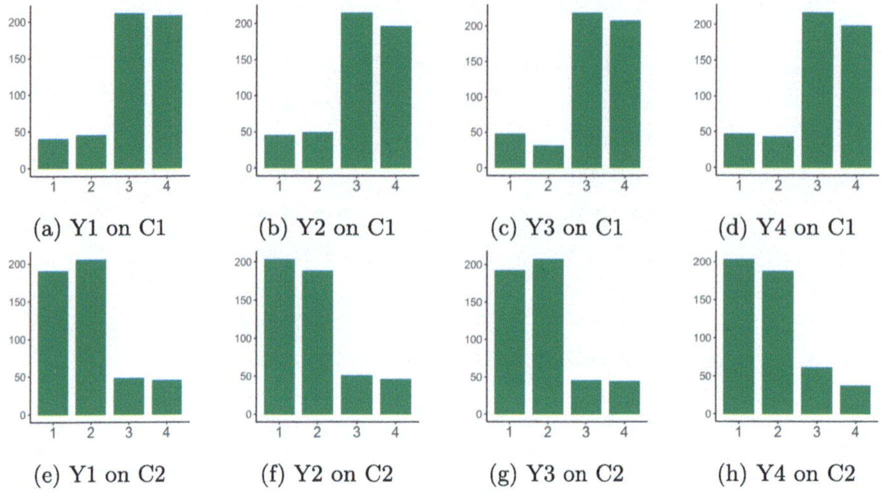

Fig. 2. Block Y: frequency plots for each variable and cluster

Table 7. Block contributions

	IntraC1	IntraC2	InterC1C2
Block X	33.808	33.312	44.054
Block Y	33.061	33,746	34,758
Block Z	33,130	32,941	21,186

3.4 Comparison of Partitions

In order to assess the agreement among the three clustering procedures, PAM on the raw ordinal data, CFA based clustering (CFA-bc), and Ordinal Approach based clustering (OA-bc), and to verify their alignment with the known ground truth partition, the Adjusted Rand Index (ARI) was computed. Table 8 reports the pairwise ARI between the three methods and the ARI of each method with respect to the ground-truth labels.

From Table 8, it is evident that all pairwise ARI values exceed 0.94, indicating very high agreement among the three clustering partitions. Moreover, each of the three partitions closely matches the known ground-truth clusters, signal that all three procedures almost perfectly recover the "true" partition of the data. We conclude that our proposed approach yields a clustering solution that is fully consistent with traditional methods and the known ground truth, while still providing the additional interpretability of per-block contributions without prior feature reduction.

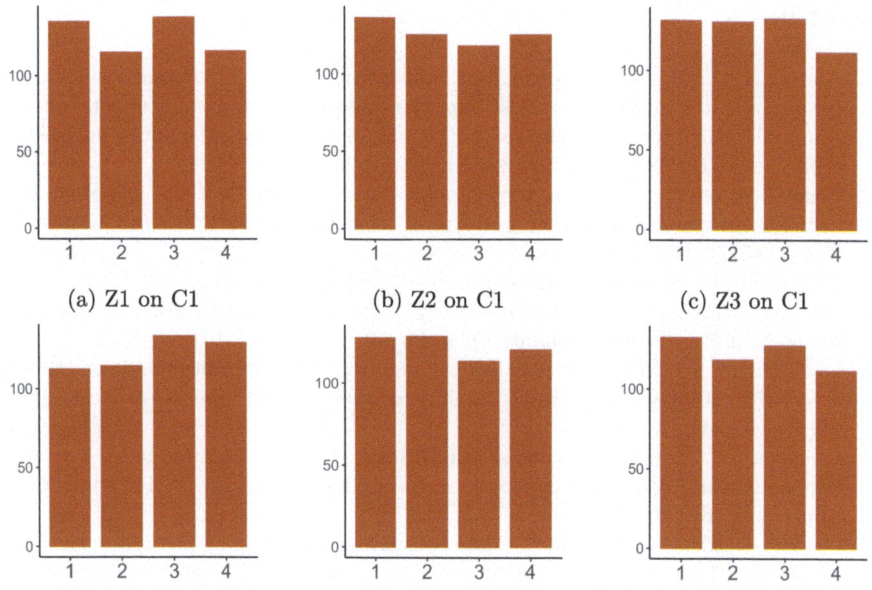

(a) Z1 on C1 (b) Z2 on C1 (c) Z3 on C1

(d) Z1 on C2 (e) Z2 on C2 (f) Z3 on C2

Fig. 3. Block Z: frequency plots for each variable and cluster

Table 8. ARI comparing the proposed method against CFA-based and Ordinal-Approach clustering, alongside each method's ARI with respect to the known ground-truth partition.

Method	CFA–bc	OA–bc	Ground Truth
Proposed PAM	0.9447	0.9604	0.9486
CFAâĂŞbc	—	0.9521	0.9643
OAâĂŞbc	—	—	0.9722

4 Conclusions

The application of our methodology has produced, in conclusion, results consistent with those of traditional approaches, remaining true to the purpose established during the construction of the sample dataset. The main advantages that we can highlight as strengths are: preserving the overall variability of the available data, avoiding the introduction of bias or distortions, and also maintaining the ordinal nature of the data, which undergo only the normalization process.

However, these conclusions can't be generalized since the methodology was tested only on one example dataset. For this reason, a simulation study is in the works to reach more robust evidence.

References

1. Abdullah, M., Ahmed, R., Altun, Y.: Comparison between factor analysis and cluster analysis to determine the most important affecting factors for students' admission and their interests in the specializations: A sample of salahaddin university-erbil. Zanco J. Pure Appli. Sci. **34**(s6), 12–23 (2022)
2. Bédard, A., et al.: Confirmatory factor analysis compared with principal component analysis to derive dietary patterns: a longitudinal study in adult women. J. Nutr. **145**(7), 1559–1568 (2015)
3. Delvecchio, F.: Scale di misura e indicatori sociali. Cacucci, Bari (1995)
4. Denk, T., Christensen, H.: How to classify political cultures? a comparison of three methods of classification. Quality Quantity **50**, 177–191 (2016)
5. Di Cimbrini, T.: Welfare or politics? the identity of italian mutual aid societies as revealed by a latent class cluster analysis of their annual reports. Account. Hist. **20**(3), 310–341 (2015)
6. D'Souza, C.: Game meats: consumption values, theory of planned behaviour, and the moderating role of food neophobia/neophiliac behaviour. J. Retailing Consumer Serv. **66**, 102,953 (2022)
7. Gaylord-Harden, N., Gipson, P., Mance, G., Grant, K.: Coping patterns of African American adolescents: a confirmatory factor analysis and cluster analysis of the children's coping strategies checklist. Psychol. Assess. **20**(1), 10 (2008)
8. Goertz, G.: Social Science Concepts: A User's Guide. Princeton University Press (2006)
9. Goertz, G.: Social Science Concepts and Measurement: New and Completely Revised Edition. Princeton University Press (2020)
10. Gründler, K., Krieger, T.: Should we care (more) about data aggregation? Euro. Econ. Rev. **142**, 104,010 (2022)
11. Gucciardi, G.: Measuring the relative development and integration of eu countries' capital markets using composite indicators and cluster analysis. Rev. World Econ. **158**(4), 1043–1083 (2022)
12. Higuchi, A., Maehara, R.: A factor-cluster analysis profile of consumers. J. Bus. Res. **123**, 70–78 (2021)
13. Kovács, T., Ko, A., Asemi, A.: Exploration of the investment patterns of potential retail banking customers using two-stage cluster analysis. J. Big Data **8**(1), 1–25 (2021). https://doi.org/10.1186/s40537-021-00529-4
14. Landaluce-Calvo, M., Gozalo-Delgado, M.: Proposal for a dynamic composite indicator: application in a comparative analysis of trends in the eu member states towards the europe 2020 strategy. Soc. Indic. Res. **154**, 1031–1053 (2021)
15. Mazur-Wierzbicka, E.: Measurement of progress in the environmental area: Poland against the countries of the european union. Int. J. Environ. Res. Public Health **20**(1), 563 (2022)
16. Młynarzewska-Borowiec, I.: Implementation of the europe 2020 sustainable growth priority in eu member states: assessment for the period 2010–2020. In: Financial Stability, Economic Growth and Sustainable Development, pp. 229–250. Routledge (2024)
17. Munck, G., Verkuilen, J.: Conceptualizing and measuring democracy: evaluating alternative indices. Comp. Pol. Stud. **35**(1), 5–34 (2002)
18. Prosser, M., Trigwell, K.: Confirmatory factor analysis of the approaches to teaching inventory. Br. J. Educ. Psychol. **76**(2), 405–419 (2006)

19. Rakhmawan, S.: Work from home in the era of covid-19 pandemic: the analysis and impact of it. East Java Econ. J. **5**(2), 221–244 (2021)
20. Slattery, M.: Analysis of dietary patterns in epidemiological research. Appl. Physiol. Nutr. Metab. **35**(2), 207–210 (2010)
21. Teorell, J., Coppedge, M., Lindberg, S., Skaaning, S.E.: Measuring polyarchy across the globe, 1900–2017. Stud. Comp. Int. Dev. **54**(1), 71–95 (2019)
22. Vettori, G., Bigozzi, L., Miniati, F., Vezzani, C., Pinto, G.: Identifying pre-service teachers' profiles of conceptions of learning: a cluster analysis. Soc. Psychol. Educ. **22**(5), 1131–1152 (2019)

Gender Stereotypes and Barriers in STEM: A Bayesian Statistical Analysis of Perceptions and Challenges

Rossella Duraccio[1], Maria Iannario[1], Claudia Tarantola[2(✉)], and Roberta Varriale[3]

[1] University of Naples Federico II, via L. Rodinó, 22, Naples, Italy
[2] University of Milan, Via Conservatorio 7, Milan, Italy
claudia.tarantola@unimi.it
[3] Sapienza University, Piazzale Aldo Moro, 5, Rome, Italy

Abstract. The study examines gender stereotypes in STEM fields using advanced statistical techniques within a Bayesian framework. The analysis relies on a survey collected from Italian participants in 2021, investigating perceptions of STEM subjects and professional careers. By employing supervised learning for predictive analysis based on a multilevel ordinal regression model, we aim to enhance the understanding of the barriers women face in STEM, including biases and stereotypes that shape career expectations and opportunities. Additionally, we estimate the marginal effects of key predictors to quantify the impact of factors such as gender, age, education, and workplace environment on perceptions of STEM. This approach not only enhances statistical methodologies but also provides insights into real-world social challenges.

Keywords: Bayesian Analysis · Gender Parity · Marginal effect · Multilevel Logit Model · Social Challenges · STEM

1 Introduction

STEM (Science, Technology, Engineering, and Mathematics) education plays a crucial role in both Italy and globally, as technological innovation and evolving labor market demands increasingly drive companies to seek candidates with scientific and technological expertise. STEM careers are generally associated with higher salaries, job security, and strong professional opportunities. However, despite these advantages, women remain significantly under-represented in STEM fields [3]. Recent estimates suggest that, at the current pace, it may take as long as 258 years to achieve gender parity in STEM employment [10].

This gender imbalance is not simply the result of personal choice but reflects a complex interaction of structural barriers and cultural norms. Women in STEM frequently report workplace discrimination, with nearly half indicating such experiences. According to the European Institute for Gender Equality

A. D'Ambrosio et al. (Eds.): CLADAG-VOC 2025, SCDAKO, pp. 132–143, 2025.
https://doi.org/10.1007/978-3-032-03042-9_12

(EIGE), 47% of those affected are Italian. Additionally, women in STEM roles tend to earn less than their male counterparts in equivalent positions, further exacerbating gender inequality. Deep-rooted stereotypes continue to discourage women from pursuing STEM careers, reinforcing the perception of these fields as male-dominated and unwelcoming, and thereby limiting female participation and advancement.

One key obstacle is the challenge of work-life balance. Women often bear a disproportionate share of caregiving and household responsibilities, which limits their ability to engage in the demanding nature of many STEM roles. Many employers lack adequate flexibility and support systems to accommodate individuals managing these dual responsibilities. The shortage of female role models and mentors in senior STEM positions further discourages young women from entering the field and advancing professionally.

However generally, what most strongly influences the under-representation of women in STEM is the pivotal role played by gender stereotypes. From an early age, cultural and educational biases lead to the internalization of science and technology as male domains. A recent study involving Italian high school students and teachers [20] revealed that gender stereotypes continue to shape students' self-perceptions and teachers' expectations, thus strengthening gendered educational and professional pathways. In Italy, these dynamics persistently influence students' choices, contributing to ongoing disparities in access to and success within STEM careers [17].

Broader international research confirms that students' preferences for STEM careers are shaped by an intricate interplay of social, cultural, and disciplinary factors [5]. These can either amplify or mitigate gendered perceptions of science and technology [16]. For instance, males consistently outperform females in mathematics, physics, chemistry, and computer science, while females show a slight advantage in biology [4,17]. These performance gaps—real or perceived—contribute to gendered patterns in subject interest and career choice.

These stereotypes not only undermine confidence in mathematical and technical abilities among girls and women but also feed into institutional barriers, such as biased hiring practices and limited access to mentorship or leadership roles. The so-called 'leaky pipeline'—the gradual attrition of women from STEM pathways at each stage of education and career—can thus be seen as the cumulative result of persistent cultural narratives and structural inequities [19].

This study aimed to explore gender stereotypes within STEM fields, using data collected through a survey conducted from November 20 to December 11, 2021. A total of 1003 individuals from various regions of Italy participated, with 77% of the respondents being female. The survey was carried out as part of the activities of the Laboratory for Statistical Data Analysis at the University of Federico II.

Data collection was based on a snowball sampling strategy, initiated through the involvement of university student associations, which facilitated the dissemination of an online questionnaire among a wide and diverse group of participants. This approach allowed for the engagement of individuals across different academic backgrounds and regions, thus enriching the variety of perspectives included in the study.

The questionnaire included several sections: demographic and personal information, opinions about STEM subjects, and perceptions about people working in STEM.

The survey data are complex due to their hierarchical structure, the nature of the variables, and potential biases. Responses are nested within contextual factors such as region of residence, parental education, and female employment rates, creating dependencies that require a multilevel modeling approach. Additionally, voluntary participation introduces self-selection bias, with a higher proportion of female respondents potentially influencing results. Social desirability bias may also affect responses, as participants might select socially acceptable answers rather than their true beliefs. Furthermore, relationships among variables are highly interdependent, as perceptions of STEM careers are shaped by multiple interacting factors, including gender, education, and socio-economic background.

To analyze the data, we employ a multilevel ordinal logit model within a Bayesian framework, which allows us to account for the hierarchical structure of responses while uncovering underlying patterns in gender stereotypes and their influence on women's participation in STEM. The study demonstrates the critical role of statistical science in modern data analysis by applying these techniques to real-world data, offering novel insights into gender disparities in STEM fields. The remainder of the paper proceeds as follows. Section 2 describes the data and survey design. Section 3 provides a brief overview of the methodology, while Sect. 4 presents the results, followed by a discussion. Finally, Sect. 5 concludes the paper and suggests directions for future research.

2 Data and Survey Design

The questionnaire was structured into several sections to capture different aspects of gender stereotypes in STEM

- *Demographics and personal information*: gender, age, and region of residence.
- *Opinions regarding STEM subjects*: perceptions of scientific disciplines and career aspirations.
- *Perceptions about individuals working in STEM*: beliefs about gender roles and representation in STEM fields.

The selected item under investigation and the corresponding covariates are theoretically informed by the existing literature on gender stereotypes and obstacles in STEM and have been validated in previous research. Their inclusion was based on theoretical and empirical relevance, reflecting established findings that students' beliefs and expectations about STEM subjects and careers are shaped by gendered socialization processes, educational contexts, and dominant cultural narratives [3,5,20].

In particular, constructs such as 'STEM belonging' [5] (see Autoval in Table 1), perceived competence [16], and stereotypical attributions of scientific talent [18] (see Internalized image of scientist in Table 1) have been shown to play a crucial role in explaining gender differences in STEM participation and persistence. The design of the questionnaire therefore reflects a multidimensional approach, incorporating both attitudinal and contextual variables, in line with grounded theories of stereotype internalization and expectancy-value models of achievement motivation [21]. Respondents rated their agreement with key statements using a 5-point Likert scale (1 = strongly disagree, 5 = strongly agree). The primary focus of this contribution is gender bias in STEM, assessed through a key question selected for in-depth analysis. Specifically, the ordinal variable under examination is:

> "Men tend to have a greater inclination toward studying scientific subjects."

A summary of key respondent characteristics, including demographic attributes, educational background, and attitudes toward STEM, is presented in Table 1.

3 Multilevel Ordinal Logit Model Within a Bayesian Framework

The ordinal regression model is motivated by assuming an underlying latent continuous variables that is related to the observed ordinal response through the threshold concept [9].

The analysed data are present in a hierarchical structure yielding a multilevel framework [7,8,15]. Therefore, in this analysis, we propose a multilevel ordinal logistic model to estimate the odds of being in a particular stage k as

$$\ln\left(\frac{P(Y \leq k|X\beta + Z\mu)}{P(Y > k|X\beta + Z\mu)}\right) = \tau_k - (X\beta + Z\mu) \tag{1}$$

where $X\beta$ captures the *global (population-level) effects* of the covariates, $Z\mu$ represents the *group-level (random) effects*, allowing for variability across groups, and τ_k are the threshold parameters for the ordinal response.

In the analysis we define the following grouping levels: (i) region of residence (20 levels); (ii) parental education level (4 levels); and (iii) female employment rate by region (3 levels: low, medium, high).

Therefore, in this framework we distinguish between two key sources of variability. Between-group variability which is captured by the random effects variance σ_u^2, and within-group variability σ_e which refers to the residual variation within each group, capturing individual-level differences that are not explained by the model.

When a logit link is used, as in the present model, the individual-level residual variance (commonly referred to as the within-group variance) is fixed to $\pi^2/3 \approx$

Table 1. Characteristics of survey respondents.

Characteristic	Percentage (%)
Gender	
Female	77
Male	23
Generation	
Generation Z (1997 or later)	58
Born before 1997	42
Region of Residence	
Northern Italy	35
Central Italy	40
Southern Italy	25
Female Employment Rate in Region	
Low	30
Medium	45
High	25
Parents' Educational Qualifications (Parent_edu)	
No formal education	5
Middle/High school diploma	45
Bachelor's/Master's degree	40
Ph.D.	10
Being confident in doing science (Autoval)	
Yes	53
No	26
Do not know	21
Role of Teachers in STEM Interest	
Effective	70
Not Effective	30
Internalized Image of Scientists	
Typically male	50
Typically female	10
Any gender	40
Work-Family Balance Perception	
Agree that balance is not important	55
Disagree	45

3.29 due to the logistic distribution assumption for the latent variable. This constraint ensures identifiability of the model parameters [1]. As a result, only the between-group variability—captured through group-level random effects—is estimated, enabling a meaningful decomposition of the total variance across hierarchical levels.

In this study, we adopt a Bayesian approach, as it is particularly well-suited when the number of groups is small or when dealing with multiple hierarchical levels—both of which characterize our case study. Combining the likelihood with the priors distributions on the model parameters, that we assume to be independent, we obtain the corresponding posterior distribution

$$P(\boldsymbol{\tau}, \boldsymbol{\beta}, \sigma_u^2, \sigma_e^2 \mid Y) \propto P(Y \mid \tau, \boldsymbol{\beta}, \sigma_u^2, \sigma_e^2) P(\tau) P(\boldsymbol{\beta}) P(\sigma_u^2). \tag{2}$$

Here $P(\boldsymbol{\tau})$ and $P(\boldsymbol{\beta})$ are the priors on the thresholds and fixed effects; $P(\sigma_u^2)$ is the prior on the between-group variability.

In our setting, we use a weakly informative prior on the coefficients $\boldsymbol{\beta}$ and on the intercepts τ_k.

$$\boldsymbol{\beta} \sim \mathcal{N}(\tilde{\beta}, 100)$$
$$\tau_k \sim \mathcal{N}(\tilde{\tau_k}, 100)$$

where $\tilde{\beta}$ and $\tilde{\tau_k}$ are the maximum likelihood estimates (MLE) of a baseline cumulative link model obtained using the `polr` function in the `MASS` package in R, [22].

For between-group variances, we apply Student's t-priors with 3 degrees of freedom. This choice ensures wide tails, allowing for greater robustness to outliers while maintaining a proper distribution with finite mean and variance. Here, the mean and variance are set—without loss of generality to 0 and 2.5, respectively.

In addition, to quantify the proportion of total variance attributable to group-level structure, we computed the Intraclass Correlation Coefficient (ICC) for each grouping factor using the following expression:

$$\text{ICC} = \frac{\sigma^2_{\text{between}}}{\sigma^2_{\text{between}} + \pi^2/3} \tag{3}$$

Since the prior are not conjugate to the model we use Markov Chain Monte Carlo techniques to simulate from the corresponding posterior distribution.

Model estimation has been performed using the Stan (`brms` R package, [2]). A Hamiltonian MCMC sampling was conducted with four chains, each with 2000 iterations and 200 bun-in. Standard convergence diagnostics such as graphical examinations of traceplot and calculation of the Gelman and Rubin $\hat{R}$ statistic of [6] are employed.

We also compute the marginal effect of covariates, with an example for the analysis of fixed effects being:

$$\frac{\partial P(Y = k|X, Z)}{\partial x_k} = \beta_k \left(\Lambda(\tau_k - X\beta - Z\mu) - \Lambda(\tau_{k-1} - X\beta - Z\mu)\right) \tag{4}$$

where $\Lambda(\cdot)$ denotes the cumulative logistic function.

Bayesian estimates of the marginal effect estimates are obtained via ergodic means from the posterior samples. More precisely, we draw samples from the posterior distributions of β_k, τ_k, and μ, computing the corresponding derivatives for each sample and averaging over them. While we use the logistic function for robustness reasons, other link functions can also be explored as alternatives. Further methodological developments can be found in the manuscript by [12].

4 Results and Discussion

We present the posterior estimates from the Bayesian multilevel ordinal logit model applied to the STEM perception data. As already mentioned, due to the voluntary nature of participation, the sample is self-selected and not random. Therefore, response rate cannot be precisely estimated, but post-stratification weights could be considered for future analyses.

To evaluate potential multicollinearity issues in the model, we examined the posterior correlation matrix of the fixed-effect coefficients. As expected in models that include interaction terms, some moderate to high correlations emerged between structurally related parameters - in particular, between main effects and their corresponding interactions. These correlations reflect the model structure and are not indicative of problematic collinearity.

Importantly, all diagnostic criteria confirmed the reliability of the posterior estimates: the Markov chains showed good convergence, the effective sample sizes were sufficiently high, and posterior distributions were unimodal and well-behaved. These results indicate that, although multicollinearity arises naturally due to the inclusion of interaction terms, it does not compromise inference. The Bayesian framework, supported by weakly informative priors, offers additional robustness by regularizing parameter estimates and enhancing the stability and interpretability of the results. While estimation remains stable, caution is warranted when interpreting individual coefficients due to the inherent dependencies introduced by interaction terms.

The predictor of the estimated model includes *Internalized Image of Scientists*, *Being confident in doing science*, *Role of Teachers in STEM Interest*, *Generation, Gender* for capturing the global effects and *Parent_edu*, *Work-Family Balance Perception*, *Region of residence* for assessing the group-level (random) effects.

The main results on the regression parameters β, on which we focus our interpretation in this contribution, are in Table 2. A preliminary assessment of parameter significance was conducted using the HDI+ROPE decision rule [13, 14].

The gender-related covariates (e.g., Internalized Image of Scientists: *male*, Gender: *female*) suggest that male respondents are more likely to identify with the image of a scientist and to perceive themselves as naturally inclined toward studying scientific subjects, compared to female respondents.

Table 2. Regression Coefficients

Parameter	Estimate	Est.Error	l-95% CI	u-95% CI	ESS
$\hat{\tau}_1$	−0.10	1.12	−2.44	2.10	2109
$\hat{\tau}_2$	0.85	1.12	−1.51	3.06	2121
$\hat{\tau}_3$	2.10	1.13	−0.29	4.30	2132
$\hat{\tau}_4$	3.82	1.14	1.43	6.04	2180
Internalized Image of Scientists *male*	1.75	0.61	0.59	2.96	3924
Internalized Image of Scientists *female*	0.03	0.88	−1.67	1.86	3324
Autoval YES	0.07	0.41	−0.73	0.87	2769
Autoval NO	0.09	0.38	−0.64	0.83	2910
Role of Teacher (effective)	−0.16	0.13	−0.42	0.08	10209
Generation before 1997	0.47	0.25	−0.02	0.97	5035
Gender *female*	−0.34	0.35	−1.02	0.34	2865

Generations born before 1997 exhibit a positive effect likely driven by institutional restrictions, differences in career paths, or varying attitudes toward innovation and change. The role of teachers, instead, presents a negative coefficient. This may reflect a no differential impact of teacher encouragement depending on the student's gender or pre-existing attitudes, potentially by fostering more inclusive environments and encouraging equal opportunities. A particularly striking finding emerges when examining cluster effects (whose parameters are not reported here for simplicity): in regions with average to high female employment rates, women are more likely to perceive STEM as an exclusive field or predominantly suited to men. This counterintuitive trend may be driven by the fact that increased female employment is often concentrated in sectors that offer better work-family balance, reinforcing traditional career choices.

The estimated ICCs suggest that most of the variability in STEM-related perceptions is explained by individual-level characteristics (see Table 3). However, certain group-level factors—such as family occupational background and region of residence—exhibit non-negligible between-group variance, indicating the presence of contextual effects.

Table 3. Intraclass Correlation Coefficients (ICC) for random grouping factors

Group	ICC (median)	95% CI Low	95% CI High
Parent_edu	0.007	0.00001	0.149
family_work	0.114	0.025	0.521
Reg_resid	0.104	0.00016	0.807

The results of this study are inherently shaped by the specific cultural and societal contexts in which it was conducted. Variations in how terms such as ‘scientist’ or ‘gender’ are interpreted across different regions or populations can impact the generalizability of the findings beyond the studied group.

To illustrate these dynamics, in Fig. 1 we present a graph depicting the marginal effects for gender and age groups, categorized by generation, gender, and perceptions of the image of a scientist. The marginal effects by gender indicate that gender significantly influences outcomes, with women exhibiting slightly lower probabilities of engaging in certain behaviors compared to men. This suggests that gendered perceptions and societal expectations may play a role in shaping decision-making processes.

Across all graphical representations, we observe greater uncertainty in the first response category (strong disagreement), likely due to smaller sample sizes in this group. Additionally, the frequency of responses in the strongly agree category remains relatively low across all three cases presented.

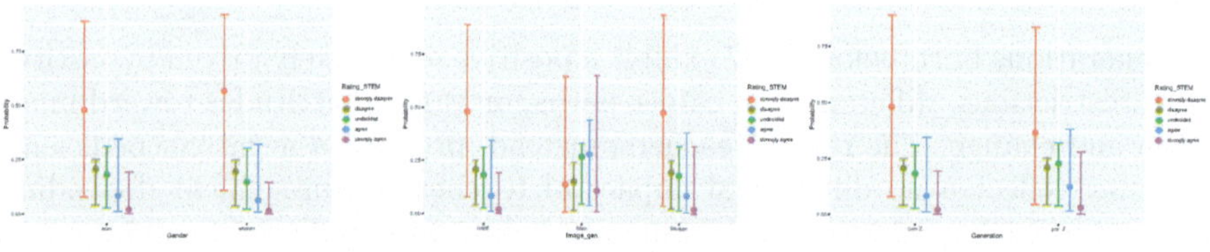

Fig. 1. Marginal effects illustrating the relationship between the predictors and the outcome variable, highlighting the impact of key factors such as generation, gender, and perception of a scientist’s image.

5 Conclusion and Future Work

This study provides a statistical framework for analyzing gender disparities in STEM, leveraging a Bayesian multilevel ordinal model to account for the hierarchical nature of survey data. Our results highlight the persistence of gender stereotypes and structural barriers that influence women’s participation in STEM fields. Notably, men are more likely to perceive themselves as naturally inclined toward scientific subjects, reinforcing gender biases from an early stage. Moreover, regional differences in female employment rates suggest that broader socio-economic factors shape gender perceptions in STEM, further emphasizing the need for targeted interventions.

This study provides valuable insights into the factors influencing perceptions and behaviours across different generations and genders. Our analysis demonstrates that generation plays a significant role, with individuals born before 1997 showing a moderate positive effect in comparison to other groups. Additionally, the role of gender reveals notable differences, suggesting that societal perceptions and biases may influence the way individuals engage with various professional fields, particularly in the context of science. While this study offers valuable insights into the dynamics of generation, gender, and the perception of scientists, there are several limitations to consider. Moreover, as already mentioned, due to the voluntary nature of participation, the sample is self-selected and not random. Therefore, the response rate cannot be accurately estimated, but post-stratification weights could be considered for future analyses. The sample size for certain groups, particularly in extreme response categories, is relatively small, leading to increased uncertainty in the analysis. This highlights the need for an analytical approach based on response styles to better understand and interpret these variations. A larger and more diverse sample could provide a more comprehensive understanding of the factors at play. In addition, the measures used to capture perceptions of scientists and related stereotypes may not fully encompass the complexity of how these perceptions are formed or how they influence behavior. More detailed or qualitative data could provide additional nuance. Despite these limitations, the findings offer important starting points for further research and highlight key areas for addressing gender bias and stereotypes in the scientific field. From a policy perspective, for instance, the findings underline the urgency of addressing gender stereotypes through structural changes in education and workplace environments. Strategies to mitigate stereotypes and promote inclusion should focus on:

- Increasing the visibility of female role models in STEM education and professional fields
- Encouraging gender-sensitive STEM education from early schooling stages.
- Promoting mentorship programs that provide guidance and networking opportunities for young women in STEM
- Implementing workplace policies that support work-life balance for both gender.

Despite the relatively large size (N = 1003), the sample is not representative due to its non-random nature and overrepresentation of women (77%). This limitation may influence generalizability. In future work, post-stratification techniques or sampling adjustments could help address these issues.

Future research should explore longitudinal data to track changes in gender perceptions over time and investigate the effectiveness of targeted interventions in reducing STEM-related stereotypes. Finally, methodological aspects, such as the development of ordinal superiority measures as already developed in a univariate context [11], constitute an element under study.

Acknowledgments. Maria Iannario and Claudia Tarantola gratefully acknowledge the support of the Marie Skłodowska-Curie Actions under the European Union's Horizon Europe research and innovation program for the Industrial Doctoral Network on Digital Finance (acronym: DIGITAL, project no. 101119635) and PRIN 2022 - CUP: E53C24002270006.

References

1. Agresti, A.: Analysis of Ordinal Categorical Data, 2nd edition. J. Wiley & Sons, Hoboken (2010)
2. Bürkner, P.C.: brms: an R package for Bayesian multilevel models using Stan. J. Stat. Softw. **80**(1), 1–28 (2017)
3. Charles, M., Bradley, K.: Indulging our gendered selves? Sex segregation by field of study in 44 countries. Am. J. Sociol. **114**(4), 924–976 (2009)
4. Cheryan, S., Ziegler, S.A., Montoya, A.K., Jiang, L.: Why are some STEM fields more gender balanced than others? Psychol. Bull. **143**(1), 1–35 (2017)
5. Dost, G.: Students' perspectives on the 'STEM belonging' concept at A-level, undergraduate, and postgraduate levels: an examination of gender and ethnicity in student descriptions. IJ STEM Ed **11**, 12 (2024)
6. Gelman, A., Rubin, D.B.: Inference from iterative simulation using multiple sequences. Stat. Sci. **7**, 457–472 (1992)
7. Gelman, A., Hill, J.: Data Analysis Using Regression and Multilevel/Hierarchical Models. Cambridge University Press (2006)
8. Hedeker, D., Gibbons, R.D.: Longitudinal Data Analysis. Wiley-Interscience (2006)
9. Hedeker, D., Gibbons, R.D.: A random-effects ordinal regression model for multilevel analysis. Biometrics **50**(4), 933–944 (1994)
10. Holman, L., Stuart-Fox, D., Hauser, C.E.: The gender gap in science: How long until women are equally represented? PLOS Biology **16**(4) (2018)
11. Iannario, M., Kateri, M., Tarantola, C.: Modelling scale effects in rating data: a Bayesian approach. Qual. Quant. **58**, 4053–4071 (2024)
12. Iannario, M., Tarantola, C.: Methodological Advancements in Bayesian Multilevel Models for Ordinal Data. Manuscript (2025)
13. Kruschke, J.: Doing bayesian data analysis: A tutorial with r, JAGS, and stan. Academic Press (2014)
14. Kruschke, J.K., Liddell, T.M.: The bayesian new statistics: Hypothesis testing, estimation, meta-analysis, and power analysis from a bayesian perspective. Psychonomic Bull. Rev. **25**, 178–206 (2018)
15. Natarajan, R., Kass, R.E.: Reference bayesian methods for generalized linear mixed models. J. Am. Stat. Assoc. **95**(449), 227–237 (2000)
16. Mansour, N.: Exploring the impact of social, cultural, and science factors on students' STEM career preferences. Res. Sci. Educ. **55**, 641–668 (2025)
17. Priulla, A., D'Angelo, N., Attanasio, M.: An analysis of Italian university students' performance through segmented regression models: gender differences in STEM courses. Genus **77**, 11 (2021)
18. Smeding, A.: Women in Science, Technology, Engineering, and Mathematics (STEM): an investigation of their implicit gender stereotypes and stereotypes' connectedness to math performance. Sex Roles **67**, 617–629 (2012)
19. Stoet, G., Geary, D.C.: The gender-equality paradox in science, technology, engineering, and mathematics education. Psychol. Sci. **29**(4), 581–593 (2018)
20. Sulla, F., Agueli, B., Lavanga, A., Logrieco, M.G.M., Fantinelli, S., Esposito, C.: Analysis of the development of gender stereotypes and sexist attitudes within a group of Italian high school students and teachers: a grounded theory investigation. Behav Sci (Basel). **15**(2), 230 (2025)

21. Wang, M.-T., Degol, J.: Gender gap in science, technology, engineering, and mathematics (STEM): Current knowledge, implications for practice, policy, and future directions. Educ. Psychol. Rev. **29**(1), 119–140 (2017)
22. Venables, W.N., Ripley, B.D.: Modern Applied Statistics with S, Fourth edition. Springer, New York (2002). ISBN 0-387-95457-0

Trajectory Reconstruction in Muon Scattering Tomography Using Two-Component Mixture Modelling

Marta Ferrari(✉), Alessandra R. Brazzale, and Giovanna Menardi

Department of Statistical Sciences, University of Padova, Via C. Battisti 241, 35121 Padua, PD, Italy
marta.ferrari.9@phd.unipd.it,
{alessandra.brazzale,giovanna.menardi}@unipd.it

Abstract. Muon tomography is a technique that aims to reconstruct the internal composition of an unknown volume by analyzing the trajectories of muons as they traverse the volume. The reference method proposed by Schultz et al. (2007) models the scattering angles and displacements of the observed muons using a Gaussian distribution and employs the Expectation-Maximization (EM) algorithm for maximum likelihood estimation. We propose an extension of this model which uses a mixture of two Gaussian distributions to better account for the heavy tails observed in the distribution of the scattering of muon tomography data. The proposed model is fitted using the stochastic EM (SEM) algorithm. An application to simulated data from an experiment aimed at evaluating the wear level in the inner walls of an insulating tube, is given.

Keywords: Mixture Model · Muon Tomography · Scattering · Stochastic EM

1 Introduction

1.1 Background and Motivation

Muons are subatomic particles in the lepton family, naturally produced when cosmic rays interact with Earth's upper atmosphere. These particles constantly shower the Earth at a rate of approximately 1 muon per cm^2 per minute. Due to their higher mass compared to electrons, muons possess strong penetrating power, enabling them to travel through several meters of dense material before being absorbed or decaying. Muon Scattering Tomography (MST) leverages this property to reconstruct three-dimensional images of the internal structure of complex or inaccessible volumes by analyzing the trajectories of these particles. Introduced by [1], MST has proven effective in diverse fields: in volcanology for imaging volcanic interiors and forecasting eruptions; in archaeology for non-invasive exploration of ancient structures; in security for detecting radioactive or illicit substances with minimal radiation risk; and in medicine as a potential

A. D'Ambrosio et al. (Eds.): CLADAG-VOC 2025, SCDAKO, pp. 144–155, 2025.
https://doi.org/10.1007/978-3-032-03042-9_13

alternative to electrons and protons in radiotherapy. One of the key advantages of this technique is its non-invasive and non-destructive nature, enabling the exploration of unknown volumes without causing damage or posing risks to living beings.

MST is based on the principle that, as muons pass through a material, they interact primarily with atomic nuclei via the Coulomb force, which causes them to slow down and deviate from their original trajectory. As a muon travels through a material, it undergoes multiple Coulomb scattering events, each resulting in a small deflection. The total deviation a muon experiences depends on the material it passes through. Materials with a high atomic number Z, such as tungsten and uranium, cause significant deflections in the muon's path, while medium-Z materials like steel and iron, and low-Z materials like carbon and aluminum, lead to smaller deviations. The goal of tomography is to study these scattering events and use the resulting data to estimate the material properties that the muons have traversed.

To assess the scattering distribution, detectors are placed on opposite sides of the volume being examined. These detectors track the location and angle of the muons. By analyzing these measurements, it is possible to determine how much each muon has been deflected from its original trajectory, thereby inferring the density of the material it has passed through. In this work, we aim to build upon the foundational model developed by [2] for multiple scattering and the estimation procedure used to reconstruct muon trajectories within a three-dimensional volume. After overviewing the geometry underlying muon scatter tomography (Sect. 1.2), we present the reference Gaussian model of [2] (Sect. 2.1). To account for the heavier tails of the distribution of the scattering events, we rather describe the scattering as a unimodal Gaussian mixture (Sect. 2.2). The model is estimated by maximum likelihood using a stochastic version of the Expectation Maximization (SEM) algorithm (Sect. 3). Since our model intrinsically includes a double latent structure, caused by both the mixture component and the unknown trajectory of the muons, a stochastic step needs to be added. The procedure is then illustrated within an application to simulated data (Sect. 4).

1.2 The Underlying Geometry

Let the volume under consideration be partitioned into M cubic voxels, each with side length s (measured in cm), which we shall assume to be small enough for the voxel being of homogeneous material (Fig. 1). Two sets of detectors, which are assumed to record measurements without error, track the positions $(X^{\text{IN}}, Y^{\text{IN}}, Z^{\text{IN}})$ and $(X^{\text{OUT}}, Y^{\text{OUT}}, Z^{\text{OUT}})$, as well as the angles $(\theta_x^{\text{IN}}, \theta_y^{\text{IN}})$ and $(\theta_x^{\text{OUT}}, \theta_y^{\text{OUT}})$ (measured in radians) at the entrance and exit of the investigated volume, respectively. The straight trajectory of a muon between $(X^{\text{IN}}, Y^{\text{IN}}, Z^{\text{IN}})$ and $(X^{\text{OUT}}, Y^{\text{OUT}}, Z^{\text{OUT}})$ is the sum of the path lengths L_j traveled by the muon in each voxel j, with $j = 1, \ldots, M$,

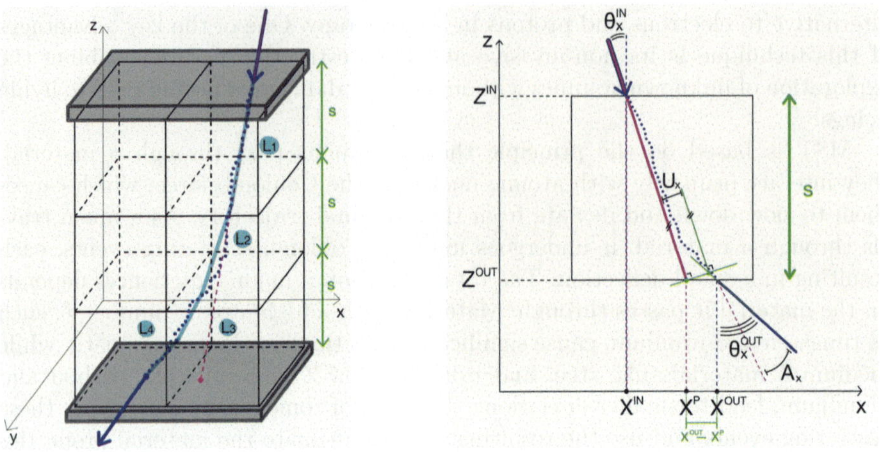

Fig. 1. Left: Scattering of a muon in a volume divided in 6 voxel. The real trajectory is represented in blue, the pink trajectory is the one the muon would pass if it was not deflected. Right: Graphical representation of the projection onto the xz-plan of a trajectory of a muon in a voxel, with indication of the involved quantities.

$$\sum_{j=1}^{M} L_j = \sum_{j=1}^{M} s\sqrt{1+\tan^2\theta_{x,j}^{\mathrm{IN}}+\tan^2\theta_{y,j}^{\mathrm{IN}}} = s\sum_{j=1}^{M} l_{xy,j},$$

where $\theta_{x,j}^{\mathrm{IN}}$ and $\theta_{x,j}^{\mathrm{OUT}}$ represent the muon's position at the entrance and exit of voxel j. The scattering angles $A_{x,j}$ and $A_{y,j}$ in voxel j are defined as

$$A_{x,j} = \theta_{x,j}^{\mathrm{OUT}} - \theta_{x,j}^{\mathrm{IN}}, \quad A_{y,j} = \theta_{y,j}^{\mathrm{OUT}} - \theta_{y,j}^{\mathrm{IN}}.$$

The displacements in the x and y directions for voxel j are given by

$$U_{x,j} = (X_j^{\mathrm{OUT}} - X_j^P)\cos(\theta_{x,j}^{\mathrm{IN}})l_{xy,j}\frac{\cos(A_{x,j}+\theta_{x,j}^{\mathrm{IN}})}{\cos(A_{x,j})},$$

$$U_{y,j} = (Y_j^{\mathrm{OUT}} - Y_j^P)\cos(\theta_{y,j}^{\mathrm{IN}})l_{xy,j}\frac{\cos(A_{y,j}+\theta_{y,j}^{\mathrm{IN}})}{\cos(A_{y,j})},$$

where (X_j^P, Y_j^P, Z_j^P) is the projected location of the muon's exit from voxel j if its trajectory had not been deflected, while $(X_j^{\mathrm{OUT}}, Y_j^{\mathrm{OUT}}, Z_j^{\mathrm{OUT}})$ is the scattered location.

The total scattering measures for the muon across the entire volume hence are

$$A_x = \sum_{j=1}^{M} A_{x,j}, \quad A_y = \sum_{j=1}^{M} A_{y,j},$$

and

$$U_x \approx \sum_{j=1}^{M}(U_{x,j} + T_j A_{x,j}), \quad U_y \approx \sum_{j=1}^{M}(U_{y,j} + T_j A_{y,j}), \tag{1}$$

where $T_M = 0$ and $T_j = \sum_{h=j+1}^{M} L_h$ for each $j \leq M - 1$. Scattering angles need to be assumed small enough to justify approximating the muon trajectories in (1) as straight lines. Each scattering measure is assumed to be independent between voxels and coordinates. If this condition is not met, higher-order terms in the displacement calculations must be taken into account, as these cannot be ignored.

2 A Two-Component Mixture Model for Scattering

2.1 The Baseline Model

To describe the probability density function of the scattering angle $A_{x,j}$ and displacement $U_{x,j}$ of a muon traversing a voxel j of homogeneous material with density λ_j at coordinate x, [2] propose the bivariate Gaussian model

$$H_{x,j} = (A_{x,j}, U_{x,j})^\top \sim N_2(0, \Sigma_{x,j}), \tag{2}$$

where $\text{vec}(\Sigma_{x,j}) = (\sigma^2_{A_{x,j}}, \sigma_{A_{x,j}U_{x,j}}, \sigma_{A_{x,j}U_{x,j}}, \sigma^2_{U_{x,j}})$. Assuming a unitary muon speed, and for a given momentum p of the muon, $\sigma^2_{A_{x,j}}$ can be approximated by

$$\sigma^2_{A_{x,j}} \cong \frac{15^2}{p^2}\frac{L_j}{r_j}, \tag{3}$$

where r_j is the radiation length of the material. The quantity of interest is the scattering density λ_j, which is specific to the voxel material and expressed in mrad$^2/cm$. It represents the mean square scattering angle of muons with nominal momentum p_0 passing through a unit depth of that material

$$\lambda_j = \left(\frac{15}{p_0}\right)^2 \frac{1}{r_j}.$$

Thus, setting $p_r^2 = (p_0/p)^2$, expression (3) can be re-written as

$$\sigma^2_{A_{x,j}} = \frac{15^2}{p^2}\frac{L_j\lambda_j p_0^2}{15^2} = L_j\lambda_j\left(\frac{p_0}{p}\right)^2 = L_j\lambda_j p_r^2.$$

Using similar arguments, [2] derive also the expressions for the remaining two terms in $\Sigma_{x,j}$, namely

$$\sigma^2_{U_{x,j}} = \frac{1}{3}L_j^2\sigma^2_{A_{x,j}} = \frac{1}{3}L_j^3\lambda_j p_r^2 \quad \text{and} \quad \sigma_{A_{x,j}U_{x,j}} = \frac{1}{2}L_j\sigma^2_{A_{x,j}} = \frac{1}{2}L_j^2\lambda_j p_r^2.$$

Some algebraic manipulation shows that, accounting for the geometry of the involved objects, $\Sigma_{x,j}$ becomes

$$\Sigma_{x,j} = \lambda_j \begin{bmatrix} L_j & \frac{L_j^2}{2} \\ \frac{L_j^2}{2} & \frac{L_j^3}{3} \end{bmatrix} p_r^2 = \lambda_j B_j p_r^2. \tag{4}$$

Analogous expressions hold for the angular scattering and displacement $H_{y,j} = (A_{y,j}, U_{y,j})^\top$ along the coordinate y, which are further assumed to be independent of their counterpart (2).

When a muon crosses a volume partitioned into M voxels, the angular scatterings and displacements $H_{x,j}$ and $H_{y,j}$ of the single voxels are not directly observed, and thus need be treated as hidden variables. We only observe the total angular scattering and displacement of a muon occurring across all M voxels along the x and y-axes, $D_x = [A_x, U_x]^\top$ and $D_y = [A_y, U_y]^\top$. As derived in [2], the overall scattering for a muon is described by

$$(D_x, D_y)^\top \sim N_4\left(0, \begin{bmatrix} \Sigma_x & 0 \\ 0 & \Sigma_y \end{bmatrix}\right),$$

with $\Sigma_x = \Sigma_y = p_r^2 \sum_{j=1}^{M} \lambda_j W_j$ and

$$W_j = \begin{bmatrix} L_j & \frac{L_j^2}{2} + L_j T_j \\ \frac{L_j^2}{2} + L_j T_j & \frac{L_j^3}{3} + L_j^2 T_j + L_j T_j^2 \end{bmatrix}. \tag{5}$$

2.2 Extending the Baseline Model

As recognized by [3], the scattering distribution of a muon crossing a homogeneous material typically exhibits heavier tails than a Gaussian distribution. To account for the tail events, whose omission could affect the estimate of possible deviation angles, we propose an alternative model for the scattering along the x-axis of a muon crossing a voxel j of homogeneous material with density λ_j,

$$H_{x,j} = (A_{x,j}, U_{x,j})^\top \sim \tau N_2\left(0, \Sigma_{x,j}\right) + (1-\tau) N_2\left(0, K\Sigma_{x,j}\right).$$

Here, $\Sigma_{x,j}$ is set as in (4) and $K > 1$ inflates the variability of the second component to capture the more extreme scattering events. Since the distribution of muon scattering angles is known to be approximately Gaussian only within its central 98% quantile range [2], we set $\tau = 0.98$ accordingly. The second component models the heavier tails, representing the remaining 2% of events in the scattering distribution. The same holds along the y coordinate.

For convenience, we denote the latent variables that describe the mixture component generating a scattering event by $Z_{x,j}, Z_{y,j} \sim Be(1, \tau)$. Then, conditional on $Z_{x,1}, \ldots, Z_{x,M}, Z_{y,1}, \ldots, Z_{y,M}$, it can be shown that the overall observed scattering now distributes as

$$(D_x, D_y)^\top \mid Z_{x,1}, \ldots, Z_{x,M}, Z_{y,1}, \ldots, Z_{y,M} \sim N_4\left(0, \begin{bmatrix} \tilde{\Sigma}_x & 0 \\ 0 & \tilde{\Sigma}_y \end{bmatrix}\right), \tag{6}$$

where, extending the results by [2], the covariance matrix $\tilde{\Sigma}_x$ proves to be

$$\tilde{\Sigma}_x = p_r^2 \sum_{j=1}^{M} \left(K^{1-Z_{x,j}} \lambda_j W_j\right),$$

with W_j given in (5). The same arguments give an analogous expression for the covariance matrix $\tilde{\Sigma}_y$.

3 Model Estimation via Stochastic EM

3.1 The Likelihood Function

Suppose we have a sample of n independently observed muon trajectories, characterized by their angular scatterings and displacements along the x- and y-axes. Let these displacements be denoted as $d_x = (d_{x_1}, \dots, d_{x_n})$ and $d_y = (d_{y_1}, \dots, d_{y_n})$, where

$$d_{x_i} = [a_{x_i}, u_{x_i}]^\top, \quad \text{and} \quad d_{y_i} = [a_{y_i}, u_{y_i}]^\top.$$

Furthermore, define $h_x = (h_{x_1}, \dots, h_{x_n})$, $z_x = (z_{x_1}, \dots, z_{x_n})$, with

$$h_{x_i} = (h_{x_i,1}, \dots, h_{x_i,M}), \qquad z_{x_i} = (z_{x_i,1}, \dots, z_{x_i,M})$$

for $i = 1, \dots, n$ and with analogous definitions for h_y and z_y. The primary objective is to estimate the vector of scattering densities $\lambda = (\lambda_1, \dots, \lambda_M)$, which plays a crucial role in specifying the covariance matrices $\tilde{\Sigma}_x$ and $\tilde{\Sigma}_y$. Estimating λ enables the visual representation of the material composition within the examined volume.

To estimate the model via Maximum Likelihood (ML), and due to the latent nature of vectors h_x, h_y, z_x, z_y, we are required to define the complete data log-likelihood function for the n muons

$$\begin{aligned}
&\ell_c(\lambda | d_x, d_y, h_x, h_y, z_x, z_y) = \\
&= \sum_{i=1}^{n} \sum_{j=1}^{M} \Big\{ z_{x_i,j} \Big[\log \tau + \log \phi_1(h_{x_i,j}) \Big] + (1 - z_{x_i,j}) \Big[\log(1-\tau) + \log \phi_2(h_{x_i,j}) \Big] \\
&\quad + z_{y_i,j} \Big[\log \tau + \log \phi_1(h_{y_i,j}) \Big] + (1 - z_{y_i,j}) \Big[\log(1-\tau) + \log \phi_2(h_{y_i,i}) \Big] \Big\}, \qquad (7)
\end{aligned}$$

where $j = 1, \dots, M$ indexes the voxels, and $\phi_1(h_{x_i,j})$, $\phi_2(h_{x_i,j})$, $\phi_1(h_{y_i,j})$ and $\phi_2(h_{y_i,j})$ are shorthand notations for the bivariate centered normal densities that define the bivariate mixture model governing displacements along the x- and y-axes within a single voxel.

It is important to note that only the total angular scattering and displacement measurements—d_x and d_y—are observed after each muon crosses the entire volume. The scattering events within individual voxels, $h_{x_i,j}$ and $h_{y_i,j}$, remain unobserved, as do the latent variables $z_{x_i,j}$ and $z_{y_i,j}$ for all i and j. To obtain the likelihood function for the model, these unobserved variables would need to be integrated out. However, due to the presence of missing data, expression (7) is more suitably maximized using the Expectation-Maximization (EM) algorithm.

3.2 The EM Algorithm

The EM algorithm is an iterative method for finding maximum likelihood estimates of parameters in the presence of incomplete or latent data [4]. The underlying idea is to maximize the expected value of the complete likelihood, with respect to the distribution of missing data. To this aim, an E-step computes the expected value given the current parameter estimates, and is then alternated with an M-step, where parameters are updated by maximizing the expected likelihood obtained in the E-step. The process repeats until convergence.

In the following, the two steps are detailed for a single iteration identified by the index $(q+1)$.

E-step. Compute the auxiliary function

$$\begin{aligned} Q(\lambda, \lambda^{(q)} |\, d_x, d_y) &= \mathbb{E}_{H_x,H_y,Z_x,Z_y|d_x,d_y,\lambda^{(q)}}[\ell_c(\lambda|d_x, d_y, H_x, H_y, Z_x, Z_y)] \\ &= \mathbb{E}_{H_x,Z_x|d_x,\lambda^{(q)}}[\ell_c(\lambda|d_x, H_x, Z_x)] + \mathbb{E}_{H_y,Z_y|d_y,\lambda^{(q)}}[\ell_c(\lambda|d_y, H_y, Z_y)] \\ &= \sum_{i=1}^{n} \Big\{ \mathbb{E}_{Z_{x_i}|d_{x_i},\lambda^{(q)}} \Big[\mathbb{E}_{H_{x_i}|Z_{x_i},d_{x_i},\lambda^{(q)}}[\ell_c(\lambda|d_x, H_x, Z_x)] \Big] \\ &\quad + \mathbb{E}_{Z_{y_i}|d_{y_i},\lambda^{(q)}} \Big[\mathbb{E}_{H_{y_i}|Z_{y_i},d_{y_i},\lambda^{(q)}}[\ell_c(\lambda|d_y, H_y, Z_y)] \Big] \Big\}, \end{aligned} \tag{8}$$

that is, the expected value of the complete data log-likelihood (7) with respect to the conditional distribution $(Z_x, Z_y, H_x, H_y \mid d_x, d_y, \lambda^{(q)})$ of the hidden variables.

Computing the expectation terms in (8) over all possible latent variable configurations —$Z_{x_i} = (Z_{x_i,1}, \ldots, Z_{x_i,M})$ and $Z_{y_i} = (Z_{y_i,1}, \ldots, Z_{y_i,M})$ for $i = 1, \ldots, n$—is, in practice, computationally intractable, as it involves a total number of terms growing as $2n \times 2^M$. To address this challenge, we developed a stochastic variant of the EM algorithm, known as the Stochastic EM (SEM), firstly proposed by [5], which introduces an additional step, known as the S-step, at each iteration. In this step, values for the latent variables Z_{x_i} and Z_{y_i}, $i = 1, \ldots, n$, are sampled from their respective posterior distributions

$$p(Z_{x_i} = z_{x_i} \mid d_{x_i}) \quad \text{and} \quad p(Z_{y_i} = z_{y_i} \mid d_{y_i}). \tag{9}$$

Several simulation algorithms can be used in this context, including Gibbs Sampling, Metropolis–Hastings, and Approximate Bayesian Computation. See, e.g. [6,7]. These techniques facilitate efficient estimation by approximating the missing data distribution, thus enabling the practical implementation of the EM framework in complex models.

Trial and error let us opt for an Independent Metropolis–Hastings sampler. This Markov chain Monte Carlo (MCMC) technique requires defining a proposal distribution that determines the pace at which the chain jumps. We went for an independent proposal distribution for the values of z_{x_i} and z_{y_i}, for $i = 1, ..., n$, as given below for iteration $(q+1)$.

S-step. Repeat, until convergence, the following steps.

1. Simulate vectors $z^*_{x_i} \sim Bin(M, \tau)$ and $z^*_{y_i} \sim Bin(M, \tau)$.

2. Compute

$$\alpha_x = \min\left\{1, \frac{p\left(Z_{x_i} = z_{x_i}^{(q+1)}\right) p\left(Z_{x_i} = z_{x_i}^* \mid d_{x_i}\right)}{p\left(Z_{x_i} = z_{x_i}^*\right) p\left(Z_{x_i} = z_{x_i}^{(q+1)} \mid d_{x_i}\right)}\right\}$$

$$= \min\left\{1, \frac{\frac{p\left(Z_{x_i}=z_{x_i}^{(q+1)}\right) p\left(d_{x_i}|Z_{x_i}=z_{x_i}^*\right) p\left(Z_{x_i}=z_{x_i}^*\right)}{p\left(d_{x_i}\right)}}{\frac{p\left(Z_{x_i}=z_{x_i}^*\right) p\left(d_{x_i}|Z_{x_i}=z_{x_i}^{(q+1)}\right) p\left(Z_{x_i}=z_{x_i}^{(q+1)}\right)}{p\left(d_{x_i}\right)}}\right\}$$

$$= \min\left\{1, \frac{p\left(d_{x_i} \mid Z_{x_i} = z_{x_i}^*\right)}{p\left(d_{x_i} \mid Z_{x_i} = z_{x_i}^{(q+1)}\right)}\right\}$$

for the x-axis and an analogous calculation for α_y regarding the y-axis.

3. Assign to $z_{x_i}^{(q+1)}$ the new value $z_{x_i}^{(*)}$ with probability α_x and to $z_{y_i}^{(q+1)}$ the new value $z_{y_i}^{(*)}$ with probability α_y. Otherwise, keep $z_{x_i}^{(q+1)}$ and $z_{y_i}^{(q+1)}$.

After convergence, compute the expected value of the auxiliary function (8) using the samples $z_{x_i}^{(q+1)}$ and $z_{y_i}^{(q+1)}$, for $i = 1, ..., n$, produced above.
For another application of Metropolis–Hastings sampler to SEM, see e.g. [8].

The simulation step precedes the expectation step, after which the maximization step is carried out.

M step. Maximize the auxiliary function (8) to yield

$$\lambda^{(q+1)} = \arg\max_{\lambda} Q(\lambda, \lambda^{(q)} \mid d_x, d_y).$$

In Appendix A, we will show that

$$Q(\lambda, \lambda^{(q)} \mid d_x, d_y) \propto \sum_{j=1}^{M} Q_j(\lambda_j, \lambda^{(q)};\ d_x, d_y).$$

Since each term in the summation of the auxiliary function that involves λ_j appears only in $Q_j(\lambda_j, \lambda^{(q)};\ d_x, d_y)$ for a specific index j, we can maximize $Q(\lambda, \lambda^{(q)};\ d_x, d_y)$ in λ_j by setting its gradient with respect λ_j to zero and solving for λ_j. This yields the iterative update formula for step $q + 1$,

$$\lambda_j^{(q+1)} = \frac{1}{2n_j} \sum_{i:L_{i,j} \neq 0} \frac{\tilde{G}_{x_i,j}\left(\lambda^{(q)}, z_{x_i,j}^{(q+1)}\right) + \tilde{G}_{y_i,j}\left(\lambda^{(q)}, z_{y_i,j}^{(q+1)}\right)}{2},$$

where $\tilde{G}_{x_i,j}$ and $\tilde{G}_{y_i,j}$ are defined in Appendix A, expression (11).

The three steps of the stochastic EM (SEM) algorithm—S-step, E-step, and M-step—are repeated iteratively until convergence.

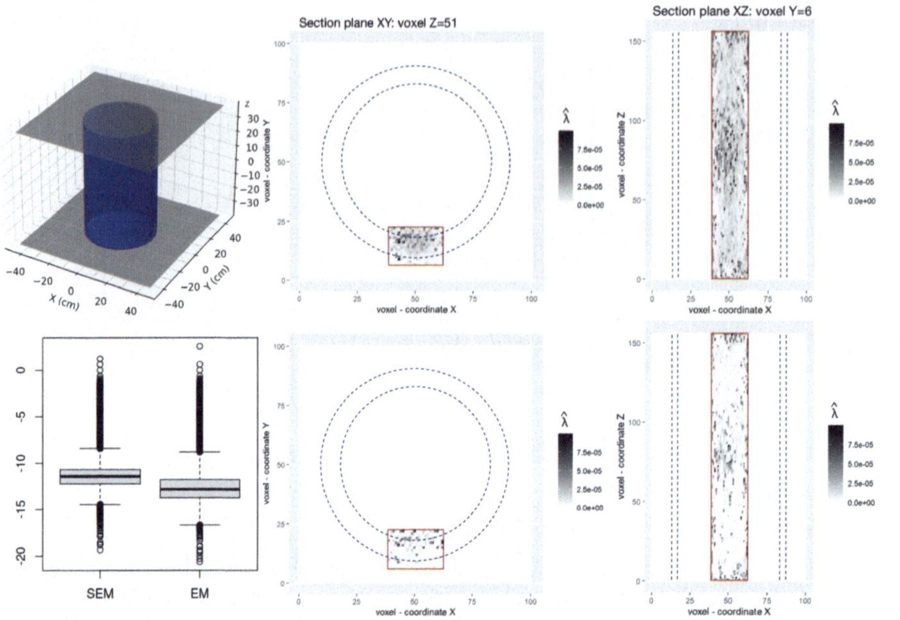

Fig. 2. Representation of the insulated pipe under analysis (top left), empirical log-distribution of the positive estimated λ values (bottom left), and corresponding voxel-wise heatmaps of the estimates obtained via SEM (top) and EM (bottom), on the xy and xz planes of the selected subregion (outlined in red).

4 An Application

The procedure described so far is illustrated, and compared with the baseline Gaussian model [2], using a simulated dataset from an experiment conducted at the Department of Modern Physics, University of Cantabria. The experiment aimed to estimate the wear level of a 2 cm thick steel gas pipe—insulated with rock wool and wrapped in aluminum—by reconstructing the materials and their thicknesses. To maintain computational feasibility, the analysis is restricted to measurements of the total scattering of 10,000 muons passing through a subregion partitioned into about 60,000 cubic voxels with a side length of 0.5 cm.

Reconstructing the material composition of the layers forming the insulated pipe is particularly challenging. Reducing the voxel size, indeed, increases the number of voxels as well as the number of muons required, since the smaller the voxels, the lower the probability that each will be traversed by at least one muon. Moreover, although the chosen voxel size ensures that approximately 25% of the voxels intersect the pipe (and therefore have a non-zero scattering density), there is no guarantee that the material within each voxel is homogeneous. Finally, even though the exact density of the materials forming the layers is not known—both steel and rock wool are composite materials—the range of plausible values for the λ's is low in magnitude resulting in a difficult distinction even between voxels occupied by the pipe and truly empty ones.

In light of these considerations, it is ultimately a success that both the SEM and the EM algorithms are able to detect the presence of the pipe—see the reconstructed sectional images in Fig. 2. The same figure also compares the log-distributions of the estimated λ values, further confirming that the estimates via SEM are generally higher, that is, the SEM algorithm identifies the filled voxels with a greater accuracy. However, the reconstructions remain relatively coarse, and the estimated λ values are low.

The computational cost of both algorithms grows substantially with the number of muons and voxels. For SEM, efficiency is strongly affected by the choice of sampling strategy in the stochastic step: methods like the Gibbs sampler, which require updating each voxel crossed by a muon, lead to long iteration times. In our case study, one SEM iteration takes about 35 s with the Metropolis–Hastings algorithm, but around 20 min with Gibbs sampling. While SEM typically converges in substantially fewer iterations—around 1,250 in our application compared to 2,066 for EM—its higher per-iteration cost may offset this advantage.

5 Final Remarks

Due to technical limitations and available computational resources, extensive testing of alternative algorithms was not feasible. However, several potential improvements could enhance the proposed method. These include adjusting the relationship between the variances of the two mixture components, exploring alternative mixture models and simulation algorithms, refining the SEM algorithm, developing an estimation strategy for K within the model, and incorporating corrections for detector errors and numerical distortions.

The reference methodology assumes minimal scattering effects, meaning the muon's true trajectory does not significantly deviate from the straight-line path between its entry and exit points within the volume. Instead of assuming a predefined trajectory, the proposed method reconstructs it layer by layer, parallel to the xy-plane. By integrating information on the muon's entry and exit angles and positions, this approach allows for a more flexible and accurate trajectory estimation.

Acknowledgments. *The publication was produced with funding from the Italian Ministry of University and Research under the Call for Proposals related to the scrolling of the final rankings of the PRIN 2022 call* (Project title "Latent variable models and dimensionality reduction methods for complex data" – Project No. 20224CRB9E, CUP C53C24000730006 – PI Prof. Paolo Giordani – Grant Assignment Decree No. 1401 adopted on 18.9.2024 by MUR).

Appendix

A The E-Step in Detail

From the second row of (8), it is evident that the two expectations related to scattering in the x- and y-coordinates can be computed separately. The following steps are evaluated for a generic coordinate, omitting the subscript.
Let's work out the expected value of

$$\begin{aligned}
&\mathbb{E}_{H|d,z^{(q+1)},\lambda^{(q)}}\left[\ell_c(\lambda|d,H,z^{(q+1)})\right] \qquad (10)\\
&=\sum_{j=1}^{M}\sum_{i:L_{i,j}\neq 0}\mathbb{E}_{H_{i,j}|d_i,z_i^{(q+1)},\lambda^{(q)}}\left\{\left[z_{i,j}^{(q+1)}\left(\log\tau-\frac{1}{2}\log|\Sigma_{i,j}|-\frac{1}{2}H_{i,j}^T\Sigma_{i,j}^{-1}H_{i,j}\right)\right]+\right.\\
&\left.\quad+(1-z_{i,j}^{(q+1)})\left(\log(1-\tau)-\frac{1}{2}\log|K\Sigma_{i,j}|-\frac{1}{2}H_{i,j}^T(K\Sigma_{i,j})^{-1}H_{i,j}\right)\right\}\Bigg\},\\
&=\begin{cases}\sum_{j=1}^{M}\sum_{i:L_{i,j}\neq 0}-\log\lambda_j-\dfrac{\mathbb{E}\left(H_{i,j}^TB_{i,j}^{-1}H_{i,j}\mid d_i,z_i^{(q+1)},\lambda^{(q)}\right)}{2\lambda_j p_{r,i}^2} & \text{if } z_{i,j}^{(q+1)}=1\\
\sum_{j=1}^{M}\sum_{i:L_{i,j}\neq 0}-\log\lambda_j-\dfrac{\mathbb{E}\left(H_{i,j}^TB_{i,j}^{-1}H_{i,j}\mid d_i,z_i^{(q+1)},\lambda^{(q)}\right)}{2K\lambda_j p_{r,i}^2} & \text{if } z_{i,j}^{(q+1)}=0.\end{cases}
\end{aligned}$$

To compute both expectations, $K^{1-z_i^{(q+1)}}$ is used in the subsequent steps, addressing both cases, $z_i^{(q+1)}=1$ and $z_i^{(q+1)}=0$. Since both $H_{i,j}$ and D_i are Gaussian, the conditional distribution $H_{i,j}|d_i,z_i^{(q+1)},\lambda^{(q)}$ is Gaussian [9],

$$H_{i,j}|d_i,z_i^{(q+1)},\lambda^{(q)}\sim N_2\left(\tilde{\Sigma}_{D_iH_{i,j}}^T\tilde{\Sigma}_i^{-1}d_i,\ \left(K^{1-z_i^{(q+1)}}\Sigma_{i,j}\right)-\tilde{\Sigma}_{D_iH_{i,j}}^T\tilde{\Sigma}_i^{-1}\tilde{\Sigma}_{D_iH_{i,j}}\right),$$

where $\tilde{\Sigma}_{D_iH_{i,j}}$ is the covariance matrix of $H_{i,j}|d_i,z_i^{(q+1)},\lambda^{(q)}$. The distribution $H_{i,j}^TB_{i,j}^{-1}H_{i,j}|d_i,z_i^{(q+1)},\lambda^{(q)}$ is a quadratic form. Let $C=B_{i,j}^{-1}$ and $X=H_{i,j}|d_i,z_i^{(q+1)},\lambda^{(q)}$, then the expectation of X^TCX is

$$\mathbb{E}_{X^TCX}=\text{tr}(C\Sigma_X)+\mu_X^TC\mu_X,$$

where μ_X and Σ_X are the mean and covariance matrix of X. Therefore, using the following identity

$$\tilde{\Sigma}_{D_iH_{i,j}}B_{i,j}^{-1}\tilde{\Sigma}_{D_iH_{i,j}}^T=W_{i,j}(K^{1-z_{i,j}^{(q+1)}}p_{r,i}^2\lambda_j)^2,$$

the expectation of the quadratic form can be computed, knowing that $B_{i,j}$ and $\tilde{\Sigma}_i$ are symmetric and that $\text{tr}(AB)=\text{tr}(BA)$,

$$\begin{aligned}
\mathbb{E}\left(H_{i,j}^TB_{i,j}^{-1}H_{i,j}\mid d_i,\lambda^{(q)},z_i^{(q+1)}\right)&=\text{tr}\Big[B_{i,j}^{-1}\left(K^{1-z_i^{(q+1)}}\Sigma_{i,j}\right)+\\
&-B_{i,j}^{-1}\tilde{\Sigma}_{D_iH_{i,j}}^T\tilde{\Sigma}_i^{-1}\tilde{\Sigma}_{D_iH_{i,j}}\Big]+d_i^T\Sigma_i^{-1}\left(\tilde{\Sigma}_{D_iH_{i,j}}B_{i,j}^{-1}\tilde{\Sigma}_{D_iH_{i,j}}^T\right)\Sigma_i^{-1}d_i.
\end{aligned}$$

This simplifies to

$$K^{1-z_{i,j}^{(q+1)}}p_{r,i}^2\left\{2\lambda_j^{(q)}+(\lambda_j^{(q)})^2K^{1-z_{i,j}^{(q+1)}}p_{r,i}^2\left[d_i^T\Sigma_i^{-1}W_{i,j}\Sigma_i^{-1}d_i-\text{tr}\left(\Sigma_i^{-1}W_{i,j}\right)\right]\right\}.$$

Next, the following quantity is defined to simplify the notation

$$\tilde{G}_{i,j}\left(\lambda^{(q)}, z_{i,j}^{(q+1)}\right) = \frac{1}{K^{1-z_{i,j}^{(q+1)}} p_{r,i}^2} \mathbb{E}\left(H_{i,j}^T B_{i,j}^{-1} H_{i,j} \mid d_i, z_{i,j}^{(q+1)}, \lambda^{(q)}\right), \quad (11)$$

which simplifies the further expressions. At this point, the expectation in (10) can be rewritten as

$$\mathbb{E}_{H|d,z^{(q+1)},\lambda^{(q)}}\left[\ell_c(\lambda|d, H, z^{(q+1)})\right] \sum_{j=1}^{M} \left[-n_j \log \lambda_j - \frac{1}{2\lambda_j} \sum_{i:L_{i,j}\neq 0} \tilde{G}_{i,j}\left(\lambda^{(q)}, z_{i,j}^{(q+1)}\right)\right].$$

Finally, the auxiliary function in (8) is

$$Q(\lambda, \lambda^{(q)}|d_x, d_y) \propto \sum_{j=1}^{M} \left[-n_j \log \lambda_j - \frac{1}{2\lambda_j} \sum_{i:L_{i,j}\neq 0} \tilde{G}_{x_i,j}\left(\lambda^{(q)}, z_{x_i,j}^{(q+1)}\right) + \right.$$
$$\left. -n_j \log \lambda_j - \frac{1}{2\lambda_j} \sum_{i:L_{i,j}\neq 0} \tilde{G}_{y_i,j}\left(\lambda^{(q)}, z_{y_i,j}^{(q+1)}\right)\right] \propto \sum_{j=1}^{M} Q_j\left(\lambda_j, \lambda^{(q)}; d_x, d_y\right).$$

References

1. Borozdin, K.N., et al.: Radiographic imaging with cosmic-ray muons. Nature **422**(6929), 277 (2003)
2. Schultz, L.J., et al.: Statistical reconstruction for cosmic ray muon tomography. IEEE Trans. Image Process. **16**(8), 1985–1993 (2007)
3. Lynch, G. R., Dahl, O. I.: Approximations to multiple coulomb scattering. Nucl. Instrum. Methods Phys. Res. B (1990)
4. Dempster, A.P., Laird, N.M., Rubin, D.B.: Maximum likelihood from incomplete data via the EM algorithm. J. R. Stat. Soc. Ser. B **39**(1), 1–22 (1977)
5. Celeux, G., Diebolt, J.: The SEM algorithm: a probabilistic teacher algorithm derived from the EM algorithm for the mixture problem. Comput. Stat. Q. **2**, 73–82 (1985)
6. Robert, C. P., Casella G.: Monte Carlo statistical methods. 2nd ed. Springer texts in statistics. Springer, New York (2004)
7. Sisson, S.A., Fan, Y., Beaumont, M.: Handbook of Approximate Bayesian Computation, 1st edn. CRC Press, New York (2018)
8. Kuhn, E., Lavielle M.: Coupling a stochastic approximation version of EM with an MCMC procedure. ESAIM: Probab. Stat. 8, pp. 115–131 (2004)
9. Johnson, R.A., Wichern, D.W.: Applied multivariate statistical analysis, 6th edn. Pearson Prentice Hall, Upper Saddle River, N.J (2007)

Revealing the Nature of Italian Life Expectancy: A Comparative Study of ARIMA Models Using COVID-19 Shock

Girolamo Franchetti and Massimiliamo Politano(✉)

Economics and Statistics Department, University of Naples Federico II, Naples, Italy
{girolamo.franchetti,politano}@unina.it

Abstract. This study investigates how alternative ARIMA model specifications can be used to infer the underlying trend structure—deterministic or stochastic—of the life expectancy at birth time series for the Italian population over the period 19742024. By comparing two ARIMA(1,d,0) models and two ARIMA(1,d,1) models, each estimated with and without a deterministic trend component, we aim to assess not only forecast accuracy but also the capacity of each model to capture the structural dynamics of the series, particularly in the presence of exogenous shocks. The COVID-19 outbreak in 2020 is treated as a structural shock and serves as a testing ground for evaluating model adaptability and long-run behavior. Our analysis employs stationarity tests, residual diagnostics, impulse response functions (IRFs), model fit statistics, and forecast error measures. Results indicate that while trend-based ARIMA models tend to provide better in-sample statistical fit, they often fail to capture the persistent deviations induced by structural breaks. In contrast, the ARIMA(1,d,1) model without a deterministic trend offers greater flexibility and superior post-shock forecasting performance. The paper concludes by proposing a structured approach to model selection under structural uncertainty, highlighting how comparative model analysis can inform our understanding of time series behavior over a given historical period.

Keywords: Life expectancy · ARIMA models · COVID.19

1 Introduction

Forecasting demographic time series, such as life expectancy, requires explicit assumptions regarding the nature of the underlying trend—in particular, whether it follows a deterministic trajectory (characterized by smooth and predictable growth) or a stochastic one (where shocks can have persistent, long-term effects) [1–5]. These assumptions are not merely technical, but fundamentally shape how we interpret temporal dynamics and respond to structural changes in

A. D'Ambrosio et al. (Eds.): CLADAG-VOC 2025, SCDAKO, pp. 156–171, 2025.
https://doi.org/10.1007/978-3-032-03042-9_14

the data [8,9,11,13]. This issue becomes especially salient when a series is subjected to significant exogenous shocks [7], which may either temporarily deviate from or permanently alter its trajectory.

Life expectancy reflects a multifactorial process influenced by health interventions, socioeconomic development, public policy, environmental factors, and, crucially, epidemiological events [12,16]. The COVID-19 pandemic in 2020 represents a unique, global, and exogenous disruption, offering an empirical setting to investigate the resilience and adaptability of different time series models [17,18].

Research Question: *Does the time series of Italian life expectancy over the period 1974–2024 behave more like a deterministic-trend process subject to temporary fluctuations, or a difference-stationary process characterized by permanent innovations? Furthermore, how do different ARIMA model specifications—with and without deterministic trends—perform in the presence of the COVID-19 shock in terms of model fit, forecasting accuracy, and dynamic adjustment?*

To address these questions, we undertake a structured econometric analysis using ARIMA models [6,13]. Our methodological strategy comprises five analytical steps:

1. **Step 1:** Examine the stationarity properties of the life expectancy series using visual diagnostics and formal unit root tests;
2. **Step 2:** Identify appropriate ARIMA model structures through ACF/PACF diagnostics;
3. **Step 3:** Estimate and compare model performance using in-sample fit statistics and residual diagnostics;
4. **Step 4:** Evaluate out-of-sample forecast accuracy, especially over the post-pandemic period (2020–2034);
5. **Step 5:** Analyze impulse response functions (IRFs) to assess each model's dynamic behavior in response to exogenous shocks.

Through this comparative framework, we aim to shed light on both empirical model performance and the deeper structural properties of the underlying demographic process. Ultimately, our findings contribute to the ongoing methodological debate on time series specification under structural uncertainty, particularly in the context of public health and demographic forecasting.

2 Data and Initial Diagnostics

The empirical object of analysis is the annual life expectancy at birth for the Italian population over the period 1974–2024.[1] The series exhibits a pronounced upward trend, consistent with improvements in medical technology, healthcare coverage, and living standards. However, a sudden and substantial decline is visible in 2020, coinciding with the onset of the COVID-19 pandemic (Fig. 1).

[1] Data were obtained from the ISTAT database: https://esploradati.istat.it/databrowser/#/en.

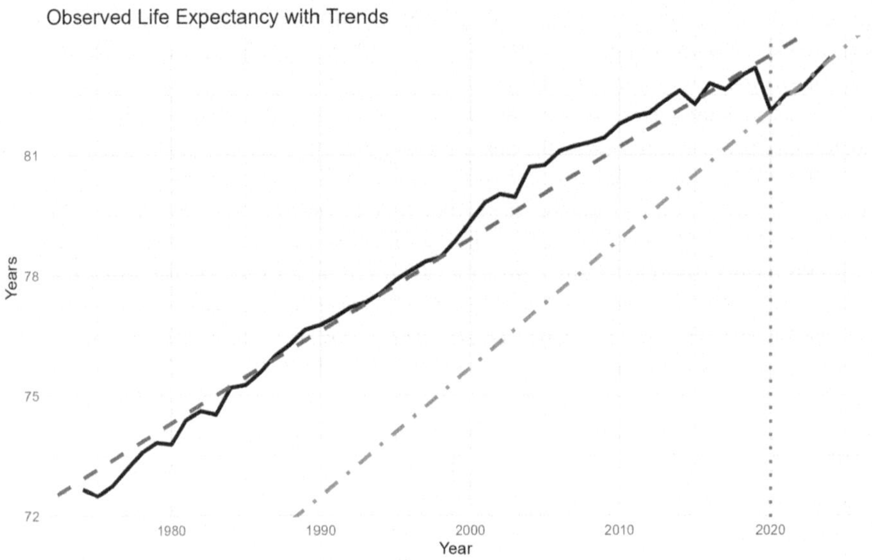

Fig. 1. Original Series with Fitted Linear Trend (1974–2024). 2020 reflects the COVID-19-induced drop.

Prior to estimating any model, it is essential to determine whether the series is stationary, and if not, whether the non-stationarity arises from a deterministic trend or from stochastic integration. These considerations inform whether the appropriate transformation is differencing, detrending, or both.

We apply two complementary statistical tests:

- The **Augmented Dickey-Fuller (ADF)** test, which evaluates the null hypothesis that the series contains a unit root (i.e., it is non-stationary);
- The **Kwiatkowski-Phillips-Schmidt-Shin (KPSS)** test, which evaluates the null hypothesis that the series is stationary around a deterministic trend (Table 1).

The ADF test fails to reject the null hypothesis of a unit root, indicating that the series is non-stationary. Concurrently, the KPSS test rejects the null of trend stationarity, implying that stationarity cannot be achieved by removing a deterministic trend. The joint interpretation of these results—in line with econometric best practices—strongly supports the presence of a stochastic trend. This finding

Table 1. Stationarity Tests: ADF and KPSS on the Original Series

Test	Test Statistic	5% Critical Value
ADF (with drift)	−1.832	−2.930
KPSS (with trend)	0.208	0.146

justifies modeling the series as integrated of order one, i.e., I(1), and applying first-order differencing as a necessary step toward achieving stationarity.

Trend Nature Derivation. Taken together, the ADF and KPSS results point to the presence of a difference-stationary process in the data, wherein shocks may induce permanent shifts in the level of the series. However, both tests have well-documented limitations: they are sensitive to sample size, lag structure, and the presence of structural breaks—a crucial factor in our case, given the abrupt decline in 2020.

As such, while these preliminary diagnostics provide a foundation for transformation strategy, they must be complemented by model-based validation. This motivates the transition to the next analytical step, where ARIMA models with different structural assumptions are estimated and evaluated based on their empirical adequacy, explanatory power, and robustness to structural change.

3 Model Estimation and Specification

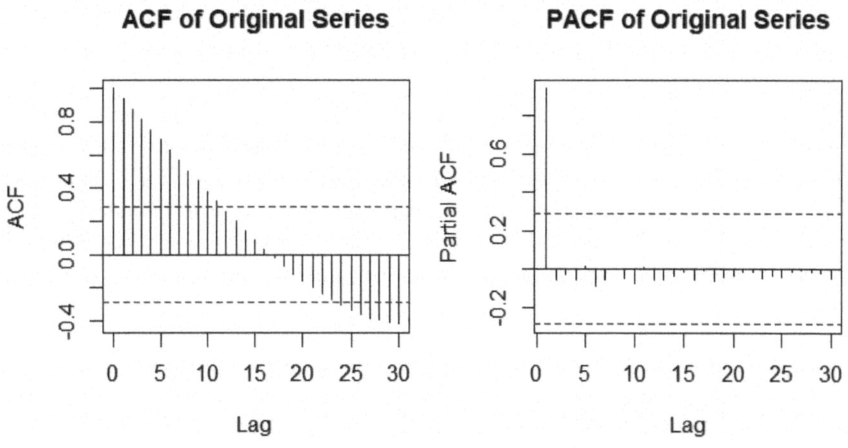

Fig. 2. ACF and PACF of the original (non-differenced) series.

Preliminary Identification via ACF and PACF. The initial model selection process begins with the analysis of autocorrelation patterns. The autocorrelation function (ACF) of the original life expectancy series displays a slowly decaying structure across the first several lags. This long memory behavior is consistent with a non-stationary series possessing a unit root. For instance, the ACF coefficients remain relatively high at lags 1 through 10 (e.g., 0.941, 0.878, 0.817), indicating strong temporal dependence (Fig. 2).

Conversely, the partial autocorrelation function (PACF) demonstrates a sharp cut-off after the first lag. The PACF at lag 1 is large and significant, while subsequent lags are statistically insignificant and fluctuate around zero. This signature is typical of an AR(1) process in the presence of differencing.

These observations are consistent with the stationarity tests discussed in Sect. 2, and they support an initial specification of an ARIMA(1,1,0) model. However, to capture richer short-term dynamics and potential measurement errors, we also consider models with a moving average (MA) component, such as ARIMA(1,1,1).

Trend Treatment and Differencing Strategy. An essential component of ARIMA modeling involves deciding how to transform a non-stationary series into a stationary one. This process depends on whether the trend is assumed to be deterministic or stochastic. Accordingly, we apply two distinct preprocessing strategies to the life expectancy data, aligned with the conceptual structure of the models under comparison.

- **Models without a deterministic trend:** In these specifications, we assume that the trend component is stochastic and best addressed through differencing. The original series is differenced once, based on the results of the ADF and KPSS tests and supported by the slow decay pattern in the ACF. First-order differencing is thus employed to remove the stochastic trend and achieve mean stationarity.
- **Models with a deterministic trend:** Here, we posit the existence of a linear deterministic trend in the data, which is removed prior to differencing. This is achieved via ordinary least squares (OLS) regression on a time index. The residuals from this regression—interpreted as deviations from the trend—are subsequently differenced as needed. This approach is suitable when the trend is viewed as exogenous or policy-driven and not subject to random shocks.

This dual approach allows us to explore whether the life expectancy series is better modeled by an exogenous, smooth trend or an endogenous, evolving one. In both cases, the resulting transformed series are subjected to further unit root testing to verify that stationarity has been successfully achieved (Table 2).

Table 2. Number of Differences Applied for Stationarity

Model	Differencing Order
ARIMA(1,d,0) without trend	1
ARIMA(1,d,1) without trend	1
ARIMA(1,d,0) with trend	1
ARIMA(1,d,1) with trend	1

The table above confirms that all model variants ultimately require a single difference to induce stationarity. However, the interpretation of the differencing operation differs across model classes. In models without a trend, differencing captures the full stochastic evolution of the process. In trend-adjusted models, it captures fluctuations around a pre-estimated deterministic path. As we will show, these conceptual differences have significant implications for model performance, especially in the face of structural shocks.

Model Candidates and Rationale. To systematically explore the structure of the time series, we estimate and compare four ARIMA model variants. Each model differs in the presence or absence of autoregressive (AR) and moving average (MA) components, and all are specified with a first-order difference ($d = 1$) based on the unit root tests and autocorrelation diagnostics.

The general ARIMA(p, d, q) model can be written as:

$$\Phi(L)(1 - L)^d y_t = \Theta(L)\varepsilon_t,$$

where:

- y_t is the observed time series (life expectancy);
- L is the lag operator, such that $Ly_t = y_{t-1}$;
- $(1 - L)^d$ denotes differencing of order d;
- $\Phi(L) = 1 - \phi_1 L - \ldots - \phi_p L^p$ is the autoregressive polynomial of order p;
- $\Theta(L) = 1 + \theta_1 L + \ldots + \theta_q L^q$ is the moving average polynomial of order q;
- ε_t is a white noise error term, with mean zero and constant variance.

Model Candidates and Rationale. Based on the preliminary diagnostics—specifically, the slow exponential decay of the autocorrelation function (ACF) and the sharp cut-off in the partial autocorrelation function (PACF) after lag 1—we focus our analysis on autoregressive integrated moving average (ARIMA) models with first-order differencing ($d = 1$), an autoregressive term of order one (AR(1)), and optionally a moving average term of order one (MA(1)).

This leads to two principal model specifications:

- **ARIMA(1,1,0):**

  $$(1 - \phi_1 L)(1 - L)y_t = \varepsilon_t$$

  This model includes a first-order autoregressive component ϕ_1 applied to the differenced series. It captures persistence in the deviations from the stochastic trend and is directly motivated by the PACF structure observed in the data.
- **ARIMA(1,1,1):**

  $$(1 - \phi_1 L)(1 - L)y_t = (1 + \theta_1 L)\varepsilon_t$$

  This more flexible specification combines both autoregressive and moving average terms, allowing the model to capture richer short-term dynamics and potentially serially correlated shocks or measurement noise.

Each of these specifications is estimated in two versions:

1. **Without deterministic trend:** The original series is differenced directly, under the assumption that the trend is stochastic (i.e., driven by a unit root process).
2. **With deterministic trend:** A linear trend is first removed via ordinary least squares (OLS) regression on time, and the residuals are subsequently differenced to isolate the stochastic component.

This results in a total of four candidate models, defined by the presence or absence of a moving average component (MA(1)) and a deterministic trend. All models assume first-order differencing ($d = 1$), as supported by unit root tests and autocorrelation diagnostics.

- **ARIMA(1,1,0) without trend:**

$$(1 - \phi_1 L)(1 - L)y_t = \varepsilon_t$$

This is a difference-stationary model with a single autoregressive term. It models stochastic trends without any deterministic component, allowing for permanent level shifts in response to shocks.
- **ARIMA(1,1,1) without trend:**

$$(1 - \phi_1 L)(1 - L)y_t = (1 + \theta_1 L)\varepsilon_t$$

This more flexible specification adds a moving average term to capture serial correlation in the innovations of the differenced series. It improves short-run fit without imposing a deterministic trend.
- **ARIMA(1,1,0) with trend:**

$$(1 - \phi_1 L)(1 - L)(y_t - \alpha - \beta t) = \varepsilon_t$$

This model assumes a deterministic linear trend $\alpha + \beta t$, which is removed before differencing. The stochastic component is then modeled as an AR(1) process.
- **ARIMA(1,1,1) with trend:**

$$(1 - \phi_1 L)(1 - L)(y_t - \alpha - \beta t) = (1 + \theta_1 L)\varepsilon_t$$

This specification combines a deterministic trend with both autoregressive and moving average components. It provides the greatest flexibility for capturing both structural trend and transitory deviations.

These models are selected not only for their theoretical consistency with the observed autocorrelation patterns, but also for their parsimony and interpretability in the presence of structural shocks. Their adequacy is evaluated using standard information criteria—namely, the Akaike Information Criterion (AIC) and Bayesian Information Criterion (BIC)—alongside residual diagnostics and out-of-sample forecasting performance. The next section further examines how these specifications differ in the treatment of trend components and in the transformations required to ensure stationarity.

4 Impulse Response and Dynamic Behavior

To analyze how each ARIMA model propagates the effect of an exogenous shock over time, we compute the impulse response function (IRF) associated with a one-time innovation. The IRF represents the expected dynamic adjustment of the process following a unit shock to the error term at time t, i.e., $\varepsilon_t = 1$, while all other future shocks are assumed to be zero.

We focus on the ARIMA(1,1,1) model, specified as:

$$(1 - \phi_1 L)(1 - L)y_t = (1 + \theta_1 L)\varepsilon_t,$$

which can be rearranged into its state-space equivalent for the differenced series $\Delta y_t = y_t - y_{t-1}$:

$$\Delta y_t = \phi_1 \Delta y_{t-1} + \varepsilon_t + \theta_1 \varepsilon_{t-1}.$$

Now, consider a one-time shock $\varepsilon_t = 1$ at time t, with $\varepsilon_{t+1} = \varepsilon_{t+2} = \ldots = 0$. The impulse response function (IRF) then traces the effect on Δy_{t+h} and, cumulatively, on y_{t+h}.

The dynamic path is obtained recursively:

$$\Delta y_t = 1 + \theta_1 \cdot 0 = 1,$$

$$\Delta y_{t+1} = \phi_1 \cdot \Delta y_t + 0 + \theta_1 \cdot 1 = \phi_1 + \theta_1,$$

$$\Delta y_{t+2} = \phi_1 \cdot \Delta y_{t+1} + 0 + \theta_1 \cdot 0 = \phi_1(\phi_1 + \theta_1),$$

$$\Delta y_{t+3} = \phi_1 \cdot \Delta y_{t+2} = \phi_1^2(\phi_1 + \theta_1),$$

$$\vdots$$

$$\Delta y_{t+h} = \phi_1^{h-1}(\phi_1 + \theta_1), \quad h \geq 2.$$

Thus, the cumulative effect on the level y_{t+h} is given by:

$$IRF(h) = \sum_{j=0}^{h} \Delta y_{t+j},$$

which shows that the autoregressive parameter ϕ_1 governs the rate at which the effect of the initial shock decays over time. When $|\phi_1| < 1$, the impact of the shock gradually diminishes but may persist for many periods, depending on the value of ϕ_1. When ϕ_1 is close to 1, the decay is slow, and the shock has near-permanent effects—characteristic of highly persistent stochastic trends.

By contrast, in models with a deterministic trend, such as:

$$(1 - \phi_1 L)(1 - L)(y_t - \alpha - \beta t) = \varepsilon_t,$$

the deterministic trend $\alpha + \beta t$ acts as an anchor, forcing the series to eventually return to a fixed growth path. Consequently, the IRF in such models tends to converge back to zero, even after large shocks, implying that structural breaks are treated as temporary.

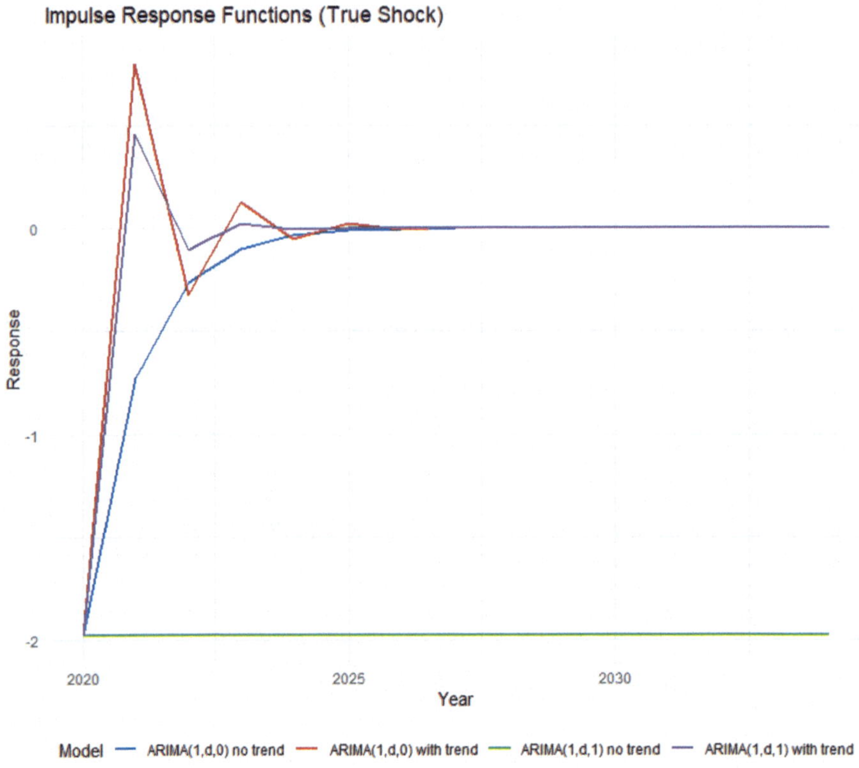

Fig. 3. Impulse Response Functions to a Negative Shock in 2020

As shown in Fig. 3, the ARIMA(1,1,1) model without trend produces a persistent yet gradually decaying IRF, consistent with the partial recovery in life expectancy observed after the 2020 pandemic shock. Conversely, models with deterministic trends exhibit rapid reversion, underestimating the long-term impact of the disruption.

In summary, the autoregressive coefficient ϕ_1 *plays a central role in shaping the post-shock trajectory of the series. The IRF provides a structural lens through which the plausibility and realism of different ARIMA models can be assessed in the presence of rare but impactful events.*

It is important to emphasize that the IRF represents the deviation from the model's expected path in the absence of the shock. In practical terms:

- For models **without a deterministic trend**, the IRF is added to the last observed level y_{t-1}, since the model assumes a stochastic trend. The post-shock trajectory is therefore:

$$y_{t+h} = y_{t-1} + \sum_{j=0}^{h} IRF(j).$$

- For models **with a deterministic trend** (e.g., linear $\alpha + \beta t$), the IRF is added to the extrapolated trend component:

$$y_{t+h} = \hat{\alpha} + \hat{\beta}(t+h) + IRF(h).$$

In this way, the impulse response function reflects the *transitory or permanent displacement* relative to the model's baseline expectation—whether that baseline is a constant (in difference-stationary models) or a time-dependent trend (in trend-stationary models).

This distinction is critical when interpreting the effect of structural shocks: in difference-stationary processes, the entire path is shifted, while in trend-stationary processes, the shock represents only a temporary deviation from the pre-defined trajectory.

5 Forecasting Performance and Model Evaluation

Beyond capturing historical dynamics, a robust time series model must deliver accurate and interpretable forecasts—especially under conditions of structural change. To this end, we evaluate the forecasting performance of each ARIMA specification over the period 20202034, with a particular focus on the response to the COVID-19 shock.

Forecast accuracy is assessed using standard performance metrics, including Mean Absolute Error (MAE), Root Mean Square Error (RMSE), and pseudo-R^2 values. The pseudo-R^2 is calculated as:

$$\text{pseudo-}R^2 = 1 - \frac{\text{MSE}_{\text{model}}}{\text{Var}(y_{\text{obs}})},$$

which compares the mean squared forecast error to the variance of the observed series, offering an intuitive measure of explanatory power over the forecast window.

To ensure that the forecasts reflect structural dynamics and not just trend extrapolation, we explicitly incorporate impulse response functions (IRFs) into the forecast generation. The 2020 shock is introduced as a one-time innovation in the error term, and each model's internal dynamics are allowed to propagate this disturbance over time. This approach embeds structural realism into the projections by simulating how each model would have reacted ex-ante to the COVID-19 shock.

Figure 4 illustrates the forecast trajectories implied by each model. Deterministic trend models project a relatively fast return to pre-shock trajectories,

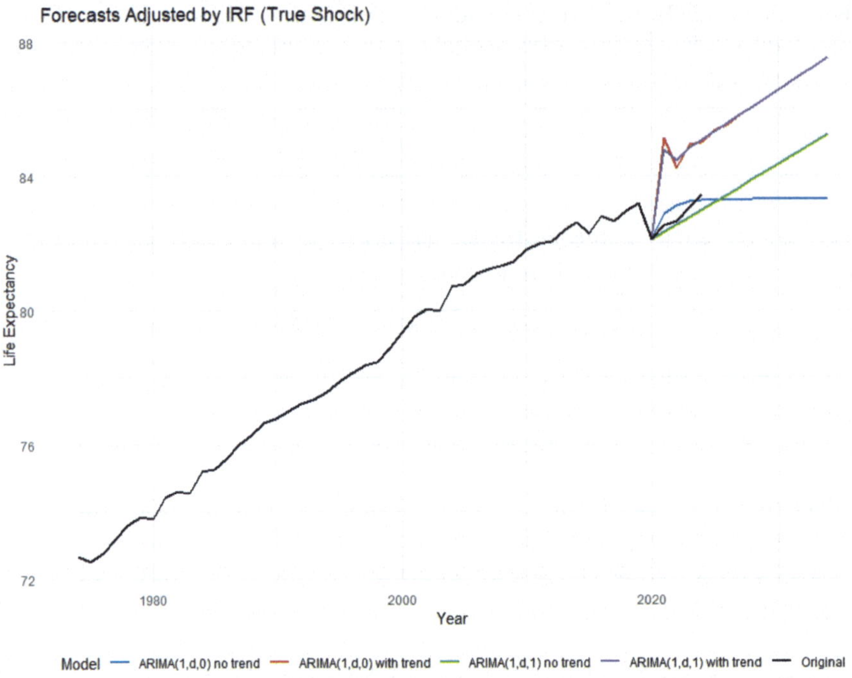

Fig. 4. Forecasting Comparison Across Model Specifications (19742034)

while stochastic-trend models—particularly **ARIMA(1,1,1) without trend**—forecast more persistent effects and a gradual recovery. These visual differences reflect fundamentally distinct assumptions about shock persistence and system memory.

The accuracy of the forecasts is summarized in Table 3, which compares predicted and observed life expectancy values over 20202024.

The **ARIMA(1,1,1) without trend** model demonstrates the best predictive accuracy across all criteria, with the lowest MAE and RMSE and a pseudo-R^2 of 0.841. This model exhibits robustness to structural breaks and aligns more closely with the actual path of life expectancy in the wake of COVID-19.

In contrast, deterministic trend models—despite their superior in-sample statistical fit—underperform in the post-2020 window. Their forecasts systematically overestimate recovery, resulting in higher prediction errors and even negative pseudo-R^2 scores. This discrepancy underscores the limitation of assuming full mean reversion in the presence of a persistent exogenous shock.

In summary, while deterministic trend models may appear statistically appealing under stable conditions, they prove less reliable in disrupted environments. More flexible specifications—especially ARIMA models with both autoregressive and moving average terms but without imposed trend—offer improved forecasting performance and greater adaptability to real-world structural changes.

Table 3. Comparison of Forecasted and Observed Life Expectancy Values (Decimal Format)

Year	Observed	AR(1,d,0) NT	AR(1,d,1) NT	AR(1,d,0) TR	AR(1,d,1) TR
2020	82.08	82.34	82.19	82.22	82.17
2021	82.50	82.57	82.43	82.66	82.49
2022	82.64	82.76	82.66	83.09	82.81
2023	83.03	82.92	82.89	83.53	83.12
2024	83.42	83.04	83.10	83.96	83.43

6 Comparative Model Evaluation

In this section, we systematically compare the four ARIMA model specifications along multiple performance dimensions, incorporating both in-sample fit and out-of-sample forecast accuracy [14]. The objective is to understand not only which model best approximates the historical data, but also which one generalizes effectively under structural instability.

6.1 In-Sample Model Fit

We begin with an evaluation of statistical fit over the estimation sample using conventional information criteria: the Akaike Information Criterion (AIC), the Bayesian Information Criterion (BIC), and the log-likelihood [10,15,19]. These metrics are presented in Table 4.

Table 4. Model Diagnostics and Performance Metrics

Model	AIC	BIC	LogLik	MAE	RMSE	Pseudo-R^2
ARIMA(1,d,1) no trend	1.03	6.45	2.49	**0.158**	**0.182**	**0.841**
ARIMA(1,d,0) no trend	22.13	25.74	−9.06	0.285	0.320	0.510
ARIMA(1,d,1) with trend	−11.83	−6.41	8.92	1.658	1.699	−12.83
ARIMA(1,d,0) with trend	**−13.21**	**−9.59**	**8.60**	1.677	1.725	−13.27

From the table, it is clear that models incorporating deterministic trends (particularly ARIMA(1,d,0) with trend) yield the best in-sample fit, as reflected in the lowest AIC and BIC values. However, this statistical superiority comes at a cost when forecasting under conditions of structural disruption.

6.2 Out-of-Sample Accuracy and Structural Robustness

When comparing forecast performance over the COVID-19 period (20202024), the models with deterministic trends exhibit marked degradation in accuracy.

Their high MAE and RMSE values, along with large negative pseudo-R^2 scores, suggest a systematic overestimation of recovery—an artifact of assuming full reversion to a pre-shock deterministic path.

Conversely, the **ARIMA(1,d,1) without trend** model achieves the lowest forecast errors and highest pseudo-R^2 among all specifications. This model accommodates the persistence of the 2020 shock while still capturing short-term dynamics, providing a more realistic depiction of the partial rebound in life expectancy observed in recent years.

6.3 Residual Diagnostics

Further support for model adequacy is provided by residual autocorrelation analysis. Figure 5 displays the ACF and PACF of residuals for all four models.

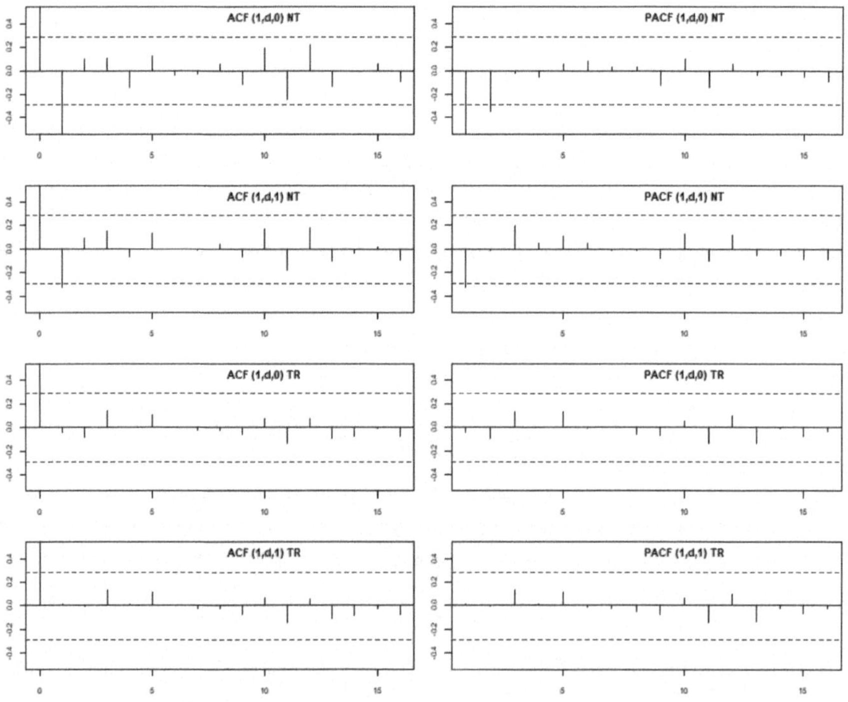

Fig. 5. ACF and PACF of Residuals from All Models

The residuals from the ARIMA(1,d,1) without trend model show no significant autocorrelation, indicating that the model successfully captures the underlying structure of the data. In contrast, the residuals from trend-based models display low-level serial dependence, suggesting that important dynamics may have been omitted or misspecified due to the imposition of a deterministic structure.

6.4 Summary of Findings

The comparative analysis reveals a fundamental trade-off between in-sample fit and structural adaptability. Models optimized for historical data—particularly those with deterministic trends—can fail to anticipate or incorporate structural changes such as pandemics. More flexible models, especially ARIMA(1,d,1) without trend, offer a better balance by accommodating both persistent level shifts and short-term corrections.

This reinforces the principle that model selection should not rely solely on information criteria or in-sample fit but must also account for theoretical plausibility and empirical robustness in the presence of shocks.

7 Conclusion

This study provides a comprehensive comparative analysis of ARIMA model specifications applied to Italian life expectancy data over the period 19742024, with a particular focus on the response to an exogenous structural shock—the COVID-19 pandemic. By evaluating model behavior across five key dimensions—trend specification, differencing strategy, dynamic adjustment, forecast accuracy, and residual diagnostics—we offer both empirical insights and methodological guidance for modeling demographic time series under uncertainty.

Our results demonstrate that model assumptions regarding the nature of the trend—deterministic versus stochastic—have profound implications for performance. While deterministic trend models (particularly ARIMA(1,d,0) with trend) achieve excellent in-sample fit, they exhibit poor forecast accuracy in the wake of structural disruption, often overestimating post-shock recovery. In contrast, the **ARIMA(1,d,1) without trend** model consistently delivers superior out-of-sample performance, lower forecast errors, and more plausible impulse responses, aligning closely with the observed partial rebound in life expectancy following the 2020 decline.

These findings emphasize the value of flexibility and parsimony in time series modeling. Models that allow for both short-run dynamics and persistent shocks—without imposing rigid deterministic structure—are better suited to capturing real-world demographic processes, especially under conditions of instability.

Beyond empirical forecasting, this work contributes to a broader theoretical understanding of demographic systems as stochastic and evolving processes. In a world increasingly shaped by non-linear shocks—pandemics, climate events, geopolitical crises—modeling frameworks must be equipped not only to fit the past but to anticipate structurally plausible futures.

In this sense, the ARIMA(1,1,1) without trend emerges not just as a statistically adequate model, but as a conceptually coherent and policy-relevant tool for demographic forecasting. It enables decision-makers to internalize uncertainty, evaluate alternative scenarios, and respond more effectively to systemic risks.

We believe these results hold relevance beyond the Italian case, offering a transferable modeling approach for national statistical agencies, health planners, and policy institutions grappling with the demographic consequences of a volatile global environment. Future work may extend this framework to multivariate models, explore regime-switching structures, or integrate high-frequency health and mortality indicators to enhance predictive power.

Ultimately, this study affirms the power of structurally grounded, empirically validated models in navigating the complexity of demographic change—and underscores the need for resilient forecasting tools in an era of uncertainty.

References

1. Alshehri, A., El-Halawany, F., Abu-Zinadah, H.: Prediction of life expectancy in saudi arabia by 2030 using arima models. J. Phys: Conf. Ser. **1978**(1), 012060 (2021)
2. ArunKumar, K.E., Kalaga, D.V., Kumar, C.M.S., Chilkoor, G., Kawaji, M., Brenza, T.M.: Forecasting the dynamics of cumulative covid-19 cases for top-16 countries using statistical machine learning models: Arima and sarima. Appl. Soft Comput. **103**, 107161 (2021)
3. Adeleye, D.A., Muhajarine, N.: Time series prediction of under-five mortality rates for Nigeria: comparative analysis of artificial neural networks, holt-winters exponential smoothing and arima models. BMC Med. Res. Methodol. **20**(1), 1–13 (2020)
4. Cerqueira, V., Torgo, L., Mozetič, I.: Evaluating time series forecasting models: an empirical study on performance estimation methods. Mach. Learn. **109**(11), 1997–2028 (2020). https://doi.org/10.1007/s10994-020-05910-7
5. Torri, T., Vaupel, J.W.: Forecasting life expectancy in an international context. Int. J. Forecasting **28**(2), 519–531 (2012)
6. Box, G.E.P., Pierce, D.A.: Distribution of residual autocorrelations in autoregressive-integrated moving average time series models. J. Am. Stat. Assoc. **65**(332), 1509–1526 (1970)
7. Christiano, L.J., Eichenbaum, M., Evans, C.L.: Monetary policy shocks: What have we learned and to what end? Handb. Macroecon. **1**, 65–148 (1999)
8. Clements, M.P., Hendry, D.F.: Economic forecasting: some lessons from recent research. Econ. Modelling **18**(2), 323–339 (2001)
9. Diebold, F.X.: Forecasting in Economics, Business, Finance and Beyond. University of Pennsylvania, 2015. Online textbook. https://www.sas.upenn.edu/~fdiebold/ForecastingBook.html
10. Diebold, F.X., Mariano, R.S.: Comparing predictive accuracy. J. Bus. Econ. Stat. **13**(3), 253–263 (1995)
11. James, D.: Hamilton. Princeton University Press, Time Series Analysis (1994)
12. Husin, W.Z.W., Aziz, N.A.b.: Box-jenkins model for forecasting malaysia life expectancy. Int. J. Acad. Res. Bus. Soc. Sci. **11**(12), 11980 (2021)
13. Liu, Z., Zhu, Z., Gao, J., Xu, C.: Forecast methods for time series data: a survey. IEEE Access **9**, 91896–91912 (2021)

14. Lütkepohl, H.: New introduction to multiple time series analysis. Springer, Heidelberg (2005)
15. Marcellino, M., Stock, J.H., Watson, M.W.: A comparison of direct and iterated multistep ar methods for forecasting macroeconomic time series. J. Econometrics **135**(1–2), 499–526 (2006)
16. Pandey, A., Chhikara, R.: Analysis of life expectancy using various regression techniques. In: 2020 2nd International Conference on Advances in Computing, Communication Control and Networking (ICACCCN), p. 9362914 (2020)
17. Perone, G.: An arima model to forecast the spread and the final size of covid-2019 epidemic in Italy. medRxiv (2020)
18. Perone, G.: Comparison of arima, ets, nnar and hybrid models to forecast the second wave of covid-19 hospitalizations in Italy. medRxiv (2020)
19. Pesaran, M.H., Schuermann, T., Vanessa Smith, L.: Forecast scenarios for global gdp using model averaging. Int. J. Forecasting **22**(4), 527–542 (2006)

Testing for Constant Central Asymmetry Between Two Copulas

Lorenzo Frattarolo(✉)

Department of Economics, University of Verona, Verona, Italy
lorenzo.frattarolo@univr.it

Abstract. We develop a test of constant central asymmetry between two copulas estimated through their empirical counterpart. We provide inference under standard assumptions for stationary time series. The tie-break bootstrap is used for calculating p-values of the proposed Cramérvon Mises test statistic. Finite sample properties are assessed with Monte Carlo experiments. We apply the testing procedure to the US portfolio industry returns during the subprime crisis.

Keywords: Two-Sample Test · Dependence Asymmetry · Radial Symmetry · Reflection Symmetry

1 Introduction

Financial markets increase their cross-sectional dependence during downturns through different mechanisms collectively called financial contagion, generating an asymmetry in dependence [8]; [1]; [6]. Dependence asymmetry is described in the following definition: a random vector has dependence asymmetry if the probability of co-movements of its components differs in the upper and lower tails of the distribution. [5] studied how the most used measures of dependence asymmetry are related, and measures based on copula central symmetry emerged as one of the most convenient. The paper also empirically investigated the presence of copula central asymmetry in US portfolio returns. More importantly, concerning the present note, the article documented that asymmetry is time-varying. This paper proposes detecting the time variation of asymmetry in the simplest possible way. We develop a two-sample test that is robust for time series dependence. We apply the test to the US industry portfolio returns of two consecutive years during the subprime crisis to understand if the procedure is sensible to systemic co-movements in the market. We provide only the results and definitions from copula theory needed in the following, but the interested reader could refer to [4,7,9]. Given a N-dimensional random vector $\mathbf{X}$ with marginal cumulative distribution function (CDF) $F_{x_i}(X_i)$, $i = 1 \ldots, N$ the copula C is the CDF of the probability integral transform (PIT) $\mathbf{U}$ of $\mathbf{X}$

$$\mathbf{U} = (U_1, \ldots, U_N) = (F_{X_1}(X_1), \ldots, F_{X_N}(X_N)).$$

A. D'Ambrosio et al. (Eds.): CLADAG-VOC 2025, SCDAKO, pp. 172–179, 2025.
https://doi.org/10.1007/978-3-032-03042-9_15

The survival copula $\bar{C}$ is the distribution of $\mathbf{1}_N - \mathbf{U}$ where $\mathbf{1}_N$ be a N-dimensional vector with all components equal to one.

A copula C is centrally symmetric if it is equal to its survival counterpart $\bar{C}$. For this reason, in the study of central symmetry is convenient to introduce the anti-symmetrizzation of the copula

$$\Delta_C(\mathbf{u}) = \frac{1}{2}\left(C(\mathbf{u}) - \bar{C}(\mathbf{u})\right)$$

Suppose now that we have a sample from $\mathbf{X}$, of size T from N-dimensional random vector $\mathbf{X}$, $\left\{\{X_{ti}\}_{t=1}^{T}\right\}_{i=1}^{N} \equiv \{\mathbf{X}_t\}_{t=1}^{T}$. We divide the sample in two subsamples $\{\mathbf{X}_t\}_{t=1}^{T_1}$ and $\{\mathbf{X}_t\}_{t=T_1+1}^{T}$ The first subsample has T_1 observation, has as copula C_1 and anti-symmetrization $\Delta_{C_1}(\mathbf{u})$. The second subsample has $T_2 = T - T_1$,copula C_2 and anti-symmetrization $\Delta_{C_2}(\mathbf{u})$. The null hypothesis we investigate is that central asymmetry does not change:

$$\mathcal{H}_0 : \Delta_{C_1}(\mathbf{u}) = \Delta_{C_2}(\mathbf{u})$$

In particular, under this hypothesis C_1 and C_2 could be different but both centrally symmetric, resulting in $\Delta_{C_1} = \Delta_{C_2} = 0$.

2 Test

A non-parametric consistent estimation of Δ_C can use the empirical copula. We denote the set A indicator as $\mathbb{I}(A)$. In addition, we define the normalized ranks $U_{T,ti} = \frac{1}{T+1}\sum_{s=1}^{T}\mathbb{I}(X_{si} \leq X_{ti})$, $t = 1, \ldots, T$ and $i \in \mathcal{N}$. With those definitions, the empirical copula and the empirical survival copula are:

$$C_T(\mathbf{u}) = \frac{1}{T}\sum_{t=1}^{T}\mathbb{I}(\mathbf{U}_{T,t} \leq \mathbf{u}), \quad \bar{C}_T(\mathbf{u}) = \frac{1}{T}\sum_{t=1}^{T}\mathbb{I}(\mathbf{1}_N - \mathbf{U}_{T,t} \leq \mathbf{u}).$$

Then, the anti-symmetrization of the empirical copula is

$$\Delta_T(\mathbf{u}) = \frac{1}{2}\left(C_T(\mathbf{u}) - \bar{C}_T(\mathbf{u})\right) \tag{1}$$

An informative test statistic for the null hypothesis $\mathcal{H}_0$, could be a functional of $\Delta_{T_1}(\mathbf{u}) - \Delta_{T_2}(\mathbf{u})$.

We investigate in this note the following Cramérvon-Mises statistic under the random measure generated by the empirical copula:

$$S_{T_1,T_2} = \int_{(0,1]^N}\left(2\frac{\Delta_{T_1} - \Delta_{T_2}}{\sqrt{\frac{1}{T_1} + \frac{1}{T_2}}}\right)^2 dC_T = \frac{1}{T}\sum_{t=1}^{T}\left(2\frac{\Delta_{T_1}(\mathbf{U}_{T_1,t}) - \Delta_{T_2}(\mathbf{U}_{T_2,t})}{\sqrt{\frac{1}{T_1} + \frac{1}{T_2}}}\right)^2 \tag{2}$$

We briefly recall some results on the weak convergence of empirical copula processes under strong mixing data. Under the assumption on copula derivatives given in [12], for strongly mixing data with mixing coefficient $\alpha_n = o\left(n^{-a}\right)$ for some $a > 0$, the empirical copula process $\mathbb{C}_T = \sqrt{T}\left(C_T\left(\mathbf{u}\right) - C_{\mathbf{U}}\left(\mathbf{u}\right)\right)$ converge weakly, in the Hoffman-Jorgensen sense, in $\ell^\infty\left([0,1]^N\right)$ the space of bounded function on the N-dimensional unit hypercube to a Gaussian field [3]:

$$\mathbb{C}_T \rightsquigarrow \mathbb{C} = \mathbb{B}_{\mathbf{U}}\left(\mathbf{u}\right) - \sum_{i=1}^{N} \frac{\partial C_{\mathbf{U}}\left(\mathbf{u}\right)}{\partial u_i} \mathbb{B}_{i,\mathbf{U}}\left(u_i\right), \tag{3}$$

where $\mathbb{B}_{\mathbf{U}}$ is a d-dimensional Brownian sheet with covariance function

$$\mathbb{C}\mathrm{ov}\left(\mathbb{B}_{\mathbf{U}}\left(\mathbf{u}\right), \mathbb{B}_{\mathbf{U}}\left(\mathbf{v}\right)\right) = \sum_{t=-\infty}^{\infty} \mathrm{Cov}\left(\mathbb{I}\left(\mathbf{U}_0 \leq \mathbf{u}\right), \mathbb{I}\left(\mathbf{U}_t \leq \mathbf{v}\right)\right) \tag{4}$$

Analogously was showed, in [5], that the empirical survival copula process $\bar{\mathbb{C}}_T = \sqrt{T}\left(\bar{C}_T\left(\mathbf{u}\right) - \bar{C}\left(\mathbf{u}\right)\right)$ weakly converges towards a Gaussian field :

$$\bar{\mathbb{C}}_T \rightsquigarrow \bar{\mathbb{C}} = \bar{\mathbb{B}}_{\mathbf{U}}\left(\mathbf{u}\right) - \sum_{i=1}^{N} \frac{\partial \bar{C}_{\mathbf{U}}\left(\mathbf{u}\right)}{\partial u_i} \bar{\mathbb{B}}_{i,\mathbf{U}}\left(u_i\right), \tag{5}$$

where $\bar{\mathbb{B}}_{\mathbf{U}}$ is a d-dimensional Brownian sheet with covariance function

$$\mathbb{C}\mathrm{ov}\left(\bar{\mathbb{B}}_{\mathbf{U}}\left(\mathbf{u}\right), \bar{\mathbb{B}}_{\mathbf{U}}\left(\mathbf{v}\right)\right) = \sum_{t=-\infty}^{\infty} \mathrm{Cov}\left(\mathbb{I}\left(\mathbf{1}_N - \mathbf{U}_0 \leq \mathbf{u}\right), \mathbb{I}\left(\mathbf{1}_N - \mathbf{U}_t \leq \mathbf{v}\right)\right) \tag{6}$$

With arguments analogous to the one used in [11] if $\min\left(T_1, T_2\right) \to \infty$, in such a way that $T_1/(T_1+T_2) \to \lambda \in [0,1]$ one can show that

$$\left(2\frac{\Delta_{T_1}\left(\mathbf{U}_{T_1,t}\right) - \Delta_{T_2}\left(\mathbf{U}_{T_2,t}\right)}{\sqrt{\frac{1}{T_1} + \frac{1}{T_2}}}\right)^2 \tag{7}$$

converges weakly to

$$\sqrt{1-\lambda}\left(\mathbb{C}_1 - \bar{\mathbb{C}}_1\right) + \lambda\left(\mathbb{C}_2 - \bar{\mathbb{C}}_2\right) \tag{8}$$

where the underscore i, i=1, 2 means Gaussian processes relative to the copula C_i. Using the continuous mapping theorem, the proposed test statistic S_{T_1,T_2}, under the null, converge weakly to

$$\mathbb{S}_\lambda = \int \left(\sqrt{1-\lambda}\left(\mathbb{C}_1 - \bar{\mathbb{C}}_1\right) + \lambda\left(\mathbb{C}_2 - \bar{\mathbb{C}}_2\right)\right) d\mathbb{C},$$

and diverge under the alternative.

Analogous line of reasoning can be used to show that test statistics $S^{*,[m]}_{T_1,T_2}$, $m = 1, \dots, M$ computed on Tie-Break bootstrap [13] samples converge to M independent copies of $\mathbb{S}_\lambda$. Using bootstrap replicates we approximate P-values in the standard way [c.f. 5,13]:

$$\hat{P} = \frac{1}{M} \sum_{m=1}^{M} \mathbb{I} \left(S^{*,[m]}_{T_1,T_2} > S_{T_1,T_2} \right). \tag{9}$$

3 Simulation Study

This section uses simulations to study the finite sample properties of the proposed constant copula central symmetry test. In all the experiments performed, the number of bootstrap replicates is M = 250, and the estimated rejection probabilities are computed using 100 Monte Carlo independent replicates. Each figure in the section presents different dimensions of the random vector $N \in \{2, 10, 25, 50\}$. The simulation study uses a recent asymmetric generalization of a score-driven copula factor model introduced in [10] with factors distributed according to a Skew-t distribution with common asymmetry parameter

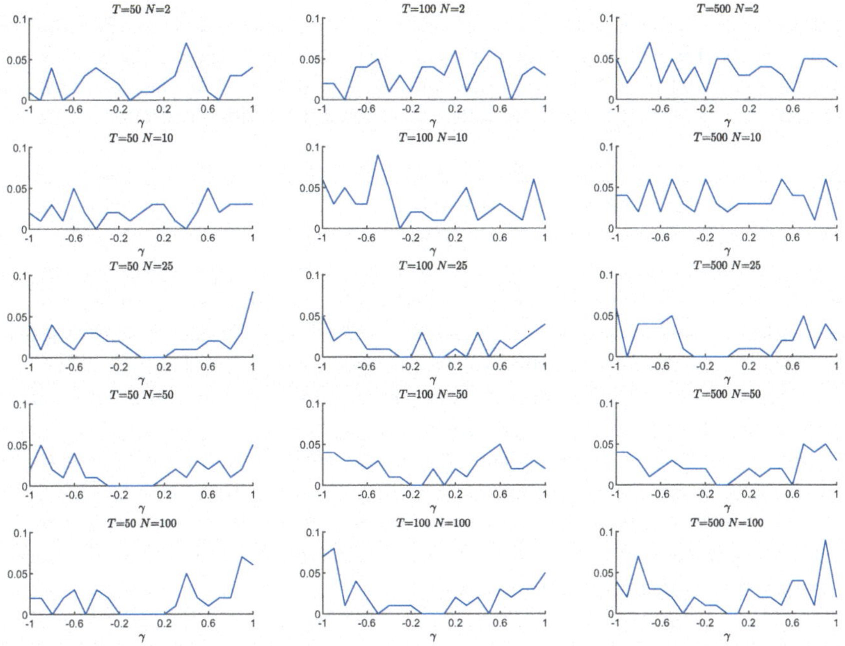

Fig. 1. Percentages of rejection at 5% significance level in blue, as estimated from 1000 replicates of the Tie-Break Bootstrap, for the tests based on S_n under the Skew-t score driven factor model and different levels of asymmetry, but equal asymmetry for both samples γ.

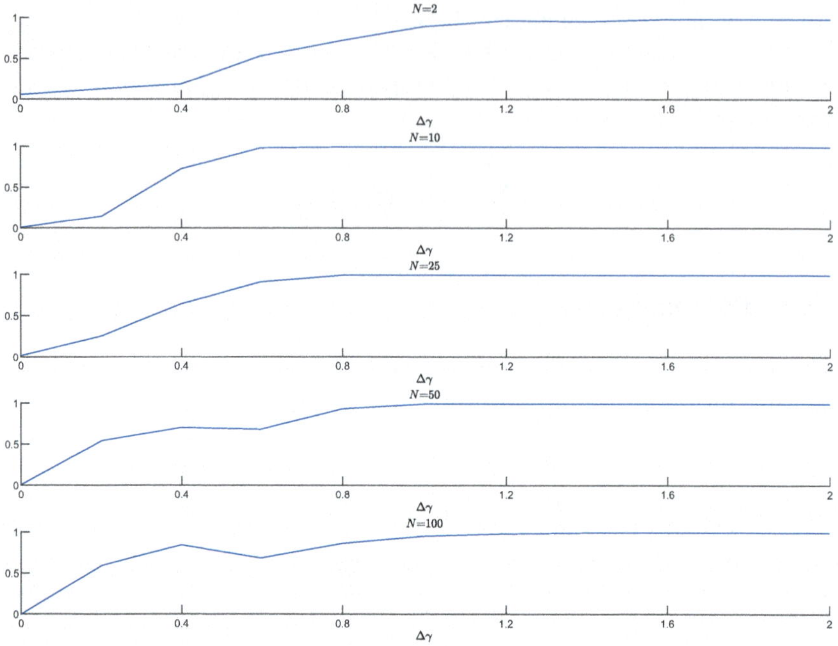

Fig. 2. Percentages of rejection at 5% significance level in blue, as estimated from 1000 replicates of the Tie-Break Bootstrap, for the tests based on S_n under the Skew-t score driven factor model and $T = 500$ observations. The first half of the sample has an asymmetry parameter γ, the second half $-\gamma$ resulting in a difference $\Delta\gamma$.

$\gamma \in [-1, 1]$. We vary γ in steps of 0.1 and choose a number of groups equal to 2 for $N \leq 10$ and 5 in the other cases. The rest of the parameter values are from the simulation study in [10]. Here and in the empirical application section, the block length follows the heuristic proposed in [3]. In addition, in the simulation we consider only the case $\lambda = 1/2$ so $T_1 = T_2 = T/2$ since in the empirical section the test will be performed on two consecutive years of daily returns.

To investigate the finite sample size, the copula factor model is the same in both samples. Results are reported in Fig. 1, for different numbers of observations $T \in \{50, 100, 500\}$, and show overall size in line with the nominal level size, but a more conservative test when increasing the number of time series.

To study the power of the tests based on S_{T_1,T_2} the copula factor model has asymmetry parameter $\gamma \in [0, 1]$ in the first subsample and $-\gamma$ in the second subsample resulting in a difference $\Delta\gamma$.
The results are reported in Fig. 2, for $T = 500$, where as expected power is an increasing function of $\Delta\gamma$.

4 Empirical Application

In this section, we study the copula central asymmetry of equity portfolio returns composed of US stocks. We use a benchmark dataset publicly available on Kenneth French website[1]. We use the daily returns of 30 portfolios chosen by industry[2]. In Fig. 3, we show the test performed on data from 1 January 1995 to 31 December 2023 and test for central symmetry of the joint copula of the 30 portfolios separately for each pair of two consecutive years in the sample. The choice of the period allows us to understand if the test can detect the subprime crisis.

The test rejects the null (5% level) at the beginning of the subprime crisis (2006-2007) and less strongly (10% level) at its end (2009-2010) then again reacts during the European sovereign debt crisis (2012-2013). The last rejection

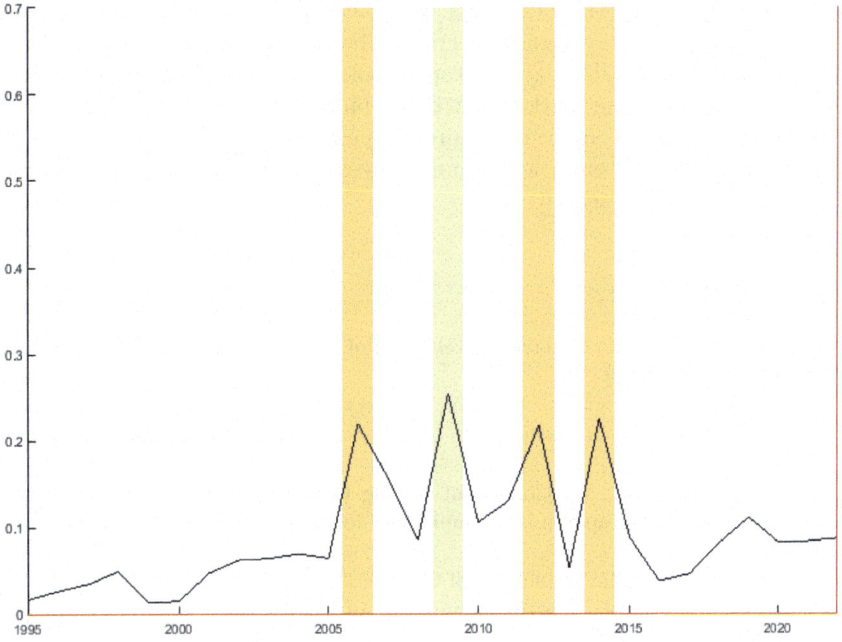

Fig. 3. Test Statistics(black line) and constant copula central symmetry test for portfolios based on 30 industries. The procedure is performed separately for two consecutive years in the sample. Red bars are rejections at the significance levels 1%, orange bars are rejections at the significance level 5%, and yellow bars are rejections at the significance level 10%. Dates refer to the first of the two consecutive years on which the test is performed.

---

[1] http://mba.tuck.dartmouth.edu/pages/faculty/ken.french/data_library.html.

[2] A detailed description of the dataset is available on the Kenneth French website.

at (5%) during 2014-2015 could only partially be explained by market reaction to the slowdown of the Chinese economy in 2015.

5 Conclusion

The detection of dependence asymmetry is essential in the quantitative assessment of financial markets. We extend to a two-sample framework, a previously developed test [5], in the time series context, based on distributional distance, using a novel, robust bootstrap framework introduced in [13]. This extension has as a null hypothesis a constant copula central asymmetry in the two samples. A small simulation study with a state-of-the-art copula factor model DGP shows reliable statistical size and power. The power of the test allows application of the procedure to pairs of consecutive years using daily returns. In a limited empirical example, using 30 US portfolios based on industrial classification, we show that the test reacts at the beginning and the end of the subprime crisis and during the European Sovereign debt crisis. The test rejects the null only in one additional case (2014-2015). Those results look promising, if the objective is to detect instabilities in the market, but developing a proper methodology for the suggested change point analysis requires the extension of the theorems of this paper to the sequential empirical copula process [see [2] and reference therein]. This is left for future research.

References

1. Ang, A., Chen, J.: Asymmetric correlations of equity portfolios. J. Financ. Econ. **63**(3), 443–494 (2002)
2. Bucher, A., Kojadinovic, I., Rohmer, T., Segers, J.: Detecting changes in cross-sectional dependence in multivariate time series. J. Multivar. Anal. **132**, 111–128 (2014)
3. Bücher, A., Ruppert, M.: Consistent testing for a constant copula under strong mixing based on the tapered block multiplier technique. J. Multivariate Anal. **116**, 208–229 (2013)
4. Durante, F., Sempi, C.: Principles of Copula Theory. CRC Press (2015)
5. Frattarolo, L.: Copula central asymmetry of equity portfolios. arXiv preprint arXiv:2501.00634 (2025)
6. Hong, Y., Jun, T., Zhou, G.: Asymmetries in stock returns: statistical tests and economic evaluation. Rev. Financial Stud. **20**(5), 1547–1581 (2007)
7. Joe, H.: Dependence Modeling with Copulas. Chapman & Hall/CRC Monographs on Statistics & Applied Probability. Taylor & Francis (2014)
8. Longin, F., Solnik, B.: Extreme correlation of international equity markets. J. Financ. **56**(2), 649–676 (2001)
9. Roger, B.: Nelsen, 2nd edn. An introduction to copulas. Springer Series in Statistics. Springer, New York (2006)
10. Oh, D.H., Patton, A.J.: Dynamic factor copula models with estimated cluster assignments. J. Econometrics **237**(2, Part C), 105374 (2023)

11. Rémillard, B., Scaillet, O.: Testing for equality between two copulas. J. Multivar. Anal. **100**(3), 377–386 (2009)
12. Segers, J.: Asymptotics of empirical copula processes under non-restrictive smoothness assumptions. Bernoulli **18**(3), 764–782 (2012)
13. Seo, J.: Tie-break bootstrap for nonparametric rank statistics. J. Bus. Econ. Stat. **42**(2), 615–627 (2024)

Exploring Student Mobility Trajectories in Higher Education During the COVID-19 Pandemic

Vincenzo Giuseppe Genova[1(✉)], Maria Prosperina Vitale[2], Giuseppe Giordano[2], and Giancarlo Ragozini[3]

[1] University of Palermo, Department of Economics, Business, and Statistics, Palermo, Italy
vincenzogiuseppe.genova@unipa.it

[2] University of Salerno, Department of Political and Social Studies, Salerno, Italy

[3] University of Naples Federico II, Department of Political Sciences, Naples, Italy

Abstract. The contribution compares student mobility trajectories in higher education before and during the COVID-19 pandemic, focusing on a regional perspective. Italy is used as a case study, which is characterised by persistent territorial inequalities, with traditional one-way mobility trajectories from Southern regions towards Northern and Central areas. Drawing on cohort data from the Italian Student National Archive, complex network data structures are extracted and analysed through community detection algorithms and regression models to reveal how the pandemic has shifted internal mobility flows, thereby altering long-established patterns. The main findings highlight a change in internal student mobility during the pandemic. Specifically, Northern universities maintained strong attractiveness nationwide, though slightly less than before the pandemic. The analysis also uncovers new mobility patterns towards the Central regions, reflecting the emergence of new educational routes. Furthermore, the persistent attractiveness of STEM programmes is confirmed.

Keywords: Educational inequalities · COVID-19 pandemic · Community detection algorithm · Territorial divide

1 Introduction

Concerning the transition from high school to university, empirical studies at the international and intranational levels [2,23,34] have shown disparities in student mobility flows between geographical areas, highlighting that demographic, economic, institutional, and social aspects influence individual educational choices. Push-pull factors affect students' decisions, such as personal motivations, family context, institution-specific resources, and socioeconomic characteristics [6,30].

Within this framework, Italy stands out as a notable case study due to the traditional one-way mobility trajectories from the South to the Centre and North

A. D'Ambrosio et al. (Eds.): CLADAG-VOC 2025, SCDAKO, pp. 180–191, 2025.
https://doi.org/10.1007/978-3-032-03042-9_16

macro-areas [3,14,15,19,39,41]. These one-way flows also reflect the broader South-to-North migration trend [10,26].

Such mobility choices are related to several factors, including the opportunity to move to an area offering a better quality of life and more job opportunities [17,27]; the attractiveness and reputation of universities [4,11,13,16]; the pursuit of long-term social mobility, often intertwined with family trajectories [28,29]; and students' characteristics and motivations [37].

The propensity to move to another region for higher education may also be affected by external conditions. Specifically, the global COVID-19 pandemic has exacerbated social and economic inequalities across higher education institutions, with profound implications for student access to educational opportunities. This health crisis has reshaped the map of international student mobility [7,12,45] and led to substantial shifts in intranational flows within countries [5,30,35]. In the Italian context, different effects of the pandemic on university settings have been described [1]. However, to the best of our knowledge, the effects of the pandemic on internal university mobility trajectories in Italy have not yet been thoroughly examined.

This study investigates changes in internal student migration due to the COVID-19 pandemic. It aims to reveal how the pandemic triggered shifts in student migration from a national perspective, modifying the established geography of mobility. Drawing on data from the Italian Student National Archive of the Ministry of Education and Research [32] both before and during the pandemic, the study seeks to map the mechanisms driving the reproduction of inequalities in Italian student mobility. Complex data structures are defined and analysed through community detection algorithms and regression models to reveal how the pandemic, considered as a natural experiment [31], has reshaped internal mobility flows and disrupted long-established patterns.

The remainder of the paper is organised as follows. Section 2 describes the data collection. Section 3 discusses the methods used to analyse student cohorts and the main findings. The final section provides some concluding remarks.

2 Data

The data are extracted from the administrative database MOBYSU.IT[1], which includes data on student cohorts enrolled in Italian universities since the academic year 2010–2011. This database contains information about students'

[1] Database MOBYSU IT. Mobilità degli Studi Universitari in Italia. Research protocol MUR - Universities of Cagliari, Palermo, Siena, Torino, Sassari, Firenze, Cattolica, and Napoli Federico II, Scientific Coordinator: Massimo Attanasio (UNIPA), Data Source: ANS-MUR/CINECA. The data have been processed under the research protocol for the study "From high school to the job placement: analysis of university careers and university mobility from Southern to Northern Italy" among the Ministry of University and Research, the Ministry of Education and Merit, the University of Palermo as the lead institution, and the INVALSI Institute. The reference researchers are Giancarlo Ragozini and Maria Prosperina Vitale.

demographic characteristics, high school backgrounds, and choices of bachelor's and master's programmes in higher education. These individual-level microdata make it possible to track students' entire academic careers.

In this study, cohorts of first-year students enrolled in Italian universities in the academic years 2018–2019, 2019–2020, 2020–2021, and 2021–2022 are extracted and analysed to determine the main patterns of student mobility trajectories between Italian territorial macro-areas before and during the COVID-19 pandemic from a regional perspective. Information on students' regions and provinces of origin, the universities they attended, and their educational programmes is considered.

The educational programmes, provided by public and private institutions in Italy (excluding online universities), are classified into the 10 fields included in the International Standard Classification of Education (ISCED-F 2013)[2]. These fields are also grouped into two categories: science, technology, engineering, and mathematics (STEM) and non-STEM. This latter classification allows us to explore the presence of educational disparities in students' decisions to move towards particular universities, considering the attractiveness of STEM or non-STEM programmes [18,19,40].

The data analysed cover the 107 Italian provinces, 78 public and private universities, and 10 ISCED fields. The number of first-year students in the Italian university system varied over the four academic years, ranging from 283,416 in 2018–2019 to 317,790 in 2021–2022.

3 Methods and Results

From the student cohort data, several data structures can be derived and further investigated by network analysis tools [14,15,24–26,38]. These range from simple unipartite weighted directed graphs, where geographical units are nodes and student flows represent the weights of links, to more complex structures, such as multipartite graphs [21,22], in which *students*, *regions*, *provinces*, *universities*, and *educational programmes* can be considered simultaneously.

We adopt a multipartite network structure to analyse students, regions, provinces, universities, and educational programmes. The procedure described in Genova et al. [24] is used to simplify, normalise, and extract statistically significant links from the derived bipartite networks. These networks are first analysed using community detection to identify clusters by applying the Infomap algorithm [8,20], a method suitable for studying patterns in network structures, instead of modularity optimisation approaches that focus on topological properties [9,33]. Flow-based methods such as Infomap are more effective at highlighting significant features and uncovering communities with similar patterns from

[2] The 10 fields are: Education; Arts and Humanities; Social Sciences, Journalism, and Information; Business, Administration, and Law; Natural Sciences, Mathematics, and Statistics; Information and Communication Technologies; Engineering, Manufacturing, and Construction; Agriculture, Forestry, Fisheries, and Veterinary; Health and Welfare; and Services.

student mobility data. Infomap (the *map equation*) leverages the relationship between community detection and minimising the length (*codelength*) of a random walker's trajectory across a network. The partition with the shortest path length most accurately reflects the community structure in bipartite network data.

We also estimate a log-linear model to assess the presence of structural breaks in student mobility trajectories among Italian regions before and during the pandemic, as well as the factors affecting these patterns. In the model, the outcome variable is the number of students moving from the $i-th$ region to the $j-th$ university at time t, denoted as $N_{ijt} \sim P(\mu_{ijt})$, where $P(\cdot)$ represents the Poisson distribution.

$$\begin{aligned} N_{ijt} = \beta_0 + \beta_{1i}\text{Region}_{it} + \beta_2\text{Students}_{it} + \beta_3\text{STEM}_{jt} + \beta_4 D_p \\ + \delta_{1i}(D_p \times \text{Region}_{it}) + \delta_2(D_p \times \text{STEM}_{jt}) + \epsilon_{ijt} \end{aligned} \quad (1)$$

In Eq. 1, to capture potential geographical disparities in mobility patterns, we include the categorical variable Region_{it} to represent the 20 Italian regions of residence. As mobility is also affected by the size of the student population, we include Students_{it}, the number of students obtaining a high school diploma in region i at time t.

We also include a binary variable for STEM and non-STEM programmes to distinguish students enrolled in STEM fields from those in other disciplines. Specifically, STEM is equal to 1 for students enrolled in programmes in *Natural Sciences, Mathematics, and Statistics*, *Engineering, Manufacturing and Construction*, *Information and Communication Technologies*, and 0 otherwise. To account for variations associated with the COVID-19 pandemic, we introduce the dummy variable D_p, which takes the value 0 for the academic years 2018–2019 and 2019–2020 and 1 for 2020–2021 and 2021–2022. This variable captures the impact of the pandemic by identifying shifts in mobility trends following the onset of the crisis.

Finally, to examine potential heterogeneous effects of the pandemic, we include two interaction terms: one between D_p and Region_{it} to assess whether the impact of the crisis on student mobility varied across different regions; and one between D_p and STEM programmes to capture whether this phenomenon affects mobility decisions.

Considering the pandemic as a natural experiment [31], and assuming that the different cohorts are homogeneous with respect to student characteristics (at least across the four consecutive academic years), the proposed model can be interpreted as a difference-in-differences model following the formulation of [43]. In this case, we are not comparing treated *vs* non-treated subjects but rather analysing the effects of a treatment (the pandemic) that affects the entire population, with respect to the different categories into which the population can be divided. Within this framework, the two interaction coefficients δ_{1i} and δ_2 assess and quantify the difference-in-differences effects [43]. Significant coefficients imply a change (before and during the pandemic) in the propensity for student mobility related to their region of residence and the chosen field of study – changes that differ from the average shift affecting the entire population.

3.1 Insights from Communities and Regression Model

Table 1 shows the communities identified by the Infomap algorithm for the four student cohorts under analysis. Notably, there is a reduction in the number of communities in 2019–2020, indicating a stabilisation of mobility patterns and confirming the results reported in Genova et al. [24], where the phenomenon was analysed until 2017–2018.

An increase in the number of communities is observed during the pandemic period, suggesting both the emergence of new routes and a shift in mobility patterns favouring the Central region, which had been less popular among students before the pandemic. These communities appear internally coherent, as indicated by the relative saving codelength of student cohorts. This Infomap metric reflects how efficiently the network's structure can be encoded based on observed mobility flows. A lower codelength indicates that movements are concentrated within well-defined communities, suggesting structured and recurrent behavioural patterns. In contrast, higher values suggest more fragmented flows.

During the health crisis, all public and private universities in Italy provided the option to attend online courses. This shift in teaching strategies in response to COVID-19 restrictions may have reduced disparities in mobility choices [44]. Before the pandemic, university mobility was often limited to students who could afford the associated costs. During the pandemic, many students were able to enrol in "brand" universities in the Central and Northern regions without bearing the living costs their families would have faced with physical relocation. This phenomenon may have given rise to a form of *artificial mobility*, masking traditional routes from Southern to Central and Northern regions.

Table 1. Infomap results for student cohorts

Cohort	Communities	Codelength	Relative saving codelength
2018–2019	16	0.90	84.08%
2019–2020	10	1.18	79.74%
2020–2021	17	1.14	80.90%
2021–2022	13	0.99	83.10%

A detailed analysis of the clustering results obtained for the first student cohort in 2018–2019 (Fig. 1 – Upper) and the last one in 2021–2022 (Fig. 1 – Lower) reveals the mobility patterns of preferential attractiveness towards Northern universities and a difference between scientific (STEM) and humanistic (non-STEM) disciplines.

In the pre-pandemic cohort, two clusters emerge around attractive universities in Northern Italy, defining the primary trajectories of student mobility across the country. In particular, one large cluster includes universities that attract students in a variety of fields such as *Humanities*, *Social Sciences*, *Business*, and *Engineering*, most of which are located in the North. Apulian provinces act as a bridge between the Polytechnic University of Turin (*Engineering*) and the University of Ferrara (*Science*), highlighting cross-regional mobility flows. Several smaller clusters are also observed to play a central role in facilitating mobility across regional boundaries.

The results for 2021-2022 reveal changes in mobility patterns during the pandemic. Four main clusters of universities are identified, showing a dichotomy between scientific (STEM) and humanistic (non-STEM) fields of study, with a persistent flow of students moving from the South to Northern regions and within the North itself [36]. Interestingly, a large cluster continues to attract students in *Humanities*, *Social Sciences*, and *Sciences*, mainly in Northern Italy. Note that during the pandemic, new small clusters appeared around universities such as Urbino, which attracts students from Rimini and Forlì-Cesena, and Pisa, which draws students from La Spezia, facilitating cross-boundary regional mobility.

Furthermore, the shift to online learning strategies in university settings during COVID-19 may have partially disrupted traditional mobility patterns, favouring a more inclusive university system. These results also suggest that, while certain mobility trends persisted, the pandemic may have reinforced the separation between different fields of study through a form of artificial mobility.

Considering the regression model findings (Table 2), the pandemic led to an overall reduction in mobility flows (N_{ijt} in Eq. 1), as indicated by the significant and negative coefficient of the dummy variable D_p (-0.064). This effect is pronounced in Northern Italian regions such as Lombardy, Veneto, and Piedmont. These regions, heavily affected by COVID-19 [42], experienced reduced student mobility. The increase in student mobility flows in regions such as Sardinia and Umbria likely reflects a form of artificial mobility facilitated by online teaching strategies adopted by universities during the pandemic. This shift allowed students to enrol at universities located in other regions, particularly those from disadvantaged family backgrounds who previously faced economic barriers to relocating for higher education.

Regarding the choice of STEM degree programmes, a positive and significant effect (0.453) is observed. Although the interaction term for the pandemic period ($D_p \times STEM$) is not statistically significant, the consistently positive effect suggests that STEM programmes continued to attract students during the health crisis. These findings are related to the network cluster results discussed above. Furthermore, a strong polarisation of mobility patterns is indicated, driven by scientific fields and highlighting the continued attractiveness of STEM disciplines.

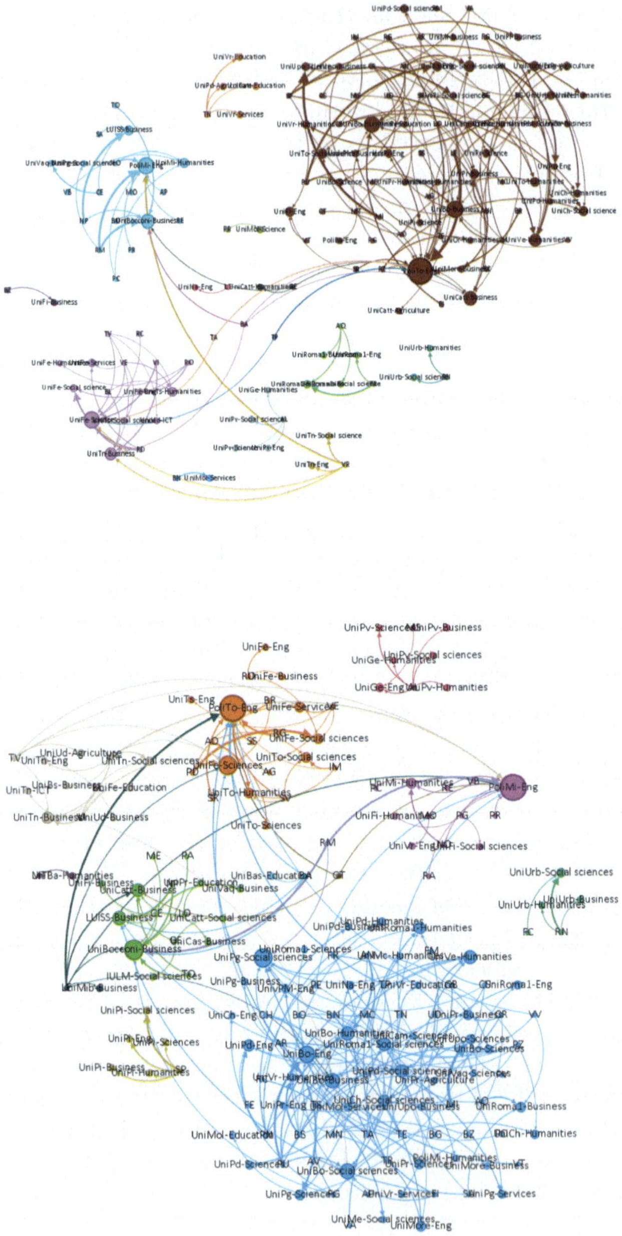

Fig. 1. Community structures of bipartite graphs for the student cohorts in 2018–2019 (Upper) and 2021–2022 (Lower). The size of the nodes is proportional to the indegree.

Table 2. Estimated coefficients of the log-linear model in Eq. 1, predicting the number of students moving from Italian region of residence i to enrol at university j (N_{ijt}).

Coefficients	Estimate	Std. Error	z value	p-value
Intercept	2.880	0.015	196.300	**<0.000**
Basilicata	2.750	0.016	169.816	**<0.000**
Calabria	2.550	0.014	185.382	**<0.000**
Campania	2.686	0.014	194.371	**<0.000**
Emilia-Romagna	2.930	0.015	192.386	**<0.000**
Friuli-Venezia Giulia	2.941	0.020	150.387	**<0.000**
Lazio	2.810	0.015	184.193	**<0.000**
Liguria	3.040	0.020	177.514	**<0.000**
Lombardy	2.421	0.013	188.750	**<0.000**
Marche	2.940	0.016	186.486	**<0.000**
Molise	2.497	0.024	104.334	**<0.000**
Piedmont	3.001	0.014	210.632	**<0.000**
Apulia	1.668	0.014	122.626	**<0.000**
Sardinia	2.568	0.022	114.358	**<0.000**
Sicily	2.359	0.013	184.701	**<0.000**
Tuscany	2.854	0.017	169.977	**<0.000**
Trentino-Alto Adige/South Tyrol	2.871	0.020	146.573	**<0.000**
Umbria	2.743	0.022	125.476	**<0.000**
Aosta Valley	2.414	0.035	68.185	**<0.000**
Veneto	1.895	0.014	136.766	**<0.000**
D_p	−0.064	0.020	−3.219	**<0.000**
Students	0.0001	<0.000	399.531	**<0.000**
Stem	0.453	0.007	61.184	**<0.000**
Basilicata:D_p	0.029	0.029	1.009	0.313
Calabria:D_p	−0.019	0.026	−0.726	0.468
Campania:D_p	−0.035	0.026	−1.344	0.179
Emilia-Romagna:D_p	−0.066	0.028	−2.385	**0.017**
Friuli-Venezia Giulia:D_p	−0.020	0.032	−0.623	0.533
Lazio:D_p	0.045	0.028	1.602	0.109
Liguria:D_p	−0.032	0.030	−1.049	0.294
Lombardy:D_p	−0.482	0.024	−19.770	**<0.000**
Marche:D_p	0.011	0.028	0.397	0.691
Molise:D_p	0.057	0.038	1.482	0.138
Piedmont:D_p	−0.103	0.027	−3.834	**<0.000**
Apulia:D_p	−0.064	0.024	−2.710	**0.007**
Sardinia:D_p	0.104	0.036	2.918	**0.004**
Sicily:D_p	−0.208	0.025	−8.451	**<0.000**
Tuscany:D_p	0.041	0.029	1.396	0.163
Trentino-Alto Adige/South Tyrol:D_p	0.033	0.032	1.032	0.302
Umbria:D_p	0.104	0.035	2.958	**0.003**
Aosta Valley:D_p	0.033	0.053	0.618	0.536
Veneto:D_p	−0.278	0.023	−11.869	**<0.000**
$STEM$:D_p	0.005	0.010	0.471	0.638

Pseudo-R^2: 0.413

4 Discussion and Concluding Remarks

This study examines four student cohorts before and during the COVID-19 pandemic to assess whether it exacerbated existing territorial inequalities in Italian student mobility routes. The main findings reveal notable shifts in internal student mobility that could disrupt established geographical trends. Specifically, a comparison of student cohorts during the pandemic with those from previous years highlights changes in mobility trends. There is a slight decrease in incoming flows to Northern regions, although Central and Northern institutions continue to attract students from Southern Italy. The increase in the number of communities during the pandemic suggests the emergence of new mobility routes towards the Central regions. A marked polarisation is also observed, largely influenced by the attractiveness of STEM fields. However, these trends may reflect a form of *artificial mobility*, driven by the online teaching strategies adopted by universities during the health crisis. The regression model results support this finding, indicating reduced mobility, particularly in Northern Italian regions that were significantly impacted by COVID-19.

By focusing on how the pandemic has reshaped student mobility patterns in Italy, policymakers can more effectively address regional disparities and formulate strategies to improve access to higher education. Identifying the key determinants of student mobility trajectories during a crisis is essential for designing policies that respond to students' evolving needs and for overcoming educational inequalities in the post-pandemic context. A comparison with more recent data would help assess the long-term impacts of the pandemic on mobility trends.

Acknowledgments. The contribution is partially supported under the National Recovery and Resilience Plan (NRRP), Mission 4, Component 2, Investment 1.1, Call for tender No. 104 published on 2.2.2022 by the Italian Ministry of University and Research (MUR), funded by the European Union – NextGenerationEU– Project Title From high school to university: Assessing peers' influence in educational inequalities and performances – CUP F53D23006150006- Grant Assignment Decree No. 1060 adopted on 07/17/2023 by the Italian Ministry of Ministry of University and Research (MUR).

The study is also partially funded by the European Union - NextGenerationEU, Mission 4, Component 2, in the framework of the GRINS -Growing Resilient, INclusive and Sustainable project (GRINS PE00000018 – CUP B73C22001260006). The views and opinions expressed are solely those of the authors and do not necessarily reflect those of the European Union, nor can the European Union be held responsible for them.

References

1. Agasisti, T., Soncin, M.: Higher education in troubled times: on the impact of COVID-19 in Italy. Stud. High. Educ. **46**(1), 86–95 (2021)
2. Attanasio, M.: Students' university mobility patterns in Europe: an introduction. Genus **78**(1), 17 (2022)

3. Attanasio, M., Enea, M.: La mobilità degli studenti universitari nell'ultimo decennio in italia. In: De Santis, G., Pirani, E., Porcu, M. (eds.) Rapporto sulla popolazione. L'istruzione in Italia, pp. 43–58. Bologna, Il Mulino (2019)
4. Bacci, S., Bertaccini, B.: Assessment of the university reputation through the analysis of the student mobility. Soc. Indic. Res. **156**(2), 363–388 (2021)
5. Balletto, G., Campisi, T., Borruso, G., Meloni, I., Scappini, B.: City form, mobility and university students in post pandemic era. In: Gervasi, O., Murgante, B., Misra, S., Rocha, A.M.A.C., Garau, C. (eds.) Computational Science and Its Applications – ICCSA 2022 Workshops. ICCSA 2022. LNCS, vol. 13380, pp. 267–281. Springer, Cham (2022). https://doi.org/10.1007/978-3-031-10542-5_19
6. Baryla, E.A., Jr., Dotterweich, D.: Student migration: do significant factors vary by region? Educ. Econ. **9**(3), 269–280 (2001)
7. Bilecen, B.: Commentary: COVID-19 pandemic and higher education: international mobility and students' social protection. Int. Migr. **58**(4), 263–266 (2020)
8. Blöcker, C., Rosvall, M.: Mapping flows on bipartite networks. Phys. Rev. E **102**(5), 052,305 (2020)
9. Blondel, V.D., Guillaume, J.L., Lambiotte, R., Lefebvre, E.: Fast unfolding of communities in large networks. J. Statist. Mech. Theory Exp. **2008**(10), P10,008 (2008)
10. Bonifazi, C., Heins, F., Licari, F., Tucci, E.: The regional dynamics of internal migration intensities in Italy. Popul. Space Place **27**(7), e2331 (2021)
11. Bratti, M., Verzillo, S.: The 'Gravity' of quality: research quality and the attractiveness of universities in Italy. Reg. Stud. **53**(10), 1385–1396 (2019)
12. Cairns, D., França, T.: Managing student mobility during the COVID-19 pandemic: an immobility turn in internationalized learning? Societies **12**(4), 105 (2022)
13. Ciriaci, D.: Does university quality influence the interregional mobility of students and graduates? the case of Italy. Reg. Stud. **48**(10), 1592–1608 (2014)
14. Columbu, S., Porcu, M., Primerano, I., Sulis, I., Vitale, M.P.: Geography of Italian student mobility: a network analysis approach. Soc. Econ. Plann. Sci. **73**, 100,918 (2021)
15. Columbu, S., Porcu, M., Primerano, I., Sulis, I., Vitale, M.P.: Correction to: analysing the determinants of Italian university student mobility pathways. Genus **78**(1), 1–1 (2022)
16. Columbu, S., Porcu, M., Sulis, I.: University choice and the attractiveness of the study area: insights on the differences amongst degree programmes in Italy based on generalised mixed-effect models. Soc. Econ. Plann. Sci. **74**, 100,926 (2021)
17. Dotti, N.F., Fratesi, U., Lenzi, C., Percoco, M.: Local labour markets and the interregional mobility of Italian university students. Spat. Econ. Anal. **8**(4), 443–468 (2013)
18. D'Agostino, A., Ghellini, G., Lombardi, G.: Movers and stayers in stem enrollment in Italy: who performs better? Genus **77**, 1–19 (2021)
19. D'Agostino, A., Ghellini, G., Longobardi, S.: Out-migration of university enrolment: the mobility behaviour of Italian students. Int. J. Manpow. **40**(1), 56–72 (2019)
20. Edler, D., Bohlin, L., Rosvall, M.: Mapping higher-order network flows in memory and multilayer networks with infomap. Algorithms **10**(4), 112 (2017)
21. Everett, M.G., Borgatti, S.P.: Partitioning multimode networks. In: Advances in Network Clustering and Blockmodeling, pp. 251–265 (2019)
22. Fararo, T.J., Doreian, P.: Tripartite structural analysis: generalizing the Breiger-Wilson formalism. Soc. Netw. **6**(2), 141–175 (1984)

23. Findlay, A., Packwood, H., McCollum, D., Nightingale, G., Tindal, S.: Fees, flows and imaginaries: exploring the destination choices arising from intra-national student mobility. Glob. Soc. Educ. **16**(2), 162–175 (2018)
24. Genova, V.G., Giordano, G., Ragozini, G., Vitale, M.P.: An analytic strategy for data processing of multimode networks. In: Advances in Data Analysis and Classification, pp. 1–23 (2023)
25. Genova, V.G., Tumminello, M., Aiello, F., Attanasio, M.: A network analysis of student mobility patterns from high school to master's. Statist. Methods App. **30**(5), 1445–1464 (2021)
26. Genova, V.G., Tumminello, M., Enea, M., Aiello, F., Attanasio, M.: Student mobility in higher education: Sicilian outflow network and chain migrations. Elect. J. Appl. Statist. Anal. **12**(4), 774–800 (2019)
27. Giambona, F., Porcu, M., Sulis, I.: Students mobility: assessing the determinants of attractiveness across competing territorial areas. Soc. Indic. Res. **133**(3), 1105–1132 (2017)
28. Impicciatore, R., Panichella, N.: Internal migration trajectories, occupational achievement and social mobility in contemporary Italy. A life course perspective. Popul. Space Place **25**(6), e2240 (2019)
29. Impicciatore, R., Tosi, F.: Student mobility in italy: The increasing role of family background during the expansion of higher education supply. Res. Soc. Stratif. Mobil. **62**, 100,409 (2019)
30. Kirby, D.: Intranational university student mobility: a case study of student migration and graduate retention in eastern Canada. Can. J. Educ. Admin. Policy/Revue canadienne en administration et politique de l'éducation **197**, 48–61 (2021)
31. Mansfield, R., et al.: The impact of the COVID-19 pandemic on adolescent mental health: a natural experiment. R. Soc. Open Sci. **9**(4), 211,114 (2022)
32. MOBYSU.IT: Database MOBYSU.IT, *Mobilità degli studi universitari italiani*, Protocollo di ricerca MIUR - Università degli Studi di Cagliari, Palermo, Siena, Torino, Sassari, Firenze e Napoli Federico II, Fonte dei dati ANS-MIUR/CINECA (2016)
33. Newman, M.E., Girvan, M.: Finding and evaluating community structure in networks. Phys. Rev. E **69**(2), 026,113 (2004)
34. Prazeres, L.: International and intra-national student mobility: trends, motivations and identity. Geogr. Compass **7**(11), 804–820 (2013)
35. Recchi, E., Safi, M.: Handbook of Human Mobility and Migration. Edward Elgar Publishing (2024)
36. Rizzi, L., Grassetti, L., Attanasio, M.: Moving from north to north: how are the students' university flows? Genus **77**(1), 1–22 (2021)
37. Rondinelli, R., Policastro, V., Scolorato, C.: How student characteristics affect mobility choices at the university level: Insights from two surveys in the campania region: Some evidence from two surveys in campania region. Statistica Applicata-Italian Journal of Applied Statistics (1) (2024)
38. Santelli, F., Ragozini, G., Vitale, M.P.: Assessing the effects of local contexts on the mobility choices of university students in Campania region in Italy. Genus **78**(1), 1–25 (2022)
39. Santelli, F., Scolorato, C., Ragozini, G.: On the determinants of student mobility in an interregional perspective: a focus on Campania region. Statistica Applicata-Italian J. Appl. Statist. **1**, 119–142 (2019)
40. Tocchioni, V., Petrucci, A., Minello, A.: Chapter short-term and long-term international scientific mobility of Italian PhDs: an analysis by gender (2021)

41. Türk, U.: Socio-economic determinants of student mobility and inequality of access to higher education in Italy. Netw. Spat. Econ. **19**(1), 125–148 (2019)
42. Varrella, S.: Coronavirus (COVID-19) in Italy: Statistics and facts (2020)
43. Villa, J.M.: Diff: simplifying the estimation of difference-in-differences treatment effects. Stand. Genomic Sci. **16**(1), 52–71 (2016)
44. Vittorietti, M., Priulla, A., Genova, V., et al.: Uncovering the interplay of territorial, socioeconomic, and demographic factors in high school to university transition. In: Book of the Short Papers SIS 2023, pp. 169–174. IT (2023)
45. Yıldırım, S., Bostancı, S.H., Yıldırım, D.Ç., Erdoğan, F.: Rethinking mobility of international university students during COVID-19 pandemic. High. Educ. Eval. Dev. **15**(2), 98–113 (2021)

Density-based community detection combining structure and attribute information

Sara Geremia and Domenico De Stefano(✉)

Department of Political and Social Sciences, University of Trieste, Trieste, Italy
sara.geremia@phd.units.it, ddestefano@units.it

Abstract. In social networks, community structure arises from a complex interplay between structural features and actor attributes. This study aims to enhance density-based community detection by integrating both topological and attribute information within a unified framework. Evaluations on simulated and real-world datasets demonstrate that incorporating attribute information improves detection performance in networks with moderate to high mixing. Results suggest that the optimal combination of topological and attribute information depends on the network structure and the type of attribute data, with a balanced approach often yielding the best results. The study also highlights the importance of choosing the topology similarity measure, with the neighbor set similarity approach proving to be more robust.

Keywords: community detection · attributed networks · leader influence · homophily

1 Introduction

In social communities, individuals tend to form close relationships with a few subjects, resulting in the formation of groups. The aggregation may be driven by diverse social mechanisms. A first may be the attraction exerted by influential actors (or network leaders) determined by the heterogeneity of degree among individuals [15]. A second mechanism is the propensity for individuals with similar interests or characteristics to bond, known as homophily. Homophily on observed attributes plays a significant role in shaping the structure and dynamics of networks, as it contributes to the formation of communities where nodes with similar characteristics tend to be more connected to each other [11]. Actors cluster into groups such that ties are more dense or intense within groups than between them. This tendency has been referred to as community structure [14]. Therefore group formation in real social networks emerges from a complex interplay between the structural features of the direct or indirect neighbourhood of actors and homophily-based attachment mechanisms. This suggests that often community detection or node clustering algorithms that rely solely on one of such

A. D'Ambrosio et al. (Eds.): CLADAG-VOC 2025, SCDAKO, pp. 192–201, 2025.
https://doi.org/10.1007/978-3-032-03042-9_17

mechanisms fail to reveal the true community structure underlying an observed network. Furthermore, node characteristics (such as demographic details or user-generated content) are often available and linked with network actors; however, in standard approaches to community detection, this additional information is often neglected [3]. Node attributes can complement the network structure, improving the determination of community boundaries. Additionally, if one source of information is missing or noisy, the other can be used.

Using solely topological information can lead to inaccurate results in many real-world applications of community detection algorithms. For instance, popular algorithms based on modularity optimisation (e.g., the Louvain algorithm proposed by [2]) suffer from solution degeneracy issues, resolution limits, and variability in results [13]. Incorporating node homophily based on available node covariates can mitigate such problems, resulting in more accurate partitions. Other classes of community detection algorithms can also benefit from leveraging this information. In particular, leader-based community detection algorithms [6] that utilise attribute data tend to yield more satisfactory results (as discussed in [10]), mainly because the attraction of leader or core nodes may depend on the characteristics of neighbouring nodes.

The present study aims to enhance community detection in leader-influenced networks by considering not only the structural properties of the network but also the similarities in the attributes associated with each node. This is achieved by extending the work of [12] on the density-based clustering of networks by developing a novel approach that integrates attribute similarity among nodes. The goal is to advance the understanding of clustering techniques in social network analysis and contribute to the development of comprehensive approaches for identifying communities within social networks with attributes.

Briefly, the proposed method is well-suited to address research questions such as: how can we uncover complex community formation mechanisms, where different dynamics coexist, such as assortative behaviour based on node characteristics (i.e., homophily) and leader-follower or leader-leader interactions?

1.1 Related works

Community detection in social networks has been extensively studied, with numerous methods developed to address this complex problem [16]. While these techniques vary in approach, many assume that all nodes in the network hold equal influence, overlooking the presence of leaders. Recognising the importance of leader nodes, few studies have explored their role in defining the community structure [4,6,18]. Leader-based models operate on the principle that communities are formed around influential nodes, with other nodes gravitating toward these leaders. While these approaches often focus solely on the structural connections, node attributes can provide valuable context and support and enhance the community detection process. In Table 1 we reported the main characteristics–community formation or detection assumptions, the data input needed by the algorithm and the output type–of some community detection methods, including those proposing leader-based community detection approach and eventually adopting nodal attributes to improve clustering solution.

The approach presented in this paper is an enhanced version of the combined leader-based and density-based community detection method named Decode [12], incorporating the contribution of attribute information when available and producing non-overlapping clusters of nodes. Among these latter, two recent studies have combined topology and attribute information to enhance leader-based community detection. The aLBCD [5] method assigns leaders based on attribute information when topological data is missing, treating attributes as a complement to network structure. TALB [10] integrates attribute information at the beginning of the algorithm, merging it with topological information. This method follows Chunaev's non-fixed topology weighted approach, which has been applied in several studies on attributed networks (see Table 5 in [3]). The topology similarity between node pairs can be determined through various techniques. One common method involves comparing the sets of neighbours for each pair of nodes using established similarity measures such as Jaccard, as implemented in the CODICIL algorithm [17]. Alternatively, methods like PLCA [1] rely directly on the information embedded in the adjacency matrix.

Table 1. Caption

Method(s)	Assumption(s)	Input type	Output type
Modularity-based (e.g., [2])	Modularity	Structural	Partition
DENGRAPH [4]	Leader and Density-based	Structural	Partition
Decode [12]	Leader and Density-based	Structural	Partition
LICOD [18]	Leader-based	Structural	Overlapping
aLBCD [5]	Leader-based	Structural+Attributive	Partition
TALB [10]	Leader-based	Structural+Attributive	Partition

The authors of CESNA [19] provide a clear description of how different data types (specifically numerical, categorical, and textual) can be handled for measuring similarities using various kernel functions. Despite advances in leader- and attribute-based methods, existing approaches still face limitations. Methods that rely on attribute information often alter centrality measures distribution and treat attributes as the sole source of information when topological data are lacking. Our proposed approach identifies leader nodes based solely on topological information, preserving the structural integrity of the network. It incorporates attribute information to refine the paths connecting nodes to their assigned leaders, improving the accuracy of community boundaries. Variations in both topological and attribute similarity measures are explored to assess their impact on community detection outcomes.

2 Density-based clustering in node-attributed networks

The clustering problem in social networks, as explored in [12], can be approached by considering the concept of high-density groups and defining density as a node-wise measure. Nodes are clustered when their density exceeds a threshold and a path connects them (Algorithm 1). In our setting, we use degree centrality as the density measure, while connection paths are modified by node-attribute similarity, where both the presence/absence of links and their weights may change when considering attribute information (non-fixed topology method). The combination of the network's topological and node attribute information is achieved via a unified distance-based approach.

For each neighbour v_j of node v_i, its attribute similarity can be determined using a distance measure. The attribute similarity matrix (W_A) can be viewed as a fully connected weighted adjacency matrix. A sparser network is extracted by retaining only edges whose weights exceed the 0.75 quantile. Then, we alter the network structure by combining the original and the attribute-aware network with a weight α:

$$w_\alpha(v_i, v_j) = \alpha \cdot w_S(v_i, v_j) + (1 - \alpha) \cdot w_A(v_i, v_j),$$

where: $w_S(v_i, v_j)$ is either an element of the adjacency matrix A or an element of the Jaccard similarity matrix W_S and $w_A(v_i, v_j)$ is an element of the filtered attribute similarity matrix W_A. DeCoDe is applied on the weighted adjacency matrix W_α.

From a computational point of view, the algorithm requires to run $O((n + m)n)$ operations on a binary network, in addition to the ones needed to compute the density, which depend on the selected node-wise measure. We used degree

1: Compute the node-wise densities $\delta(v_1), \ldots, \delta(v_n)$ accounting for edge weights
2: For each node v_i, identify the incidence edge with maximum weight:
   $e_{im} = \max_{j:e_{ij} \in E} e_{ij}$
3: **for** $0 < \lambda < \max_i \delta(v_i)$ **do**
4: Determine the upper level set V_λ:
   $V_\lambda = \{v_i \in V : \delta(v_i) \geq \lambda\}$
5: Build the subgraph $G_\lambda = (V_\lambda, E_\lambda)$ where:
   $$E_\lambda = \{e_{ij}(\lambda)\}, \quad e_{ij}(\lambda) = \begin{cases} e_{ij} & \text{if } (v_i, v_j \in V_\lambda) \cap ((e_{im} = e_{ij}) \cup (e_{jm} = e_{ij})) \\ 0 & \text{otherwise} \end{cases}$$
6: Find the connected components of G_λ
7: update $e_{im} = \max_{j:e_{ij} \in E n E_\lambda} e_{ij}$
8: **end for**
9: Build the cluster tree by associating each λ with the number of connected components of G_λ
10: Identify cluster cores as the connected components of G_λ at the lowest λ level where branches represent leaves

Algorithm 1: DeCoDe of weighted networks

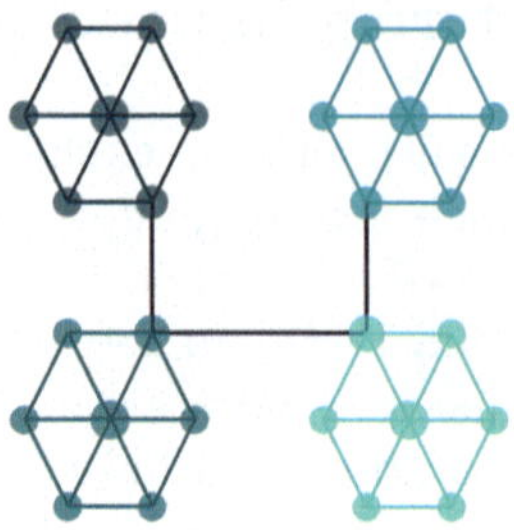

Fig. 1. Leader-based toy network. Node colors indicate true community membership, while node size reflects its degree centrality.

centrality as density measure and the procedure has proven its feasibility on networks having n in the order of the thousands.

3 Results and discussion

3.1 Simulation study

The main goal of this study is to explore the impact of attribute similarity on the performance of DeCoDe algorithm. We use a leader-based toy network (Figure 1) with varying levels of mixing. In Figure 1, in particular, the true membership reflects a community structure shaped by two social mechanisms: leader attraction and homophily-driven cohesion. Mixing is controlled by the parameter μ, defined as in LFR networks [9] which governs the fuzziness of the community structure:

$$\mu = \frac{k_i^{ext}}{k_i}$$

Here, k_i is the total degree for node v_i, and k_i^{ext} is the number of its inter-community connections. Each node shares a fraction $1 - \mu$ of its links within its community and μ with nodes in external communities.

Next, we generate homophily attribute data such that the cluster structure of attributes is the same as the community structure. Specifically, we create 2D Gaussian and Student's t-distributed attributes. Additionally, we merge four binary attribute vectors into a data matrix, maintaining the same clustering pattern. To create mixed data, we integrate Gaussian and binary features. The clustering algorithm is applied to each version of the toy network, with μ varying from 0 to 1 in increments of 0.1, and 100 replicates generated for each μ value. Performance, i.e., Normalised Mutual Information (NMI), is evaluated for different clustering results based on different α values varying from 0 to 1 in steps of 0.1 (Figure 2). When $\alpha = 0$, topology information is used only for measuring node density (i.e., detecting leaders). When $\alpha \in]0, 1[$, attributive and structural information are integrated. When $\alpha = 1$, only topology information is used, matching the original DeCoDe. Performance is also assessed using different structural similarities between node pairs, encoded in either the adjacency matrix or the Jaccard similarity matrix. Attribute similarities are calculated using appropriate measures for each data type: Euclidean distance for continuous numeric data, Simple Matching Coefficient for binary data, and Gower's distance for datasets with mixed attributes. Figure 2 displays the NMI median scores across parameters values. Results indicate that DeCoDe's performance generally decreases as the mixing parameter (μ) increases, reflecting the challenge of detecting communities with more inter-community connections. Notably, the original algorithm's NMI values drop significantly with just 0.1 mixing. Performance varies with the α parameter, where combining attribute and structural information achieves the best results in the presence of mixing. The Jaccard similarity matrix shows greater stability, particularly for binary data, suggesting that it is more robust in noisy network configurations. Thus, selecting $\alpha \neq 1$ is key to enhancing DeCoDe's performance in networks with homophily on observed attributes.

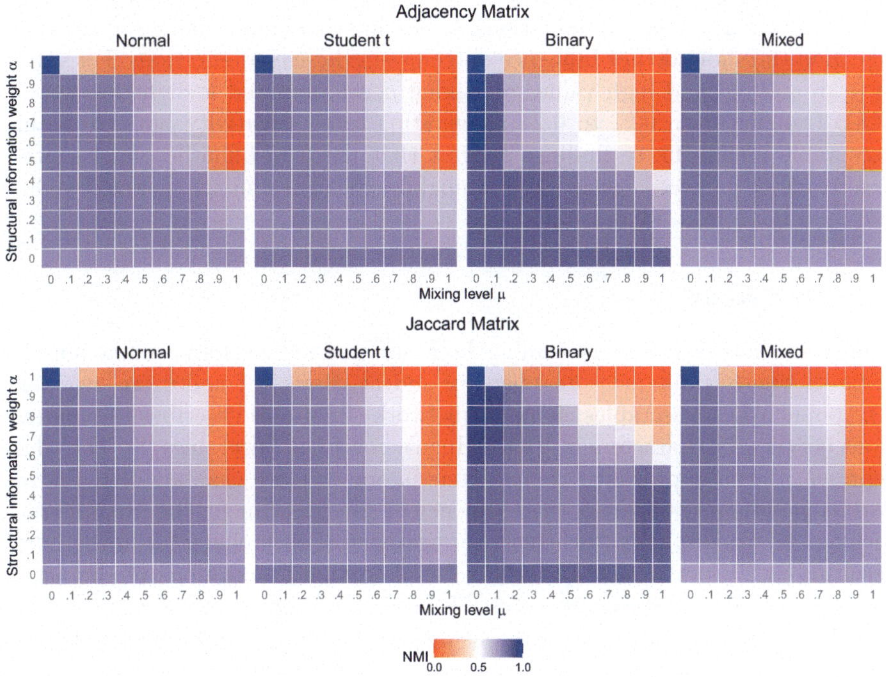

Fig. 2. Normalized Mutual Information (NMI) is evaluated for different values of weight α and mixing parameter μ. Performance is compared between adjacency-based and Jaccard-based approaches and tested across different attribute types, including Gaussian, Student's t-distributed, binary, and mixed data. High median NMI values are indicated in blue, low values are shown in red.

We also assessed the algorithm's stability, quantified using the Median Absolute Deviation (MAD) of the NMI scores across network replicates. Overall, the MAD values were low (median = 0.05; mean = 0.07), with many parameter combinations exhibiting a MAD of 0, indicating highly consistent performance across replicates. Notable variability was observed only in a few specific configurations (e.g., $\alpha > .5$ and multinomial data), where MAD values reached approximately 0.3.

To complement the simulation study, we also conducted a sensitivity analysis to evaluate the robustness of using the 0.75 quantile as threshold for sparsifying the attribute similarity matrix W_A. Specifically, we assessed how variations in the threshold impact the performance of the DeCoDe algorithm. We considered a range of quantile thresholds: from 0.5 to 0.75, with increments of 0.05. Thresholds above 0.75 were excluded, as they resulted in excessive sparsification, thus loss of meaningful relationship information. For each threshold, we evaluated clustering performance across different values of μ, α, and attribute types. The results show that DeCoDe's performance is generally stable across the considered quantile range. In particular, the MAD remained lower than 0.1 for 97.3% of parameters combinations, indicating that small changes in the sparsification threshold do not influence outcomes in these configurations. For values of α around 0.6 and high values of mixing μ, we observed a slight increase in MAD (ranging from 0.21 to 0.24), suggesting some sensitivity but still within acceptable limits. These findings suggest that the 0.75 quantile offers a practical trade-off between retaining meaningful attribute similarities and avoiding excessive noise from weak connections.

3.2 Case study

The case study involves examining how our method performs on a popular dataset, where community membership and attributes of nodes are available. The popular Les Miserables character network (Figure 3) describes the interactions between 77 characters in Victor Hugo's novel Les Miserables [8]. The edges represent the co-appearance of characters in one or more chapters of the novel, with edge weights indicating the frequency of co-appearance. The ground truth membership corresponds to the early appearance of each character in the book, resulting in 20 small communities. We used textual descriptions of characters as node attributes, which are accessible from the dataset provided in [7]. We transformed the character descriptions into a TF-IDF (Term Frequency-Inverse Document Frequency) matrix. This numerical representation allows us to incorporate the textual information on the characters into our analysis, enabling a richer understanding of the network. In Figure 3, attributive edges are highlighted in red and topology edges, representing characters' co-appearance, are shown in grey. Each block in the adjacency matrix corresponds to a chapter, and block elements indicate characters who early co-appeared in that chapter. Table 2 presents clustering performance across different α values and topology similarity matrices. Clustering performance is measured as agreement between estimated partition and the early appearance classification, evaluating NMI, Adjusted Rand Index (ARI), and the number of clusters K obtained. The attributive-only approach ($\alpha = 0$) achieves moderate performance in terms of NMI (NMI=0.56) but poor agreement in terms of ARI (ARI=0.00) and overestimates the number of clusters ($K = 30$). Balanced/topology-dominant approaches ($\alpha = 0.50/0.75$) show improvement in both NMI (0.57-0.65) and ARI (0.21-0.37) with the estimated number of clusters closer to the reference $K = 20$ in the Jaccard-based approach ($K = 18$-22). While the Jaccard-based achieves peak NMI (0.71) for the topology-only approach ($\alpha = 1$), it overestimates the number of clusters ($K = 35$). The variability observed across different approaches highlights the importance of carefully assessing how structural and attribute information can be combined for effective community detection. These findings suggest that the choice of node structural similarity measure, as well as the balance between topological and attribute-based information, can significantly impact clustering performance. This underscores the need for a tailored approach that adapts to the nature of the data, network characteristics and theoretical assumptions on the community formation mechanism to achieve meaningful and reliable community detection outcomes.

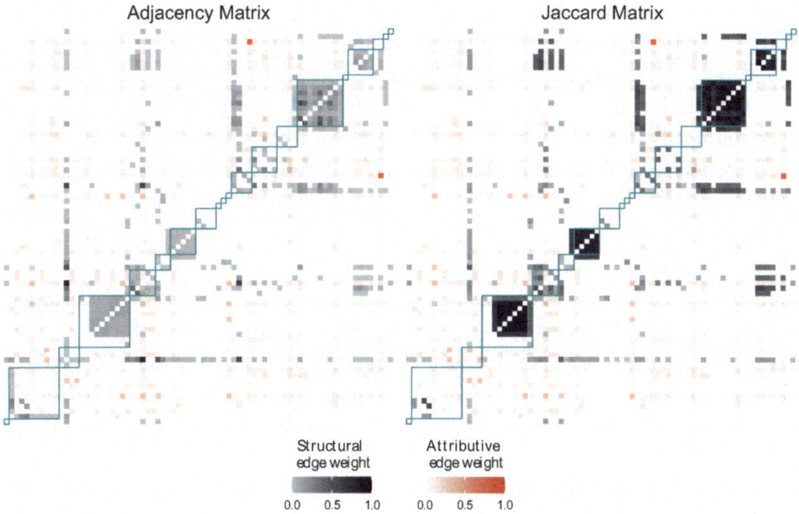

Fig. 3. Les Misérables character network. Attributive edges are shown in red and structural edges in grey. Each block in the adjacency matrix, outlined by blue lines, corresponds to a chapter, and block elements indicate characters who early co-appeared in that chapter.

Table 2. Agreement in terms of Normalised Mutual Information (NMI), Adjusted Rand Index (ARI), and number of clusters (K) between estimated and early appearance partition, across α values and topology similarity matrices.

	Adjacency Matrix			Jaccard Matrix		
α	NMI	ARI	K	NMI	ARI	K
0	0.56	0.00	30	0.56	0.00	30
0.25	0.46	0.26	7	0.49	0.05	16
0.50	0.57	0.37	7	0.64	0.21	18
0.75	0.64	0.37	7	0.65	0.24	22
1	0.65	0.34	6	0.71	0.26	35

4 Conclusion

The present study enhances density-based community detection, a particular leader-based method, integrating structural and attributive information within the DeCoDe framework. Our approach aims to identify communities in social networks where homophily and leader influence are key mechanisms. An evaluation using both simulated and real-world data suggests that integrating attribute information improves the determination of community boundaries, particularly in networks with moderate to high levels of mixing. The optimal weight parameter α varies depending on the network structure and attribute data type, with values $0 < \alpha < 1$ usually yielding superior results in the presence of mixing. This suggests the choice for a balanced approach, with α close to 0.5. The parameter was manually tuned in the experiments, and no automatic approach was proposed for its selection in real applications. The development of a data-driven strategy for optimising α remains an important direction for future research. Moreover, the choice of topology similarity measure significantly influences performance. The Jaccard-based approach demonstrated greater robustness, particularly in binary and noisy network settings. The results highlight the limitations of relying solely on structural or attribute information, particularly in complex networks with overlapping or weakly defined communities. By preserving the integrity of topological features while incorporating meaningful attribute data, our method provides a flexible framework for community detection in attributed networks.

Acknowledgements. This work has been financial supported under the National Recovery and Resilience Plan (NRRP), Mission 4, Component 2, Investment 1.1, Call for tender No. 104 published on 2.2.2022 by the Italian Ministry of University and Research (MUR), funded by the European Union - NextGenerationEU, Project Title "Models and methods for the analysis of collaboration networks", CUP J53D23011540006, Principal Investigator Domenico De Stefano.

The publication has been produced with co-funding from the European Union - Next Generation EU.

References

1. Alinezhad, E., Teimourpour, B., Sepehri, M.M., Kargari, M.: Community detection in attributed networks considering both structural and attribute similarities: two mathematical programming approaches. Neural Comput. Appl. **32**(8), 3203–3220 (2020)
2. Blondel, V.D., Guillaume, J.L., Lambiotte, R., Lefebvre, E.: Fast unfolding of communities in large networks. J. Stat. Mech: Theory Exp. **2008**(10), 10008 (2008)
3. Chunaev, P.: Community detection in node-attributed social networks: a survey. Comput. Sci. Rev. **37**, 100286 (2020)
4. Falkowski, T., Barth, A., Spiliopoulou, M.: DENGRAPH: a density-based community detection algorithm. In: IEEE/WIC/ACM International Conference on Web Intelligence (WI 2007), pp. 112–115. IEEE (2007)
5. Helal, N.A., Ismail, R.M., Badr, N.L., Mostafa, M.G.M.: An efficient algorithm for community detection in attributed social networks. In: Proceedings of the 10th International Conference on Informatics and Systems, pp. 180–184 (2016)
6. Helal, N.A., Ismail, R.M., Badr, N.L., Mostafa, M.G.M.: Leader-based community detection algorithm for social networks. WIREs Data Mining Knowl. Disc. **7**(6), e1213 (2017)
7. Kalugin, S.: Les Miserables Character Network Data (2015). https://github.com/MADStudioNU/lesmiserables-character-network
8. Knuth, D.E.: The Stanford GraphBase: A Platform for Combinatorial Computing, vol. 1. ACM Press (1993). http://www-cs-faculty.stanford.edu/~uno/sgb.html
9. Lancichinetti, A., Fortunato, S., Radicchi, F.: Benchmark graphs for testing community detection algorithms. Phys. Rev. E **78**(4), 046110 (2008)
10. Lu, D.D.: Leader-based community detection algorithm in attributed networks. IEEE Access **9**, 119666–119674 (2021)
11. Mcpherson, M., Smith-Lovin, L., Cook, J.M.: Birds of a feather: homophily in social networks. Ann. Rev. Sociol. **27**(1), 415–444 (2001)
12. Menardi, G., De Stefano, D.: Density-based clustering of social networks. J. R. Statist. Soc. Ser. A Stat. Soc. **185**(3), 1004–1029 (2022)
13. Morea, F., De Stefano, D.: Beyond one solution: a comprehensive exploration of solution space in community detection for social networks (2025). https://doi.org/10.21203/rs.3.rs-6578160/v1
14. Newman, M.E.J., Girvan, M.: Finding and evaluating community structure in networks. Phys. Rev. E **69**(2), 026113 (2004)
15. Perc, M.: The Matthew effect in empirical data. J. R. Soc. Interface **11**(98), 2014378 (2014)
16. Rosvall, M., Delvenne, J.-C., Schaub, M.T., Lambiotte, R.: Different approaches to community detection. In: Advances in Network Clustering and Blockmodeling, pp. 105–119. Wiley (2019)
17. Ruan, Y., Fuhry, D., Parthasarathy, S.: Efficient community detection in large networks using content and links. In: Proceedings of the 22nd International Conference on World Wide Web (WWW 2013), pp. 1089–1098 (2013)
18. Yakoubi, Z., Kanawati, R.: LICOD: a leader-driven algorithm for community detection in complex networks. Vietnam J. Comput. Sci. **1**(4), 241–256 (2014). https://doi.org/10.1007/s40595-014-0025-6
19. Yang, J., McAuley, J., Leskovec, J.: Community detection in networks with node attributes. In: Proceedings - IEEE International Conference on Data Mining (ICDM), pp. 1151–1156 (2013)

Bayesian Inference with Besov-Laplace Priors for Spatially Inhomogeneous Binary Classification Surfaces

Matteo Giordano(✉)

ESOMAS Department, University of Turin,
Corso Unione Sovietica 2018/bis, 10134 Turin, Italy
matteo.giordano@unito.it
https://matteogiordano.weebly.com

Abstract. In this article, we study the binary classification problem with supervised data, in the case where the covariate-to-probability-of-success map is possibly spatially inhomogeneous. We devise nonparametric Bayesian procedures with Besov-Laplace priors, which are prior distributions on function spaces routinely used in imaging and inverse problems in view of their useful edge-preserving and sparsity-promoting properties. Building on a recent line of work in the literature, we investigate the theoretical asymptotic recovery properties of the associated posterior distributions, and show that suitably tuned Besov-Laplace priors lead to minimax-optimal posterior contraction rates as the sample size increases, under the frequentist assumption that the data have been generated by a spatially inhomogeneous ground truth belonging to a Besov space.

Keywords: Bayesian nonparametric inference · frequentist analysis of Bayesian procedures · posterior contraction rates · spatially inhomogeneous functions

1 Introducion

The binary classification problem consists in predicting a binary response Y based on the value of a covariate X. Taking, without loss of generality, Y to be a 0-1 outcome, the canonical approach to this problem entails modelling Y, conditionally given X, as a Bernoulli random variable with probability of success $f(X)$. Given a labelled training dataset $(Y^{(n)}, X^{(n)}) := \{(Y_i, X_i)\}_{i=1}^n$, the central task is then to estimate the 'covariate-to-probability-of-success map' $x \mapsto f(x)$. Based on this, new unlabelled inputs X_{n+1} can be classified according to the estimate of $f(X_{n+1})$, e.g. to be of class 1 if this probability exceeds 1/2.

In the present article, we are interested in the case where the classification surface f is possibly spatially inhomogeneous, namely when it is generally flat or smooth across the domain, but it may exhibit localised high variation or even

A. D'Ambrosio et al. (Eds.): CLADAG-VOC 2025, SCDAKO, pp. 202–213, 2025.
https://doi.org/10.1007/978-3-032-03042-9_18

discontinuities. This is particularly important for a variety of applications in which there may be unknown 'critical values' of the covariates that determine rapid increases or decreases in the probability of success.

We shall consider the nonparametric Bayesian approach to this problem, endowing f with prior distributions of Besov-Laplace type. These are popular priors on function spaces that were first introduced by Lassas et al. [23], and have since then enjoyed a widespread application in imaging and inverse problems, where they empirically achieve excellent performances in the recovery of spatially inhomogeneous objects of 'bounded variation type', giving rise to edge-preserving and sparse reconstructions at the level of the maximum-a-posteriori (MAP) estimator. See e.g. [6,22,24,27,32], where many more references can be found.

Besov-Laplace priors are constructed by modelling the random coefficients in a wavelet basis expansion via independent (scaled) Laplace distributions. This offers a 'discretisation-invariant' alternative [23] to the famous 'total-variation' prior of Rudin et al. [26], and furnishes a Bayesian model for functions in the Besov scale B^{α}_{11}, $\alpha \geq 0$. These spaces measure the local variability in an L^1-sense and describe the regularity by the ℓ^1-decay of the wavelet coefficients, and are well-know to provide a mathematical characterisation for spatially inhomogeneous functions. See Sect. 2.1 for details, and [13, Section 1] for an overview on the relationship with the space of bounded variation functions. In contrast, the popular Gaussian priors, which entail L^2-type penalties, are suited to model Sobolev-regular functions with milder variations, and have been shown to perform sub-optimally in more structured recovery problems [2,4,21]. We refer the reader to Appendix E.2 in [1] for an empirical comparison of the performance of Besov-Laplace and Gaussian priors over simulated data in the white noise model with spatially inhomogeneous regression functions. Section 4.2.3 of the Ph.D. thesis [28] contains further extensive simulations results, and a nice visualisation of the superior reconstruction quality achieved by Besov-Laplace priors in the presence of spatially inhomogeneous ground truths, again in the white noise model. Alongside the aforementioned advantages, Besov-Laplace priors also maintain a favourable log-concave structure that enables computational approaches [8] and analytical study [2].

Building on the breakthroughs in the theory of the frequentist analysis of nonparametric Bayesian procedures from the last two decades [14,15,17,29,31], Agapiou et al. [2] have recently started a line of research on the performance of posterior inference with Besov-Laplace priors for the recovery of spatially inhomogeneous functions, in the large sample size limit and under the frequentist assumption that the data have been generated by a fixed 'ground truth' belonging to a Besov space. Minimax-optimal posterior contraction rates in the white noise model have been obtained by [2], later extended to drift estimation for multidimensional diffusions, density estimation and nonlinear inverse problems by [4,19,20], respectively. Further, adaptive posterior contraction rates with related hierarchical procedures have been obtained by [3,19]. We also mention the earlier related results by [5,9,25], and the recent examination by [12].

In this paper, we shall study these issues for the binary classification problem, which to our knowledge has not been considered in the context of Besov-Laplace priors and inference for spatially inhomogeneous functions. In our main result, Theorem 1, we shall show that for true classification surfaces belonging to the B^{α}_{11}-scale, suitably tuned Besov-Laplace priors attain minimax-optimal posterior contraction rates. Our proof (in Sect. 5) is based on the general posterior contraction rate theory for independent and identically distributed (i.i.d.) binary classification data, e.g. [16, Chapter 2.5], which we pursue by employing rescaled and under-smoothing Laplace priors similar to those used in [2,4,19,20]. Our result thus shows that such priors achieve optimal performance also in the binary classification problem. As a corollary, we obtain optimal convergence rates for the posterior mean estimator towards the spatially inhomogeneous true classification surface, see Corollary 1. We provide a concluding discussion in Sect. 4, with a summary of the results and an outlook on related research questions.

2 Binary Classification with Besov-Laplace Priors

2.1 Preliminaries and Notation

Throughout, we consider functions spaces defined on the d-dimensional unit cube $[0,1]^d$, $d \in \mathbb{N}$. We write $L^p([0,1]^d)$, $p \geq 1$, for the usual Lebesgue spaces on $[0,1]^d$, equipped with norm $\|\cdot\|_2$, and $\langle\cdot,\cdot\rangle_2$ for the L^2-inner product.

Let $\{\psi_{lr},\ l \in \mathbb{N},\ r = 1,\dots,2^{ld}\}$ be an orthonormal basis of $L^2([0,1]^d)$ comprising sufficiently regular, compactly supported and boundary-corrected wavelets; see e.g. [18, Chapter 4.3] for details. For $\alpha \geq 0$ and $p,q \in [1,\infty]$, let the Besov space $B^{\alpha}_{pq}([0,1]^d)$ be defined as

$$B^{\alpha}_{pq}([0,1]^d) := \left\{ w \in L^p([0,1]^d) : \|w\|^q_{B^{\alpha}_{pq}} := \sum_{l=1}^{\infty} 2^{ql(\alpha-\frac{d}{p}+\frac{d}{2})} \left(\sum_{r=1}^{2^{ld}} |\langle w,\psi_{lr}\rangle_2|^p \right)^{\frac{q}{p}} < \infty \right\},$$

where the above ℓ_p- and ℓ_q-sequence space norms are replaced by the ℓ_∞-norm if $p=\infty$ or $q=\infty$ respectively. See [18], p.370f. Such Besov scale is known to contain the traditional Sobolev and Hölder spaces: in particular, $B^{\alpha}_{22}([0,1]^d) = H^{\alpha}([0,1]^d)$ for all $\alpha \geq 0$, and $C^{\alpha}([0,1]^d) \subseteq B^{\alpha}_{\infty\infty}([0,1]^d)$ for all $\alpha > 0$ (with equality holding when $\alpha \notin \mathbb{N}$). For $p=q=1$, the spaces $B^{\alpha}_{11}([0,1]^d)$ provide a mathematical model for spatially inhomogeneous functions; in particular, the space of bounded variation functions $BV([0,1]^d)$, which is of specific interest for many applications, enjoys the (strict) inclusions $B^{1}_{11}([0,1]^d) \subset BV([0,1]^d) \subset B^{1}_{1\infty}([0,1]^d)$ (e.g. [13, Section 1]).

When there is no risk of confusion, we omit to explicitly refer to the underlying domain, e.g. writing B^{α}_{pq} for $B^{\alpha}_{pq}([0,1]^d)$. We use the symbols $\lesssim$, $\gtrsim$, and $\simeq$, for respectively one- and two-sided inequalities holding up to universal multiplicative constants. We denote by $N(\xi;\mathcal{F},\delta)$, $\xi > 0$, for the ξ-covering number of a set $\mathcal{F}$ with respect to a semi-metric δ on $\mathcal{F}$, namely the smallest number of balls of δ-radius ξ required to cover $\mathcal{F}$.

2.2 Likelihood, Prior and Posterior

Consider a random sample $(Y^{(n)}, X^{(n)}) := \{(Y_i, X_i)\}_{i=1}^n$ of binary classification data arising as

$$\begin{aligned} Y_i|X_i &\overset{\text{iid}}{\sim} \text{Bernoulli}(f(X_i)), \\ X_i &\overset{\text{iid}}{\sim} \mu, \end{aligned} \tag{1}$$

where μ is a given probability distribution on $[0,1]^d$ and $f : [0,1]^d \to [0,1]$ is some unknown probability response function. We are interested in the problem of estimating f from data $(Y^{(n)}, X^{(n)})$, whose (product) law $P_f^{(n)}$ has likelihood

$$L_n(f) := \prod_{i=1}^n f(X_i)^{Y_i}(1 - f(X_i))^{1-Y_i}. \tag{2}$$

We denote by $E_f^{(n)}$ the expectation with respect to $P_f^{(n)}$. Throughout, we will make the minimal assumption that μ is absolutely continuous with respect to the Lebesgue measure and that it possesses a bounded and bounded away from zero probability density. However, we will not pursue jointly estimating f and μ in the scenario where the latter is unknown.

We are primarily interested in the setting where f is spatially inhomogeneous, e.g. possibly flat in some areas and 'spiky' in others. A natural mathematical model for this scenario is then to take a (Borel measurable) parameter space $f \in \mathcal{F} \subseteq B^{\alpha}_{11}([0,1]^d)$ for some $\alpha \geq 0$; see Sect. 2.1 for definitions and details. We adopt the nonparametric Bayesian approach, endowing f with the following Besov-Laplace prior. These priors represent a particular instance in the class introduced by Lassas et al. [23], and have been recently shown to achieve minimax-optimal posterior contraction rates over Besov spaces in the white noise model [2] as well as in other settings [4,19,20]. In accordance with the latter references, we construct Besov-Laplace priors for classification surfaces starting from the random function, for $x \in [0,1]^d$,

$$W_n(x) = \frac{1}{n^{d/(2\alpha+d)}} \sum_{l=1}^{\infty} \sum_{r=1}^{2^{ld}} 2^{-l(\alpha-d/2)} W_{lr} \psi_{lr}(x), \quad W_{lr} \overset{\text{iid}}{\sim} \text{Laplace}, \quad \alpha > d, \tag{3}$$

where $\{\psi_{lr},\ l \in \mathbb{N},\ r = 1, \ldots, 2^{ld}\}$ is an orthonormal wavelet basis of $L^2([0,1]^d)$ as described in Sect. 2.1, and the Laplace (or double exponential) distribution has probability density $\lambda(z) = e^{-|z|}/2$, $z \in \mathbb{R}$.

The parameter α in (3) determines the (Besov-) regularity of the realisations of the random function W_n: in particular, $W_n \in C([0,1]^d) \cap B^{\alpha'}_{pq}([0,1]^d)$ almost surely for all $\alpha' < \alpha - d$ and all $p, q \in [1, \infty]$ (cf. Lemma 5.2 and Proposition 6.1 in [2]). We may then regard W_n as a random element in $C([0,1]^d)$, whose law we denote by Π_{W_n}. In accordance with the terminology of [2], we call Π_{W_n} a 'rescaled $(\alpha - d)$-regular Besov-Laplace prior'.

To construct a prior on the set of probability response functions on $[0,1]^d$, whose range equals the compact interval $[0,1]$, we proceed transforming W_n in

(3) via the logistic link $H(z) = e^z/(e^z + 1)$, $z \in \mathbb{R}$, namely taking the random function

$$f_{W_n}(x) := H[W_n(x)] = \frac{e^{W_n(x)}}{e^{W_n(x)} + 1}, \qquad x \in [0,1]^d. \tag{4}$$

We denote by Π_n the law of f_{W_n} as above, which, in slight abuse of terminology, we call a rescaled $(\alpha - d)$-regular Besov-Laplace prior for the classification surface f.

Given Π_n as above, Bayes' formula (e.g. [16], p.7) yields the posterior distribution $\Pi_n(\cdot|Y^{(n)}, X^{(n)})$ of $f|(Y^{(n)}, X^{(n)})$, which is equal to

$$\Pi_n(A|Y^{(n)}, X^{(n)}) = \frac{\int_A L_n(f) d\Pi_n(f)}{\int_{\mathcal{F}} L_n(f') d\Pi_n(f')}, \qquad A \subseteq \mathcal{F} \text{ measurable},$$

where L_n is the likelihood from (2). According to the Bayesian paradigm, $\Pi_n(\cdot| Y^{(n)}, X^{(n)})$ represents the comprehensive solution to the problem of estimating f from data $(Y^{(n)}, X^{(n)})$ arising as in model (1). It may be used to obtain point estimates (e.g. via the posterior mean) and uncertainty quantification (via credible sets).

3 Main Results

In this section, we study the (frequentist) consistency of $\Pi_n(\cdot|Y^{(n)}, X^{(n)})$, assuming binary data $(Y^{(n)}, X^{(n)})$ drawn based on some fixed 'true' unknown response probability function $f_0 \in \mathcal{F}$ to be estimated, and investigating the asymptotic concentration of the posterior distribution around f_0 when $n \to \infty$. We quantify the speed of concentration according to the usual notion of 'posterior contraction rates', that is sequences $\varepsilon_n \to 0$ such that, for large enough $M > 0$,

$$E_{f_0}^{(n)} \Pi_n \left(f : \delta(f, f_0) > M\varepsilon_n \middle| Y^{(n)}, X^{(n)} \right) \to 0$$

as $n \to \infty$. Above, δ is a distance between classification surfaces. We refer to [16] for an extensive overview on the field of the frequentist analysis of nonparametric Bayesian procedures.

Our main result establishes that the posterior distribution resulting from rescaled $(\alpha - d)$-regular Besov-Laplace priors contracts towards ground truths $f_0 \in B_{11}^{\alpha}([0,1]^d)$ at minimax-optimal rate in L^2-distance.

Theorem 1. *For fixed $\alpha > d$, let the prior Π_n be a rescaled $(\alpha - d)$-regular Besov-Laplace prior constructed as in* (4) *for W_n as in* (3)*. Let $(Y^{(n)}, X^{(n)}) = \{(Y_i, X_i)\}_{i=1}^n$ be a random sample of binary classification data arising as in* (1) *for some fixed $f = f_0 \in B_{11}^{\alpha}([0,1]^d)$ satisfying $\inf_{x \in [0,1]^d} f_0(x) > 0$. Then, for $M > 0$ large enough, as $n \to \infty$,*

$$E_{f_0}^{(n)} \Pi_n \left(f : \|f - f_0\|_2 > M n^{-\frac{\alpha}{2\alpha + d}} \middle| Y^{(n)}, X^{(n)} \right) \to 0.$$

Recall that the covariate distribution μ is assumed to posses a probability density that is bounded and bounded away from zero. In this case, the L^2-loss with respect to which the posterior contraction rates is obtained is equivalent to the $L^2(\mu)$-loss, and thus we could have formulated the statement of Theorem 1 in terms of the latter as well. The obtained rate $\varepsilon_n = n^{-\alpha/(2\alpha+d)}$ is known to be minimax-optimal for estimating in L^2-distance ground truths belonging to the space $B^\alpha_{11}([0,1]^d)$, e.g. [13].

Our result is in the spirit of the available theory for Gaussian priors in binary classification [31], for which minimax optimal posterior contraction rates in L^2-loss are obtained for spatially homogeneous functions under traditional Sobolev- or Hölder-type regularity assumptions. On the other hand, Gaussian priors are known to be unable to optimally model spatially inhomogeneous functions [4], and therefore they cannot be expected to achieve the same posterior contraction rates attained by Besov-Laplace priors in the setting at hand.

The assumption that $f_0 > 0$ over the whole domain $[0,1]^d$ crucially guarantees that the ground truth can be written as a functional composition with respect to the employed link function H. Since, by construction, the support of the prior Π_n is contained within the set of such compositions, this restriction seems unavoidable to obtain posterior consistency in the present setting. We note that the same assumption underpins the contraction rate analysis for Gaussian priors in binary classification of [31], where the prior is similarly constructed via compositions with (logistic-type) link functions. For the important case where we may have zero response probability at some $x \in [0,1]^d$, posterior contraction rates have been established using completely different priors constructions, in particular ones based on the Dirichlet process, e.g. Sect. 9.5.6 in [16]. However, we are not aware of any such result in the presence of spatially inhomogeneous classification surfaces.

Theorem 1 establishes that the posterior distribution asymptotically concentrates over L^2-balls centred around the ground truth with minimal radius (in the minimax sense). The following corollary shows that the same rate of convergence $\varepsilon_n = n^{-\alpha/(2\alpha+d)}$ is attained also by the associated posterior mean, which is hence a minimax-optimal estimator of the spatially inhomogeneous classification surface. Let $\bar{W}_n := E^{\Pi_{W_n}}[W|Y^{(n)}, X^{(n)}]$ be the mean of the posterior distribution of $W|(Y^{(n)}, X^{(n)})$, arising from prior Π_{W_n} constructed as in (4). In slight abuse of terminology, let $\bar{f}_n := f_{\bar{W}_n} = H \circ \bar{W}_n$, be the corresponding 'posterior mean' estimate of the probability response function f.

Corollary 1. *Under the assumption of Theorem 1, we have for some sufficiently large $M > 0$, as $n \to \infty$,*

$$P^{(n)}_{f_0}\left(\|\bar{f}_n - f_0\|_2 > Mn^{-\frac{\alpha}{2\alpha+d}}\right) \to 0.$$

4 Discussion

In this article, we have investigated the performance of Besov-Laplace priors for the recovery of spatially inhomogeneous binary classification surfaces. Our main result shows that, for ground truths $f_0 \in B^{\alpha}_{11}([0,1]^d)$, suitably calibrated rescaled $(\alpha - d)$-regular priors attain minimax-optimal posterior contraction rates in L^2-distance. As a corollary, we have shown that the associated posterior mean estimator is minimax-optimal. This is in line with the existing theoretical literature on Besov-Laplace priors, where similar results have been obtained in different statistical models [4,19,20].

We have focused for conciseness on non-adaptive procedures, implicitly assuming knowledge of the regularity level α of the ground truth for the construction of the optimally-contracting posterior distribution in Theorem 1. Hierarchical procedures based on Besov-Laplace priors obtained by randomising the hyperparameters in the prior construction have recently been shown to achieve adaptive posterior contraction rates in the white noise model and in density estimation [3,19]; in light of this, pursuing similar approaches for the binary classification problem is likely feasible.

In Theorem 1, minimax-optimal posterior contraction rates for ground truths $f_0 \in B^{\alpha}_{11}([0,1]^d)$ have been obtained by rescaled $(\alpha - d)$-regular Besov-Laplace priors. The same specific prior tuning, that finely exploits the interplay between rescaling and undersmoothing, has been employed in the other available results for Besov-Laplace priors [4,19,20], and marks an interesting difference with the existing literature on Gaussian priors, where minimax-optimal posterior contraction rates are typically obtained by smoothness-matching (and non-rescaled) priors, in the presence of spatially homogeneous ground truths. This is issue is extensively discussed in [19]. Recently, Dolera et al. [12] have shown, in the white noise model, that non-rescaled smoothness-matching Besov-Laplace priors optimally contract towards ground truths belonging to the Besov scale B^{α}_{11}. However, it remains unclear at present whether their result (and proof techniques) can be extended to the setting at hand.

Lastly, the computational aspects of posterior inference with Besov-Laplace priors in binary classification represent a further interesting avenue for future research. These priors naturally lend themselves to discretisation by truncating the series in (3) at some user-specified (sufficiently high) level $L \in \mathbb{N}$, giving rise to high-dimensional log-concave product distributions. Combined with the simple likelihood structure from (2), this furnishes a basis for the implementation of Markov chain Monte Carlo (MCMC) methods to sample from the resulting posterior distribution. For recent developments concerning MCMC algorithms suited Besov-Laplace priors in infinite-dimensional statistical models, we refer to [10], where more references can be found.

5 Proofs

5.1 Proof of Theorem 1

Since f_0 is strictly positive, we may write $f_0 = f_{w_0} := H \circ w_0$ for H the logistic link function and $w_0 := H^{-1} \circ f_0$. In particular, as $f_0 \in B^{\alpha}_{11}$ and H^{-1} is smooth, we have by Theorem 10 in [7] that $w_0 \in B^{\alpha}_{11}$.

The proof of Theorem 1 is based on the general theory for posterior contraction rates in i.i.d. sampling models, adapted to the case of binary classification data and nonparametric priors constructed via the logistic link function. In particular, combining Theorem 8.9 in [16] (and its proof) with Lemma 2.8 in the same reference shows that if for some sequence $\varepsilon_n \to 0$ such that $n\varepsilon_n^2 \to \infty$, some constant $C > 0$, and all $n \in \mathbb{N}$ large enough,

$$\Pi_{W_n}\left(w : \|w - w_0\|_{L^2(\mu)} \leq \varepsilon_n\right) \geq e^{-Cn\varepsilon_n^2}, \tag{5}$$

and if there exists measurable sets $\mathcal{W}_n \subseteq C([0,1]^d)$ such that

$$\Pi_{W_n}(\mathcal{W}_n^c) \leq e^{-(C+4)n\varepsilon_n^2}; \qquad \log N(\varepsilon_n; \mathcal{W}_n, \|\cdot\|_{L^2(\mu)}) \lesssim n\varepsilon_n^2, \tag{6}$$

then, for sufficiently large $M > 0$ and some $\bar{C} > 0$, as $n \to \infty$,

$$E_{f_0}^{(n)} \Pi_n\left(f_w : w \in \mathcal{W}_n,\ \|f_w - f_0\|_{L^2(\mu)} \leq M\varepsilon_n \middle| Y^{(n)}, X^{(n)}\right) \geq 1 - e^{-\bar{C}n\varepsilon_n^2}. \tag{7}$$

We proceed verifying the 'small ball probability lower bound' (5) and the 'sieve condition' (6), wherein we note that we may replace the norm $\|\cdot\|_{L^2(\mu)}$ with $\|\cdot\|_2$ since μ has a bounded and bounded away from zero probability density. Set $\varepsilon_n := n^{-\alpha/(2\alpha+d)}$, and write W_n in (3) as $W_n = (n\varepsilon_n^2)^{-1}W$, where W is the non-rescaled $(\alpha - d)$-regular Besov-Laplace random element

$$W(x) := \sum_{l=1}^{\infty} \sum_{r=1}^{2^{ld}} 2^{-l(\alpha-d/2)} W_{lr}\psi_{lr}(x), \qquad x \in [0,1]^d, \qquad W_{lr} \overset{\text{iid}}{\sim} \text{Laplace}. \tag{8}$$

Writing Π_W for the law of W, we then have

$$\Pi_{W_n}(w : \|w - w_0\|_2 \leq \varepsilon_n) = \Pi_W\left(w : \|w - n\varepsilon_n^2 w_0\|_2 \leq n\varepsilon_n^3\right)$$

so that by using the 'decentering inequality' and the sup-norm small ball probability lower bound from eq. (32) and (34) in [19] respectively, we obtain that the probability of interest is greater than

$$\begin{aligned}
e^{-\|w_0\|_{B_{11}^\alpha} n\varepsilon_n^2} \Pi_W\left(w : \|w\|_\infty \leq c_1 n\varepsilon_n^3\right) &\geq e^{-\|w_0\|_{B_{11}^\alpha} n\varepsilon_n^2} \Pi_W\left(w : \|w\|_\infty \leq c_1 n\varepsilon_n^3\right) \\
&\geq e^{-\|w_0\|_{B_{11}^\alpha} n\varepsilon_n^2 + c_2 (n\varepsilon_n^3)^{-d/(\alpha-d)}} \\
&= e^{-Cn\varepsilon_n^2},
\end{aligned}$$

for constants $c_1, c_2, C > 0$. This concludes the verification of (5). Turning to condition (6), define the sieves

$$\mathcal{W}_n := \left\{w = w^{(1)} + w^{(2)} : \|w^{(1)}\|_2 \leq R\varepsilon_n, \|w^{(2)}\|_{B_{11}^\alpha} \leq R,\ \|w\|_\infty \leq R\right\}. \tag{9}$$

Then

$$\begin{aligned}
\Pi_{W_n}(\mathcal{W}_n^c) \leq 2 - \Pi_{W_n}&\left(w = w^{(1)} + w^{(2)} : \|w^{(1)}\|_2 \leq R\varepsilon_n,\ \|w^{(2)}\|_{B_{11}^\alpha} \leq R\right) \\
&- \Pi_{W_n}(w : \|w\|_\infty \leq R),
\end{aligned}$$

which, by Lemma 9 in [19], fixing $K > C + 4$ and choosing sufficiently large $R > 0$, is smaller than

$$2e^{-Kn\varepsilon_n^2} \leq e^{-(C+4)n\varepsilon_n^2}.$$

This proves the first inequality in (6). For the second, note that since $\mathcal{W}_n \subset \{w = w^{(1)} + w^{(2)} : \|w^{(1)}\|_2 \leq R\varepsilon_n, \ \|w^{(2)}\|_{B^\alpha_{11}} \leq R\}$, using the embedding of B^α_{11} into $B^\alpha_{1\infty}$ (e.g., Theorem 3.3.1 in [30]) and the complexity bound from Theorem 4.3.36 in [18], we have

$$\log N(\varepsilon_n; \mathcal{W}_n, \|\cdot\|_2) \lesssim \log N(\varepsilon_n; \{w : \|w\|_{B^\alpha_{1\infty}} \leq R\}, \|\cdot\|_2) \lesssim \varepsilon_n^{-\frac{d}{\alpha}} = n\varepsilon_n^2,$$

concluding the verification of (6) and, in view of (7), the proof of Theorem 1.□

5.2 Proof of Corollary 1

Inspection of the proof of Theorem 1, cf. (7), shows that it holds that for some sufficiently large $M > 0$,

$$E^{(n)}_{f_0} \Pi_n \left(f_w : w \in \mathcal{W}_n, \ \|f_w - f_0\|_2 \leq M\varepsilon_n \middle| Y^{(n)}, X^{(n)} \right) \geq 1 - e^{-\bar{C}n\varepsilon_n^2},$$

where $\mathcal{W}_n$ is the sieve defined in (9). Since the latter is included in a sup-norm ball and the logistic link function H is continuous with continuous inverse H^{-1}, it then follows that

$$E^{(n)}_{f_0} \Pi_{W_n} \left(w \in \mathcal{W}_n : \|w - w_0\|_2 \leq M\varepsilon_n \middle| Y^{(n)}, X^{(n)} \right) \geq 1 - e^{-\bar{C}n\varepsilon_n^2}, \qquad (10)$$

having denoted by $\Pi_{W_n}(\cdot|Y^{(n)}, X^{(n)})$ the posterior distribution of $W|(Y^{(n)}, X^{(n)})$ arising from the prior Π_{W_n} constructed as in (3). Write shorthand $D^{(n)} := (Y^{(n)}, X^{(n)})$. For $\bar{W}_n = E^{\Pi_{W_n}}[W|D^{(n)}]$ the posterior mean, we proceed showing that, as $n \to \infty$,

$$P^{(n)}_{f_0} \left(\|\bar{W}_n - w_0\|_2 > Mn^{-\frac{\alpha}{2\alpha+d}} \right) \to 0.$$

Upon doing so, the conclusion of Corollary 1 directly follows since H is Lipschitz. We use uniform integrability arguments for Besov-Laplace priors developed in [20]. Start by noting that by Jensen's inequality

$$\|\bar{W}_n - w_0\|_2 \leq E^{\Pi_{W_n}} \left[\|W - w_0\|_2 \middle| D^{(n)} \right],$$

which, for any $M > 0$, using the Chauchy-Schwarz inequality, is upper bounded by

$$M\varepsilon_n + E^{\Pi_{W_n}} \left[\|W - w_0\|_2 1_{\{\|W - w_0\|_2 > M\varepsilon_n\}} \middle| D^{(n)} \right]$$
$$\leq M\varepsilon_n + E^{\Pi_{W_n}} \left[\|W - w_0\|_2^2 \middle| D^{(n)} \right]^{1/2} \Pi_{W_n}(w : \|w - w_0\|_2 > M\varepsilon_n | D^{(n)})^{1/2}.$$

Using the above and (10), for any $A > 0$, we can then find $M > 0$ large enough such that

$$P_{f_0}^{(n)}(\|\bar{W}_n - w_0\|_2 > 2M\varepsilon_n)$$
$$\le P_{f_0}^{(n)}\left(E^{\Pi_{W_n}}\left[\|W - w_0\|_2^2|D^{(n)}\right]^{\frac{1}{2}} \Pi_{W_n}(w : \|w - w_0\|_2 > M\varepsilon_n|D^{(n)})^{\frac{1}{2}} > M\varepsilon_n\right)$$
$$\le P_{f_0}^{(n)}\left(E^{\Pi_{W_n}}\left[\|W - w_0\|_2^2|D^{(n)}\right] e^{-An\varepsilon_n^2} > M^2\varepsilon_n^2\right) + o(1).$$

For L_n the likelihood from (2) and $C > 0$ the constant in the small ball estimate (5), define the event

$$A_n = \left\{\int_{C([0,1]^d)} L_n(f_w)/L_n(f_0)d\Pi_{W_n}(w) \ge e^{-(C+2)n\varepsilon_n^2}\right\},$$

which, by Lemma 8.10 in [16], in view of the validity of (5), satisfies $P_{f_0}^{(n)}(A_n) \to 1$ as $T \to \infty$. The second probability in the second to last display is then upper bounded by

$$P_{f_0}^{(n)}\left(\left\{\frac{\int_{C([0,1]^d)} \|w - w_0\|_2^2 L_n(f_w)/L_n(f_0)d\Pi_{W_n}(w)}{\int_{C([0,1]^d)} L_n(f_{w'})/L_n(f_0)d\Pi_{W_n}(w')} > M^2\varepsilon_n^2 e^{An\varepsilon_n^2}\right\} \cap A_n\right)$$
$$\le P_{f_0}^{(n)}\left(\int_{C([0,1]^d)} \|w - w_0\|_2^2 \frac{L_n(f_w)}{L_n(f_0)} d\Pi_{W_n}(w) > M^2\varepsilon_n^2 e^{(A-C-2)n\varepsilon_n^2}\right) + o(1).$$

By Markov's inequality and Fubini's theorem, the latter probability is smaller than

$$\frac{e^{-(A-C-2)n\varepsilon_n^2}}{M^2\varepsilon_n^2} \int_{C([0,1]^d)} \|w - w_0\|_2^2 E_{f_0}^{(n)}\left[\frac{L_n(f_w)}{L_n(f_0)}\right] d\Pi_{W_n}(w)$$
$$= \frac{e^{-(A-C-2)n\varepsilon_n^2}}{M^2\varepsilon_n^2} E^{\Pi_{W_n}}\left[\|W - w_0\|_2^2\right],$$

since by a change of measure $E_{f_0}^{(n)}[L_n(f_w)/L_n(f_0)] = E_{f_w}^{(n)}[1] = 1$. To proceed, we note that the Fernique-like theorem in Sect. 2 of [11] implies that

$$E^{\Pi_{W_n}}[\|W\|_2^2] \lesssim 1.$$

Combining the last three displays then yields that for any $A > 0$, we can find $M > 0$ large enough such that

$$P_{f_0}^{(n)}(\|\bar{W}_n - w_0\|_2 > 2M\varepsilon_n) \lesssim \frac{e^{-(A-C-2)n\varepsilon_n^2}}{M^2\varepsilon_n^2} + o(1)$$

since $w_0 \in B_{11}^\alpha \subset L^2$ as $\alpha > d$. Taking $A > (C + 2)$ and using that $\varepsilon_n = n^{-\alpha/(2\alpha+d)}$, the right-hand side tends to zero as $n \to \infty$. □

Acknowledgements. We are grateful to one anonymous referee for helpful comments that lead to an improvement of this article. This research was supported by the Italian Ministry of Education, University and Research (MIUR), "Dipartimenti di Eccellenza" grant 2023–2027, and partially by MIUR, PRIN project 2022CLTYP4.

References

1. Agapiou, S., Castillo, I.: Heavy-tailed Bayesian nonparametric adaptation. Ann. Stat. **52**(4), 1433–1459 (2024)
2. Agapiou, S., Dashti, M., Helin, T.: Rates of contraction of posterior distributions based on p-exponential priors. Bernoulli **27**(3), 1616–1642 (2021)
3. Agapiou, S., Savva, A.: Adaptive inference over Besov spaces in the white noise model using p-exponential priors. Bernoulli **30**(3), 2275–2300 (2024)
4. Agapiou, S., Wang, S.: Laplace priors and spatial inhomogeneity in Bayesian inverse problems. Bernoulli **30**(2), 878–910 (2024)
5. Arbel, J., Gayraud, G., Rousseau, J.: Bayesian optimal adaptive estimation using a sieve prior. Scand. J. Stat. **40**(3), 549–570 (2013)
6. Bioucas-Dias, J.M.: Bayesian wavelet-based image deconvolution: a GEM algorithm exploiting a class of heavy-tailed priors. IEEE Trans. Image Process. **15**(4), 937–951 (2006)
7. Bourdaud, G., Sickel, W.: Composition operators on function spaces with fractional order of smoothness. In: Harmonic analysis and nonlinear partial differential equations, RIMS Kôkyûroku Bessatsu, B26. Research in the Mathematical Sciences (RIMS), Kyoto, pp. 93–132 (2011)
8. Bui-Thanh, T., Ghattas, O.: A scalable algorithm for MAP estimators in Bayesian inverse problems with Besov priors. Inverse Probl. Imaging **9**(1), 27–53 (2015)
9. Castillo, I., Nickl, R.: On the Bernstein-von Mises phenomenon for nonparametric Bayes procedures. Ann. Statist. **42**(5), 1941–1969 (2014)
10. Chen, V., Dunlop, M.M., Papaspiliopoulos, O., Stuart, A.M.: Dimension-robust MCMC in Bayesian inverse problems. arXiv preprint arXiv:1803.03344 (2018)
11. Dashti, M., Harris, S., Stuart, A.M.: Besov priors for Bayesian inverse problems. Inverse Probl. Imaging **6**(2), 183–200 (2012)
12. Dolera, E., Favaro, S., Giordano, M.: On strong posterior contraction rates for besov-laplace priors in the white noise model. arXiv preprint arXiv:2411.06981 (2024)
13. Donoho, D.L., Johnstone, I.M.: Minimax estimation via wavelet shrinkage. Ann. Statist. **26**(3), 879–921 (1998)
14. Ghosal, S., Ghosh, J.K., van der Vaart, A.W.: Convergence rates of posterior distributions. Ann. Statist. **28**(2), 500–531 (2000)
15. Ghosal, S., van der Vaart, A.: Convergence rates of posterior distributions for non-I.I.D. observations. Ann. Statist. **35**(1), 192–223 (2007)
16. Ghosal, S., van der Vaart, A.W.: Fundamentals of Nonparametric Bayesian Inference. Cambridge University Press, New York (2017)
17. Giné, E., Nickl, R.: Rates of contraction for posterior distributions in L^r-metrics, $1 \leq r \leq \infty$. Ann. Statist. **39**(6), 2883–2911 (2011)
18. Giné, E., Nickl, R.: Mathematical foundations of infinite-dimensional statistical models. Cambridge University Press, New York (2016)
19. Giordano, M.: Besov-Laplace priors in density estimation: optimal posterior contraction rates and adaptation. Elect. J. Statist. **17**(2), 2210–2249 (2023)
20. Giordano, M., Ray, K.: Nonparametric Bayesian inference for reversible multidimensional diffusions. Ann. Statist. **50**(5), 2872–2898 (2022)
21. Giordano, M., Ray, K., Schmidt-Hieber, J.: On the inability of gaussian process regression to optimally learn compositional functions. Adv. Neural. Inf. Process. Syst. **35**, 22341–22353 (2022)

22. Kekkonen, H., Lassas, M., Saksman, E., Siltanen, S.: Random tree Besov priors–towards fractal imaging. Inverse Probl. Imaging **17**(2), 507–531 (2023)
23. Lassas, M., Saksman, E., Siltanen, S.: Discretization-invariant Bayesian inversion and Besov space priors. Inverse Probl. Imaging **3**(1), 87–122 (2009)
24. Leporini, D., Pesquet, J.-C.: Bayesian wavelet denoising: Besov priors and non-Gaussian noises. Signal Process. **81**(1), 55–67 (2001)
25. Ray, K.: Bayesian inverse problems with non-conjugate priors. Electron. J. Stat. **7**, 2516–2549 (2013)
26. Rudin, L.I., Osher, S., Fatemi, E.: Nonlinear total variation based noise removal algorithms.In: Experimental Mathematics: Computational Issues in Nonlinear Science, vol. 60. 1992, pp. 259–268, Los Alamos, NM, 1991
27. Sakhaee, E., Entezari, A.: Spline-based sparse tomographic reconstruction with Besov priors. In: Ourselin, S., Styner, M.A. (Eds.) Medical Imaging 2015: Image Processing, vol. 9413, pp. 101 – 108. International Society for Optics and Photonics, SPIE (2015)
28. Savva, A.: Adaptation in the white noise model using p-exponential priors. Ph.D. thesis, University of Cyprus (2023)
29. Shen, X., Wasserman, L.: Rates of convergence of posterior distributions. Ann. Statist. **29**, 687–714 (2001)
30. Triebel, H.: Theory of Function Spaces. Monographs in Mathematics, vol. 78. Birkhäuser Verlag, Basel (1983). https://doi.org/10.1007/978-3-0346-0416-1
31. van der Vaart, A.W., van Zanten, J.H.: Rates of contraction of posterior distributions based on Gaussian process priors. Ann. Statist. **36**(3), 1435–1463 (2008)
32. Vänskä, S., Lassas, M., Siltanen, S.: Statistical X-ray tomography using empirical Besov priors. Int. J. Tomogr. Stat. **11**, S09, 3–32 (2009)

Tracking Preference Evolution with Dynamic Multidimensional Unfolding

Giuseppe Gismondi[1(✉)], Marco Cardillo[2], and Alfonso Piscitelli[2]

[1] Department of Economics and Statistics, University of Naples Federico II, Naples, Italy
giuseppe.gismondi@unina.it
[2] Department of Agricultural Science, University of Naples Federico II, Naples, Italy

Abstract. This paper introduces a novel approach to dynamic multidimensional scaling (MDS) by incorporating the unfolding paradigm to analyze longitudinal preference data. The method addresses common limitations of traditional MDS in temporal settings, particularly when within-set dissimilarities are missing or ill-defined. By extending the unfolding framework and employing data augmentation techniques based on Kemeny-equivalent dissimilarities, the proposed methodology enables meaningful visualization of preference evolution over time. Applications to artificial data and applications to sensory evaluation and healthcare mobility data illustrate the flexibility and interpretability of the model, both in simulated and in real-world scenarios.

Keywords: Unfolding · Dynamic MDS · Longitudinal preferences

1 Introduction

Understanding how individual or group preferences evolve over time is a critical task across numerous fields, from marketing and sensory studies to political science and healthcare research. Multidimensional unfolding offers a well-established framework for modeling preference structures based on proximity data, allowing researchers to visually represent the relationship between individuals and the items they evaluate. However, the extension of the traditional unfolding techniques to longitudinal data has yet to be explored. This paper proposes a novel methodology that extends the unfolding framework to dynamic contexts, bridging the gap between static preference modeling and temporal analysis. Drawing on concepts from dynamic MDS and leveraging Kemeny-equivalent dissimilarities, we introduce a data augmentation strategy that enables the construction of a comprehensive dissimilarity matrix suitable for non-metric MDS. Our approach emphasizes interpretability by assuming temporal stability in one dimension, typically the item space, allowing clearer insights into evolving preference patterns. Through an artificial dataset and two empirical applications, we demonstrate how this method can uncover meaningful changes, or stabilities, in preferences, reinforcing its potential as a valuable analytical tool.

A. D'Ambrosio et al. (Eds.): CLADAG-VOC 2025, SCDAKO, pp. 214–225, 2025.
https://doi.org/10.1007/978-3-032-03042-9_19

2 Unfolding and the Problem of Degenerate Solutions

Multidimensional unfolding, first introduced by Coombs in 1950 [8], is a technique used to analyze two-mode preference rank data. It helps estimate two configurations A and B that represent coordinates for a group of I individuals and a set of J objects in dimension p. These are based on proximity values, typically reflecting the preferences of individuals for each object. The primary objective is to create a graphical representation where the distances between the individual points and the object points correspond to the observed distances. Unfolding is often approached as a least squares problem, aiming to minimize the raw STRESS function [17]:

$$\sigma^2(A, B, \hat{\Delta}) = \sum_{i=1}^{I} \sum_{j=1}^{J} (\hat{\delta}_{ij} - d_{ij})^2$$

that measures the fit between the observed and expected distances. Here $\hat{\Delta}$ is the $I \times J$ dissimilarity matrix, where the rows represent the judges and the columns represent the items. This matrix contains the elements $\hat{\delta}_{ij}$, which represent the observed or transformed dissimilarity between individual i and item j, while d_{ij} is instead their distance in the Euclidean space in dimension p. Unfolding is also treated as a special case of Multidimensional Scaling (MDS), where the distances within the sets of subjects and objects are not included. However, this missing information can lead to degenerate solutions in some cases, particularly in ordinal (non-metric) unfolding, where the solution might be good according to the measure of fit but at the same time graphically meaningless. The prime example of this is when all subjects are clustered at a single point, making the results not interpretable. Various attempts have been made over the years in trying to tackle this problem, and these can be classified into three distinct approaches: adapting the original data matrix, adjusting the transformation, and modifying the function to minimize. The first one involves expanding or modifying the input data. Among the proposals, we recall methods that aim to augment the data matrix by reconstructing the complete dissimilarity matrix (Lingoes [18]; Steverink et al. [22]; Van Deun et al. [23]). The second category of methods focuses on limiting the transformations to ensure both the intercept and slope remain non-zero, in an attempt to avoid degenerate solutions. Note that the intercept could also be set to zero, without problems in terms of degenerate solutions. An example of this is the ordinal-ratio method by Borg and Lingoes [4], which applies both ordinal and ratio transformations to the data simultaneously, while, to handle rank-order data, techniques like smoothed monotone regression (Heiser, [15]) and "quasi-metric" two-step transformation (Kim et al. [16]) have been suggested. The third category involves adjusting the loss function, with approaches like minimizing the Stress-2 function or using strategies that weight or penalize the loss function or transformations (Busing et al. [5]; DeSarbo & Rao, [11]). For the purpose of this paper, a method in the first category proposed by D'Ambrosio et al. [12] is chosen to perform the final analysis.

3 Longitudinal Preference Data

When analyzing longitudinal preference data, the goal is to understand how individuals preferences among a set of alternatives evolve over time. We limit our analysis to that of ranking data. Over time, these rankings may shift due to changing attitudes, acquired knowledge, new experiences, or contextual influences, making it essential to use models that account for both the temporal dynamics and the structure of ranked data. Common approaches include Bradley-Terry models, which estimate the relative worth of each item based on observed rankings, and their dynamic extensions [7], which allow item worths to change across time points. Other techniques, such as Bayesian ranking models [2] where rankings are allowed to vary in time have also gathered attention over the years. The main applications are linked with fields like marketing, food tasting, political science, and behavioral research. To the best of our knowledge, very few attempts have been made towards the visualization of time-varying preferences, and hence on the extension of MDS and unfolding analyses to this type of data. To understand the problem at hand, an overview of the existing methods is given, followed by our proposal.

3.1 Dynamic MDS

A first useful consideration to understand our approach relates to what Ambrosi and Hansohm [1] call Dynamic Mds. They describe a setting where dissimilarities are measured over T different periods on the same set of objects J. To handle the problem, they suggest looking at a super-dissimilarity matrix D that, using the notation in Cox and Cox [9], looks like this:

$$D = \begin{bmatrix} D_{11} & D_{12} & \cdots & D_{1T} \\ D_{21} & D_{22} & \cdots & D_{2T} \\ \vdots & \vdots & \ddots & \vdots \\ D_{T1} & D_{T2} & \cdots & D_{TT} \end{bmatrix}$$

where each D_{tt} is the dissimilarity matrix obtained by comparing the items at period t, whose elements δ^t_{rs}, with $r, s = 1, ...J$, indicate the dissimilarity between item r and item s at time t. The matrix $D_{tt'}$ contains the dissimilarities between objects at different time periods, so $\delta^{t,t'}_{rs}$, indicate the dissimilarity between item r at time t and item s at time t', and usually $\delta^{t,t'}_{rs} \neq \delta^{t',t}_{rs}$. Even with that, the entire matrix D remains symmetric. The main problem is that the $D_{tt'}$ are often missing, but they could be computed using the information from the D_{tt} matrices. An option could be to define:

$$\delta^{t,t'}_{rs} = \frac{1}{2}(\delta^t_{rs} + \delta^{t'}_{rs}).$$

Others are to keep them as missing or to set each of the $\delta^{t,t'}_{rr} = 0$, meaning that objects don't change over time, and the rest $\delta^{t,t'}_{rs}$ with $r \neq s$ as missing. An alternative solution is proposed by Ambrosi and Hansohm [1] which minimizes an overall stress function, obtained from the sum of the stress functions for a nonmetric MDS performed on each period, subject to a constraint that doesn't allow the T points for each item to be too distant from each other. One last proposed approach is to perform an MDS analysis separately for each period, and then use a Procrustes [14] analysis to match each individual configuration in order to obtain the average configuration.

3.2 Our Proposal

To treat the problem in the presence of both items and judges, we took inspiration from the previously cited work. In this context, not only the across-time dissimilarity matrices but also the within-set dissimilarity matrices at each period are missing. Since unfolding can be treated as a specific case of MDS, our approach consists of reducing the time-variation only on the judges' side and then computing the missing within-set dissimilarity Matrices through an unfolding procedure. This means that first, the initial data matrix is augmented, and then an MDS is performed on the augmented matrix.

To understand the procedure, start from a general problem, with I judges whose preferences on a set of J items are recorded over T time periods. Instead of working directly with the dissimilarity matrices, we work with preference data matrices. This means that in practice we have T different $I \times J$ matrices, containing the ranks given by the I judges to the J items at each time period, and let's call them with Σ_t. From these, we obtain a matrix of dimensions $(I{*}T) \times J$ by simply stacking all of the T preference matrices vertically [3]. Then we consider this as the input matrix of an unfolding procedure, call it Σ_{Tot}. So, the actual matrix of interest is:

$$\Sigma_{Tot} = \begin{bmatrix} \Sigma_1 \\ \Sigma_2 \\ \vdots \\ \Sigma_T \end{bmatrix}$$

where the judges are all the judges at all different T time periods, while the set of items is considered to be the same across all periods. From this matrix, the objective is to retrieve the following dissimilarity matrix:

$$\Delta = \begin{bmatrix} \Delta_{I*T} & \Delta_{Tot} \\ \Delta_{Tot}^T & \Delta_J \end{bmatrix}$$

where the diagonal matrices are equal to the within-sets dissimilarity matrices Δ_{I*T} and Δ_J, and the off-diagonal elements are equal to the between-sets dissimilarity matrix Δ_{Tot}. To ease the notation, we consider with $I' = I * T$ the new total number of judges, with $i = 1, ..., I'$, with J the total number of items and with $L = I' + J$ the total number of "objects" taken into consideration. For augmenting the data matrix the Kemeny-equivalent unfolding procedure proposed by D'Ambrosio et al. [12] has been chosen, to overcome the problem of degeneracies. Following part of the notation of the paper, let $\Sigma^{(i)}_{Tot}$ and $\Sigma_{Tot(j)}$ be respectively the i-th row of and the j-th column of the matrix. The augmentation is structured in the following steps:

1. Create a $J \times J$ matrix Θ containing rank vectors, with diagonal elements equal to some constant k, and all the extra-diagonal elements equal to another constant h, where $k < h$. Note with $\Theta^{(j)}$, $j = 1, ..., J$ a generic row of Θ. This represents a rank with item j as the first, and the other all tied at second place.
2. Create a $I' \times J$ rectangular matrix Z, where each element is defined as $z_{ij} = 1 - \tau_x \left(\Sigma^{(i)}_{Tot}, \Theta^{(j)}\right)$, with $i = 1, ..., I', j = 1, ...J$ and τ_x being the correlation coefficient associated with the Kemeny distance [13] between the two vectors. This matrix will then contain the between-sets dissimilarities.
3. Create a $I' \times I'$ symmetric matrix X, where each element is defined as $x_{is} = 1 - \tau_x \left(\Sigma^{(i)}_{Tot}, \Sigma^{(s)}_{Tot}\right)$, with $i, s = 1, ..., I'$. This matrix will then contain the within-judges dissimilarities.
4. Create a $J \times J$ symmetric matrix Y, where each element is defined as $y_{jf} = 1 - \tau_x \left(\Sigma_{Tot(j)}, \Sigma_{Tot(f)}\right)$, with $j, f = 1, ..., J$. This matrix will then contain the within-items dissimilarities.
5. Define the following quantities:

$$\zeta = \sqrt{\frac{JI'}{\sum_i \sum_j z^2_{ij}}} \quad \chi = \sqrt{\frac{I'^2}{\sum_i \sum_s x^2_{is}}} \quad \upsilon = \sqrt{\frac{J^2}{\sum_j \sum_f y^2_{jf}}}$$

6. Create the final dissimilarity matrix:

$$\Delta^* = \left[\begin{array}{c|c} \chi X & \zeta Z \\ \hline \zeta Z^T & \upsilon Y \end{array}\right]$$

   where $\chi X = \Delta^*_{I*T}$, $\upsilon Y = \Delta^*_J$ and $\zeta Z = \Delta^*_{Tot}$.
7. Perform a non-metric MDS on the matrix Δ^*, which will be a $L \times L$ matrix, with $L = (I * T) \times J$ whose elements are δ_{ab} with $a, b = 1, ..., L$.

An important assumption we made here is that we don't allow for the dissimilarities between the items to change over time, and in the final configuration the resulting items' locations come from the compromise between all the time periods. The reason why this choice was made resides mainly in the possible context of application. Often, panel data about preferences are the result of experiments, where items differ objectively according to some criterion and judges' preferences are studied before and after some specialization training. This is usually done to understand the efficacy of said training, so the interest lies in the change regarding the judges, and not the items. Of course, there are also experiments where the judges are already experts, which means that they don't receive further training, and someone would like to understand instead how a set of items, before and after they have gone through some changes, is perceived. But this can be seen as just a case of redefining judges and items according to the aim of the experiment. Here, the roles would be swapped, and this would work as long as the quality of the judges is high enough to rule out random rankings. More generally, this applies whenever one of the two sets of objects, judges or items, can be considered stable over time, and a change happens only in the other set. Furthermore, another argument in favor of this choice is that allowing for both the judges and items to change over time would lead to a chaotic unfolding solution. What we mean by this is that, visually, it would become difficult to asses whether a judge has changed preferences, or if the eventual shift from one period to the other mimicked the shifts in the items' positions, keeping the actual preference expressed by the judge unchanged. So, we think that treating the problem restraining one set at a time could be best also for interpretation purposes. The possibility of asymmetric dissimilarities between different time periods is not accounted for here. In the following sections, first an application on artificial data and then two examples of possible applications on real datasets are described.

4 Artificial Data

In order to check the usefulness of our approach, we constructed four different cases using simulated data with the aim of showing what the method can do when presented with different challenges. First, we generated a random sample of 20 judges from a Mallows model [20] with 5 items (a, b, c, d, e), central ranking $(5, 2, 3, 4, 1)$, and a dispersion parameter $\theta = 0.65$. This will serve as the sample in period 1 for each of the four cases. The pool of judges in the second period, instead, was selected in the following way:

1. We sampled from a Mallows model 20 judges with a reverse central ranking $(1, 4, 3, 2, 5)$ and a $\theta = 0.45$.
2. We have done a pseudo-random reassignment that switches the first with the last position for about 70% of the sample. So, on average, this reassigns rank to objects a and e.
3. We maintained the same preference structure from the first period, adding a small random noise, changing positions 2 and 3 for a small number of judges.
4. We sampled from the same exact model as in period 1.

Following the steps introduced in Sect. 3, the MDS analyses were conducted using the SMACOF algorithm [10] present in the smacof R package [19]. All the analyses were performed by setting 50 initial random start points.

4.1 Results

Here, the results are presented in Fig. 1. Labels were assigned only to the judges in the first period to understand the direction of the lines. Given that the median ranking at time 1 is always $(5, 2, 3, 4, 1)$, the sub-captions relative to each subplot show the consensus ranking at time 2.

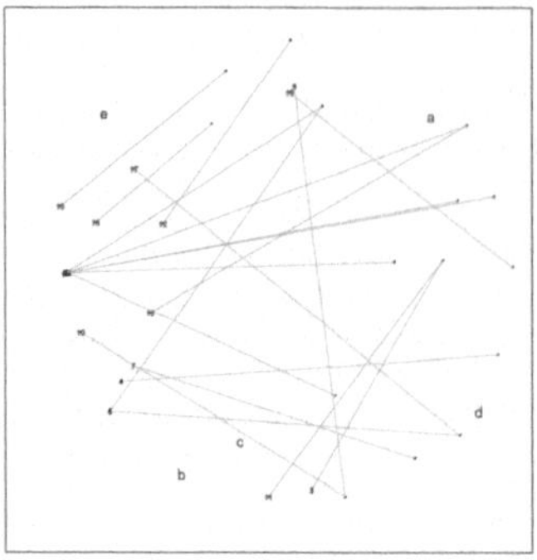

(a) Case 1: consensus (1,4,3,2,5)

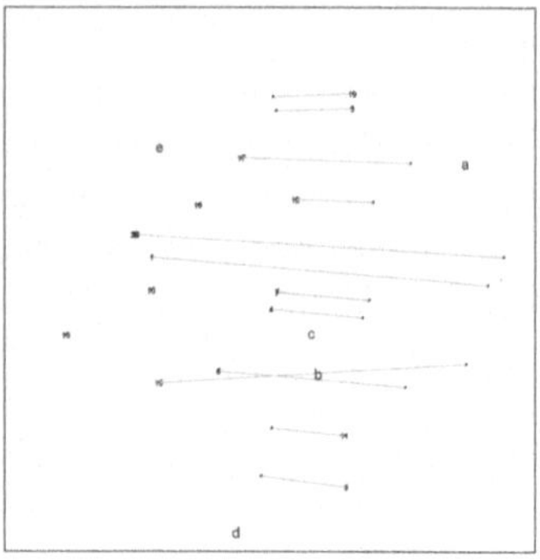

(b) Case 2: consensus (4,1,2,5,3)

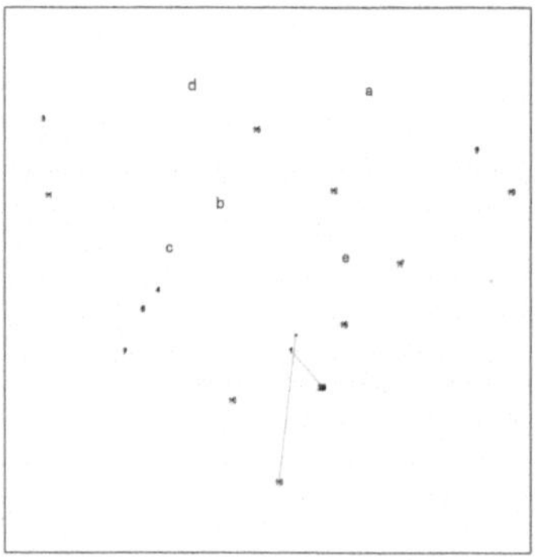

(c) Case 3: consensus (5,2,3,4,1)

(d) Case 4: consensus (5,2,3,4,1)

Fig. 1. Unfolding configurations for each case, with the associated consensus rankings for judges in the second period

In panel (a), each individual moves towards the items with a higher ranking in the second period, as expected from the new consensus ranking. In contrast, take panel (d), where the consensus does not change. Still, the figure highlights how individual preferences for the objects have changed over time, making it apparently similar to panel (a), and hinting at how looking only at the consensus ranking could be misleading. This approach is, in fact, more focused on understanding and visualizing individual shifts in preferences, while the consensus rankings can be used as an added tool to understand overall changes in the preference structure. A similar point can be made by looking at panel (c). Here, not only does the consensus remain the same, but there is a small change in the individual preferences over time. Lastly, panel (b) is probably the most interesting. As a consequence of the setup for this experiment, the trajectories are mostly horizontal, reflecting how the main direction for the change in preferences is due to the interchange between item a and item e, d also seems to be influential but to a lesser extent, while the distances from the other items remain practically the same across the two periods. So, each configuration was able to represent the situations as they were designed.

5 Real Data

Two different real-world datasets were selected to evaluate the performance of this method. In the first one, a panel of 98 assessors was asked to judge seven oils of different quality (A,B,C,D,E,F,G) before and a week after having followed a seminar. More specifically, each assessor had to assign a value from one to nine to each of the oils, where higher values mean a higher personal evaluation. This was done under a double-blind control procedure, to ensure the validity of the experiment. Among the oils, A, B, and F are the only extra-virgin olive oils and are the ones with more aromatic components, oil C is a refined olive oil which is instead more neutral, while oils D, E, and G all have some rancid defects. For the purpose of this analysis, these ratings were accordingly converted into rankings, keeping ties where present. The final dataset consists of a rectangular 196×7 matrix with each row representing the ranking assigned by assessor $i = 1, ..., 98$ at time $t = 1, 2$ to the set of oils, and that is constructed by two blocks determined by the two time periods.

In the second one, the data come from the website of the Italian Ministry of Health, which publishes annual reports on hospital admissions. One of the tables in these reports was chosen regarding interregional hospital mobility for acute care activities under ordinary hospitalization. Five years of reports have been selected, from 2011 to 2015, and each data table is a 21×21 matrix where the columns are the 21 regions of residence, and the rows represent the 21 regions of hospitalization. The 21 regions considered are: Piemonte, Valle d'Aosta, Lombardia, the autonomous province of Bolzano, the autonomous province of Trento, Veneto, Friuli Venezia Giulia, Liguria, Emilia Romagna, Toscana, Umbria, Marche, Lazio, Abruzzo, Molise, Campania, Puglia, Basilicata, Calabria, Sicilia, and Sardegna. Here the region of residence are treated as the judges while the regions of admission are the items. So, in this framework, each region expresses a preference over the possible regions of admission. This means that first we transposed the original table, and then to construct a ranking from this table it is sufficient to assign ranks to the items according to the frequencies in decreasing order. That is, for a region the item with rank one is the one with the highest absolute frequency. In the end, the resulting matrix is a 105×21 matrix, where each row represents the ranking that a region of residence $i = 1, ..., 21$ at time $t = 2011, 2012, 2013, 2014, 2015$ assigns to the set of admission regions, and is constructed by 5 blocks indicating the years.

Something to note, in both datasets, how much the items differ among themselves should not change over time. For the oils, their evaluation is based on objective characteristics, such as the concentration of phenolic compounds, level of acidity and so on, which remain the same in the two tasting sessions. Similarly, for the data about the regions of admission, in a short time span of 4 years, it is very unlikely that hospitals can make drastic changes in order to improve their medical assistance. So, keeping them fixed in the analysis allows to concentrate on how in one case the seminar changed the judges' perceptions about the oils, and in the second case to asses whether or not the decisions about hospital admissions have changed in that period.

5.1 Results

In this section, the unfolding plots, consensus rankings, and measures for the unfolding configuration for the two datasets are presented.

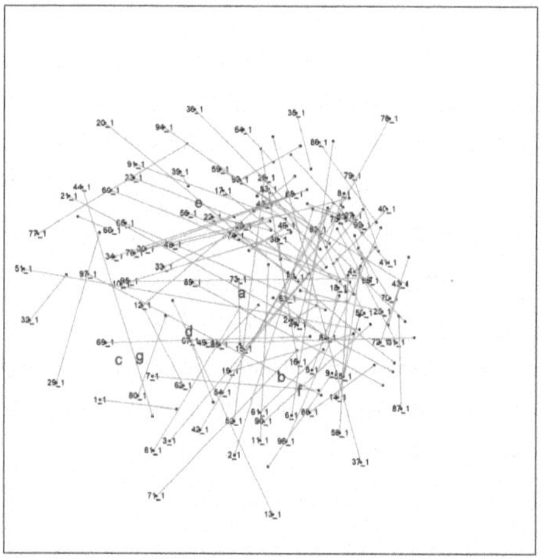

Fig. 2. Unfolding solution for the Oils dataset

The plot of the unfolding configuration in Fig. 2 for the experiment on the oils highlights a great change in preferences after the assessors attended the seminar. To understand the plot, again, labels were assigned only to the assessors before the seminar, so the direction of the line goes from the label to the corresponding dot without a label. With this in mind, the majority of the assessors move nearer to oils A and F and further from oils E, G, and C, while for the others, the results are less clear. This result is in line with what could be expected from a successful experiment; judges' preferences after the seminar tend to be more aligned with the better oils than before. A simple analysis of the consensus rankings before, $(2, 1, 3, 1, 1, 1, 2)$, and after, $(1, 2, 4, 2, 2, 2, 3)$, reveals that the main difference between the two periods lies in the position of oil A, which becomes the most preferred one. Also, the Unfolding measures in Table 1 are a sign of a decent overall configuration. The Stress measures indicate the badness of fit, and here the moderately low values are a sign of good performance. The VAF is more average. Also, the Spearman, the Kendall [6] and the TauX rank correlation coefficient, even if not too high, are still quite good.

A very different result instead is obtained from the second analysis in Fig. 3. The decisions about the admission regions seem to be quite stable over time, with each residence region not moving much with respect to the other regions and with itself in the other periods. Considering the time period analyzed and the absence of important reform or shocks, it is quite a reasonable result. It is very difficult to imagine that these kinds of decisions would change on an aggregate level in a few years, and generally that they would change at all. Here, the consensus rankings in each year are highly correlated, with a Kendall coefficient of correlation equal on average to 0.94, which further consolidates the stability of the preferences. The unfolding measures in Table 2 show a less decent overall performance, with values of Normalized Stress sufficiently low, but again a low value of the VAF. Also, the three considered correlation coefficients are more on the lower side.

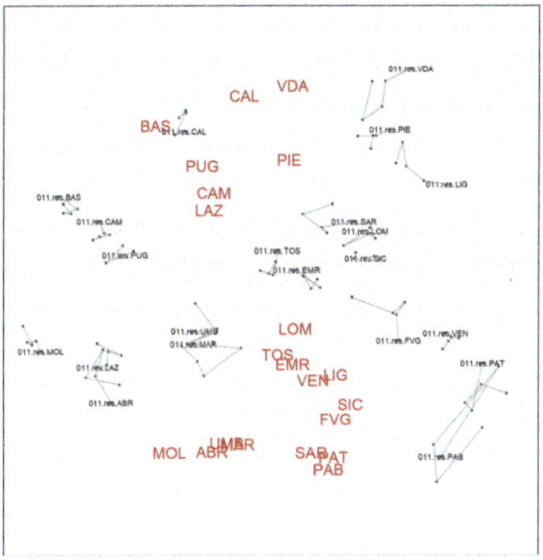

Fig. 3. Unfolding solution for the Regions dataset

The aim of the two applications was to reproduce these two different alternatives, one where preferences heavily change over time and one where the solution instead is quite static, in order to understand whether the method could distinguish between the two, also when dealing with real-world data. Furthermore, the results are consistent with what could be expected from the two situations, which is reassuring from an interpretation perspective.

Table 1. Unfolding Measures for the Oils dataset

Measure	Value
Raw Stress	135.7886
Normalized Raw Stress	0.1041
VAF	0.2959
Spearman	0.5611
Kendall	0.4068
TauX	0.3924

Table 2. Unfolding Measures for the Regions dataset

Measure	Value
Raw Stress	204.8254
Normalized Raw Stress	0.1063
VAF	0.1876
Spearman	0.4457
Kendall	0.3193
TauX	0.3054

6 Conclusions

This study presents a dynamic unfolding framework for analyzing and visualizing longitudinal preference data, offering a meaningful extension of traditional MDS approaches. This graphical tool can also be accompanied by a detailed analysis of individual differences among judges over time. By integrating unfolding principles and leveraging Kemeny-equivalent dissimilarities, the method overcomes the challenges posed by missing within-set dissimilarities and the risk of degenerate solutions. The proposed technique proves effective in distinguishing between static and evolving preference patterns, as demonstrated by the various applications. The assumption of temporal stability in one set of objects simplifies interpretation and aligns with many experimental designs, though the framework remains adaptable for alternative contexts. Future developments may consider further refinements, such as incorporating asymmetric dissimilarities or allowing for simultaneous evolution in both judges and items.

Acknowlegdments. Authors would like to thank Prof. Raffaele Sacchi for sharing the dataset of the sensory experiment on olive oils, for his suggestions and insightful remarks.

References

1. Ambrosi, K., Hansohm, J.: Ein dynamischer Ansatz zur Repräsentation von Objekten. In: DGOR: Papers of the 15th Annual Meeting/Vorträge der 15. Jahrestagung, vol. 1986, pp. 425–431. Springer, Heidelberg (1987). https://doi.org/10.1007/978-3-642-72557-9_77
2. Asfaw, D., Vitelli, V., Sørensen, Ø., Arjas, E., Frigessi, A.: Time-varying rankings with the Bayesian Mallows model. Statistics **6**(1), 14–30 (2017)
3. Bijleveld, C.C.J.H., van der Kamp, L.J.T., Mooijaart, A., van der Kloot, W.A., van der Leeden, R., van der Burg, E.: Longitudinal Data Analysis: Designs, Models and Methods. Sage Publications Ltd. (1998)
4. Borg, I., Lingoes, J.: Multidimensional Similarity Structure Analysis. Springer Science & Business Media (2012)
5. Busing, F.M.T.A., Groenen, P.J.K., Heiser, W.J.: Avoiding degeneracy in multidimensional unfolding by penalizing on the coefficient of variation. Psychometrika **70**(1), 71–98 (2005)

6. Busing, F.M.T.A.: Advances in multidimensional unfolding (Unpublished doctoral dissertation). Leiden University, Leiden, The Netherlands, Department of Data Theory (2010)
7. Cattelan, M., Varin, C., Firth, D.: Dynamic Bradley–Terry modelling of sports tournaments. J. R. Statist. Soc. Ser. C (Appl. Statist.) **62**(1), 135–150 (2013)
8. Coombs, C.H.: Psychological scaling without a unit of measurement. Psychol. Rev. **57**(3), 145–158 (1950)
9. Cox, T.F., Cox, M.A.A.: Multidimensional Scaling. CRC Press, Boca Raton (2000)
10. De Leeuw, J., Heiser, W.: Multidimensional scaling with restrictions on the configuration. Multivariate Anal. **5**, 501–521 (1980)
11. DeSarbo, W.S., Rao, V.R.: GENFOLD2: a set of models and algorithms for the GENeral unFOLDing analysis of preference/dominance data. J. Classif. **1**, 147–186 (1984)
12. D'Ambrosio, A., Vera, J.F., Heiser, W.J.: Avoiding degeneracies in ordinal unfolding using Kemeny-equivalent dissimilarities for two-way two-mode preference rank data. Multivariate Behav. Res. **57**(4), 679–699 (2022)
13. Emond, E.J., Mason, D.W.: A new rank correlation coefficient with application to the consensus ranking problem. J. Multi-Criteria Decis. Anal. **11**(1), 17–28 (2002)
14. Gower, J.C.: Generalized procrustes analysis. Psychometrika **40**(1), 33–51 (1975)
15. Heiser, W.J.: Order invariant unfolding analysis under smoothness restrictions. In: Advances in Psychology, vol. 60, pp. 3–31. Elsevier (1989)
16. Kim, C., Rangaswamy, A., DeSarbo, W.S.: A quasi-metric approach to multidimensional unfolding for reducing the occurrence of degenerate solutions. Multivariate Behav. Res. **34**(2), 143–180 (1999)
17. Kruskal, J.B.: Multidimensional scaling by optimizing goodness of fit to a nonmetric hypothesis. Psychometrika. **29**(1), 1–27 (1964)
18. Lingoes, J.C.: A general nonparametric model for representing objects and attributes in a joint metric space, pp. 277–298. Archéologie et Calculateurs. Paris, Éditions du Centre National de la Recherche Scientifique (1970)
19. Mair, P., Groenen, P., de Leeuw, J.: Multidimensional scaling using majorization: SMACOF in R. J. Stat. Softw. **102**(10), 1–47 (2022)
20. Marden, J.I.: Analyzing and Modeling Rank Data. Chapman & Hall, New York (1995)
21. Shepard, R.N.: Representation of structure in similarity data: problems and prospects. Psychometrika. **39**(4), 373–421 (1974)
22. Steverink, M., Heiser, W.J., van der Kloot, W.A.: Avoiding degenerate solutions in multidimensional unfolding by using additional distance information (2002). (Unpublished manuscript)
23. Van Deun, K., Heiser, W.J., Delbeke, L.: Multidimensional unfolding by nonmetric multidimensional scaling of Spearman distances in the extended permutation polytope. Multivariate Behav. Res. **42**(1), 103–132 (2007)

"Can You Explain That?" E2Tree, SHAP, and LIME for Interpretable Random Forests

Agostino Gnasso(✉) and Massimo Aria

Department of Economics and Statistics, University of Naples Federico II, 80126 Naples, Italy
agostino.gnasso@unina.it

Abstract. An explainability methodology for Random Forest (RF) models, called Explainable Ensemble Trees (E2Tree), has been recently introduced. E2Tree methodology enhances interpretability by transforming the decision structure of RF into a simplified, explainable tree while preserving accuracy. It achieves this by leveraging the co-occurrence of observations in RF decision paths to create a globally interpretable representation of the model's decision process. In this study, we apply three distinct explainability techniques, E2Tree, Shapley Additive Explanations (SHAP), and Local Interpretable Model-Agnostic Explanations (LIME), to analyze the decision-making process of RF models. While SHAP and LIME offer global and local perspectives, respectively, particular emphasis is placed on E2Tree for its ability to summarize the ensemble's decision logic into a single, interpretable structure. We illustrate these methods using the German Credit dataset, focusing on how variables such as loan duration, credit amount, housing status, and financial liquidity influence credit risk assessment.

Keywords: E2Tree · SHAP · LIME · Machine Learning · Classification

1 Introduction

Machine learning (ML) has become integral to decision-making in various fields, including healthcare and finance [2]. While ML models offer remarkable predictive capabilities, their reliance on complex "black-box" architectures raises concerns about transparency, trust, and ethical accountability [15]. Ensuring interpretability is crucial, particularly in high-stakes applications where human oversight is required, as emphasized by regulations like the GDPR's "right to explanation" [2]. However, defining and measuring interpretability remains challenging due to the lack of a general agreement on its formalization and evaluation [18]. Interpretable Machine Learning has its origins in early statistical models like linear regression and rule-based systems [8], with a significant shift occurring when feature importance measures in tree ensemble methods, such as

A. D'Ambrosio et al. (Eds.): CLADAG-VOC 2025, SCDAKO, pp. 226–237, 2025.
https://doi.org/10.1007/978-3-032-03042-9_20

Random Forests (RF) [4], sparked increased interest in interpretability research [6]. Existing interpretability approaches fall into two main categories: inherently transparent models (e.g., linear regression, decision trees) and post-hoc explanations for complex models (e.g., neural networks, tree ensemble methods) [14]. While tree ensemble models like RF improve predictive accuracy, they obscure individual decision paths, necessitating techniques such as permutation importance, surrogate models, and SHAP (Shapley Additive Explanations) to extract insights [12]. Despite advancements in interpretability research, challenges persist [11]. The absence of a standardized interpretability metric complicates evaluation [17], while many methods fail to quantify explanation uncertainty, reducing reliability. Feature dependencies can lead to misleading interpretations, and distinguishing correlation from causation remains problematic. Some explainability techniques lack scalability, limiting their applicability to large datasets [1]. Additionally, there is often a trade-off between accuracy and interpretability, as highly accurate models tend to be more complex and less transparent. This study employs the recently proposed E2Tree methodology [3], specifically designed to improve the interpretability and explainability of RF models. E2Tree extracts the underlying decision structure of RF, transforming it into a simplified and transparent tree representation, while preserving the model's predictive accuracy. In addition to E2Tree, we apply two well-established interpretability techniques, Shapley Additive Explanations (SHAP) [12] and Local Interpretable Model-Agnostic Explanations (LIME) [16], to provide complementary perspectives on RF decision-making, encompassing both global patterns and local instance-level explanations.

2 E2Tree in a Nutshell

To improve the interpretability and explainability of RF models while maintaining their predictive power, the E2Tree methodology has been introduced [3]. E2Tree constructs a single tree structure that encapsulates the decision-making process of the tree ensemble, offering a more understandable representation of its outcomes. This approach leverages a dissimilarity matrix, D, derived from the tree ensemble output, which quantifies the degree of discordance between pairs of observations. Specifically, D is computed by analyzing the co-occurrence frequency of observations u_i and u_j within the same terminal nodes across all trees in the ensemble. Formally, D is defined as follows:

$$D = \begin{array}{c} \\ u_1 \\ u_2 \\ u_3 \\ \vdots \\ u_i \\ \vdots \\ u_n \end{array} \begin{array}{c} \begin{array}{cccccccc} u_1 & u_2 & u_3 & & u_j & & u_n \end{array} \\ \begin{bmatrix} 0 & d_{12} & d_{13} & \cdots & d_{1j} & \cdots & d_{1n} \\ d_{21} & 0 & d_{23} & \cdots & d_{2j} & \cdots & d_{2n} \\ d_{31} & d_{32} & 0 & \cdots & d_{3j} & \cdots & d_{3n} \\ \vdots & \vdots & \vdots & \ddots & \vdots & \ddots & \vdots \\ d_{i1} & d_{i2} & d_{i3} & \cdots & d_{ij} & \cdots & 0 \\ \vdots & \vdots & \vdots & \ddots & \vdots & \ddots & \vdots \\ d_{n1} & d_{n2} & d_{n3} & \cdots & d_{nj} & \cdots & 0 \end{bmatrix} \end{array} = 1 - \begin{bmatrix} & & \\ & O_{ij} & \\ & & \end{bmatrix}$$

where d_{ij} is the dissimilarity, related to $O_{ij} = 1 - d_{ij}$.
$O_{ij} \in [0, 1]$ is the proximity measure (co-occurence) from the RF, quantifying how often two observations fall into the a terminal node across the B trees in the ensemble, indicating their similarity in the decision process. It is computed as:

$$O_{ij} = \frac{\sum_{b=1}^{B} I(u_i \wedge u_j) \cdot W_{(t)_{ij}b}}{\max(\sum_b u_i W_{(t)_i b}; \sum_b u_j W_{(t)_j b})},$$

where $I(u_i \wedge u_j)$ if both units fall in the a terminal node of tree b. The term $W_{t_{ij}b}$ serves as a local measure of node quality, that includes both u_i and u_j, for the b-th tree within the ensemble, where b ranges from 1 to B. It is computed as $W_{t_{ij}b} = \frac{\sum_{i=1}^{n_t} I(y_i = \hat{y}_t)}{n_t}$ where n_t denotes the number of observations in node t, and $\hat{y}_t$ is the predicted class label assigned by the RF model within the same node. $W_{(t)_i b}$ and $W_{(t)_j b}$ is the local measure of the individual nodes containing u_i and u_j.

3 Application on Real Data Set

This section illustrates the application of the E2Tree, SHAP and LIME methodology, on the German Credit dataset, a publicly available dataset from the UCI Machine Learning Repository. The dataset consists of 1,000 instances with 10 attributes, including financial and personal information, and is commonly used for binary classification to determine credit risk. The response variable indicates whether a loan applicant was classified as having good or bad creditworthiness based on their repayment behavior. A RF model with 1,000 decision trees is trained using the `randomForest` package in R. The dataset is partitioned into a 70%–30% train-test split [7] to ensure the external validity of machine learning models and to preserve an independent test set for potential downstream evaluations. The E2Tree framework is implemented through the `e2tree` R package. For global feature importance analysis we used SHAP (`iml` R package), while for local, instance-level interpretation of predictions we used LIME (`lime` R package). SHAP was employed to examine the average contribution and interaction effects of features across the dataset, highlighting global patterns in how the model assigns credit risk. In contrast, LIME was used to generate detailed, instance-specific explanations, helping to understand how particular feature values influenced individual classification outcomes.

3.1 E2Tree Results

The E2Tree model is constructed by generating a co-occurrence matrix O_{ij} from RF classifications. The O_{ij} matrix, compute how frequently a pair of observations appear in a RF terminal nodes. The dissimilarity matrix D is then derived as $D = 1 - O_{ij}$, forming the basis of the E2Tree structure. As shown in Fig. 1, E2Tree summarizes RF's decision structure in a single tree while maintaining interpretability. Figures 2a and 2b display the co-occurrence matrices O_{ij} and

$\hat{O}_{ij}$ as heatmaps. In both figures, rows and columns represent individual observations, ordered identically along the axes. Each cell (i, j) visualizes the co-occurrence value O_{ij} measured in RF (Fig. 2a) and $\hat{O}_{ij}$ measured in E2Tree (Fig. 2b). In the RF heatmap, darker cells indicate high co-occurrence across the ensemble, whereas lighter shades reflect weaker proximity due to averaging across multiple trees. The result is a dense and noisy matrix where most pairs show some nonzero similarity. By contrast, in E2Tree observations are either in the same node, or not. In the latter case, the co-occurrence $\hat{O}_{ij}$ in the E2Tree (Fig. 2b) becomes exactly 0, while in the RF (Fig. 2a) co-occurrence O_{ij} will (almost) never be exactly 0. This discrete structure sharpens the visual representation, revealing compact, block-diagonal structures corresponding to the tree's decision regions. Despite their different visual characteristics, both matrices encode similar proximities among observations. The refinement of E2Tree highlights dominant clusters that were only partially visible in the original RF proximity matrix. To validate the structural consistency between RF and E2Tree, the Mantel test [13] is conducted. This test assesses the null hypothesis that there is no association between the matrices derived from the RF and E2Tree models. It does so by calculating the correlation between the elements of the lower triangular portions of the two matrices and determining the significance of this correlation via random permutations. In this case, the rows and columns of one matrix are permuted 999 times to generate a null distribution of correlation values. The resulting observed Z-statistic (209.44) falls in the extreme right tail of this distribution, yielding a p-value of 0.001. This outcome provides strong statistical evidence of a significant structural similarity between the original RF ensemble and its E2Tree-based representation. The E2Tree output in Fig. 1 provides an interpretable decision tree representation of the RF model, summarizing how it predicts credit risk in the German Credit dataset. Each node represents a splitting criterion, while terminal nodes (leaves) indicate the final classification along with their probabilities. The tree follows a hierarchical structure, starting with the most decisive feature splits and refining classifications down the branches. The tree presented in Fig. 1 provides a rule-based breakdown of how the model stratifies applicants by creditworthiness, using a series of binary splits on key socio-economic variables. At the root node, 44% of applicants are classified as "good credit" and 56% as "bad credit". The first and most impactful split occurs on loan duration: applicants with durations shorter than 21 months are predicted to be bad credit in 68% of cases, while those with longer durations are associated with a 64% probability of good credit. This pattern, short durations increasing risk, is preserved across most paths and indicates a dominant main effect. Among applicants with shorter durations, the model further distinguishes based on credit amount, age, loan purpose, savings, and housing status. In particular, those with low credit amounts and younger age are frequently associated with unfavorable classifications. Several paths emphasize the combined effect of limited savings and housing status. The housing variable, which includes the category "for free", is used repeatedly as a splitter. This category represents applicants who do not bear housing costs, typically because they live with family, receive employer-provided accommodation, or inhabit a property owned by others without rent. While this may signal lower financial burden, it can also reflect reduced economic independence. In this context, the model appears to treat "for free" housing as a potential risk factor when paired with other indicators of instability, such as low savings or small credit amounts. For instance, among young applicants with little or no savings and low requested credit, the presence of rent-free housing contributes to a higher probability of bad credit classification. Conversely, in the right branch of the tree (Duration ≥ 21), more favorable outcomes are observed. Here, applicants tend to be older and request higher credit amounts. The model incorporates savings and housing status to refine its predictions. In certain branches, the combination of higher credit amount, moderate or higher savings, and longer duration leads to terminal nodes with good credit classifications. This suggests that financial capacity and stability, measured not only through income proxies but also indirectly via housing and savings, support positive assessments. Interestingly, in this more favorable context, "for free" housing is no longer associated with elevated risk and may even be interpreted positively as a sign of lower financial strain, particularly when accompanied by other favorable indicators. Age consistently appears as a splitting variable throughout the tree. In most branches, younger age corresponds to higher risk, although its effect is always modulated by other variables such as credit amount and saving behavior. This confirms the context-dependent nature of age in credit evaluation: while youth alone may not predict bad credit, it becomes a stronger risk indicator when combined with financial uncertainty or reduced autonomy. Overall, the tree structure highlights clear main effects, especially for loan duration, credit amount, age, and savings, with limited evidence of complex interactions.

3.2 SHAP Results

For global feature importance analysis we used SHAP. These methods provide complementary insights, allowing us to better understand how RF assigns credit risk classifications.

Shapley values were computed as marginal contributions of each feature to the prediction, estimated through repeated sampling of feature coalitions. The model predictions were defined in terms of class probabilities, and the target variable was excluded from the input data used in the explanation phase. Global feature importance was quantified by measuring the increase in classification error when individual features were randomly permuted, using cross-entropy as the loss function. The reported global SHAP values represent the average absolute contribution of each feature across all instances. To capture second-order effects, we also computed interaction strengths between features, which reflect how much the impact of one feature depends on the presence or value of another. Finally, local explanations were produced by analyzing individual instances, allowing us to interpret specific predictions in detail. Default parameters from the iml package were used unless stated otherwise. The SHAP analysis

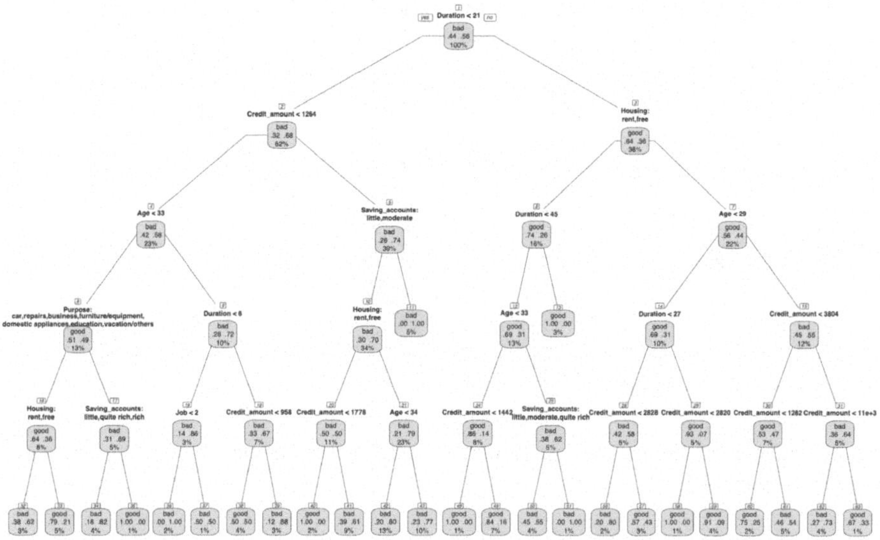

Fig. 1. Path visualization of E2Tree.

(Figs. 3 and 4) confirms that loan duration and credit amount are the most influential features in determining creditworthiness. To further explore how individual feature values contribute to model predictions, we computed and visualized SHAP values grouped by predicted class label. Figure 4 shows a SHAP summary plot where each bar represents the average contribution of a specific feature-value pair to the predicted class probability. The left and right panels display results for instances predicted as "bad" and "good" credit, respectively. The SHAP values are expressed on the probability scale, not on the log-odds scale, and indicate how much each feature shifts the predicted probability toward the corresponding class. The values were averaged across all observations within each predicted group, offering a comparative view of feature influence in different classification scenarios. It is important to note that the class labels in the panels refer to model predictions, not ground truth. Although some feature contributions may appear symmetric across panels (e.g., a positive contribution to the "good" class corresponding to a negative one for the "bad" class), these patterns are data-driven and not imposed by the method. The y-axis shows concrete feature-value combinations rather than aggregated feature names, which reflects the model's reliance on specific value ranges within features. Longer loan durations consistently increase the probability of bad credit classification, whereas shorter durations are associated with better creditworthiness. Similarly, high credit amounts contribute to a higher risk but, in some cases, align with good credit, suggesting that feature effects are context-dependent rather than absolute. Note that the SHAP summary plot (Fig. 4) illustrates how each feature influences the model's predictions. Positive values indicate an increase in the probability of the given classification, while negative values reduce it, shifting the prediction towards

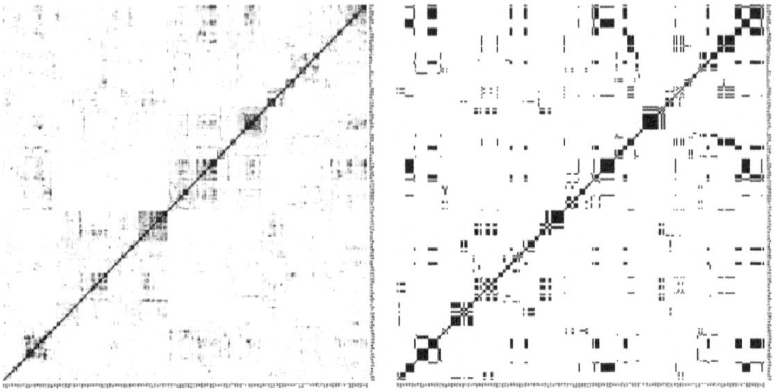

(a) (b)

Fig. 2. Heatmap of the matrix O_{ij} (panel a) and heatmap of the matrix $\hat{O}_{ij}$ estimated by E2Tree (panel b).

the opposite class. Financial liquidity also plays a crucial role, as low savings and checking account balances amplify credit risk, reinforcing the importance of available financial resources in credit evaluation. Figure 5 presents the estimated interaction strength for each feature. Interaction strength measures how much the effect of a given feature depends on the presence or value of other features. A value close to 0 indicates that the feature acts independently, i.e., its effect is mostly additive and does not vary in combination with other features. In contrast, values approaching 1 suggest a strong interaction component, meaning the feature's contribution to the prediction significantly depends on the context provided by other variables. These values are computed by comparing the model's variance explained by the feature alone versus in combination with all other features. The resulting measure is relative, in the sense that it expresses the proportion of the feature's overall effect that is due to interactions, rather than measuring the absolute strength of interaction terms. This allows for ranking features by how context-dependent their predictive contribution is. The SHAP interaction plot (Fig. 5) reveals strong dependencies between age, credit amount, and loan duration, indicating that these factors do not act independently but instead influence risk perception based on applicant profiles. Loan purpose, housing status, and financial stability exhibit moderate interaction effects, modifying risk levels depending on their combination with other variables. Conversely, sex has minimal influence, confirming that gender does not significantly impact credit assessment in this dataset.

3.3 LIME Results

To achieve instance-level interpretability, we used LIME. The LIME analysis (Fig. 6) provides a local perspective on RF's predictions, explaining how specific feature values contribute to individual classifications. For each observation to

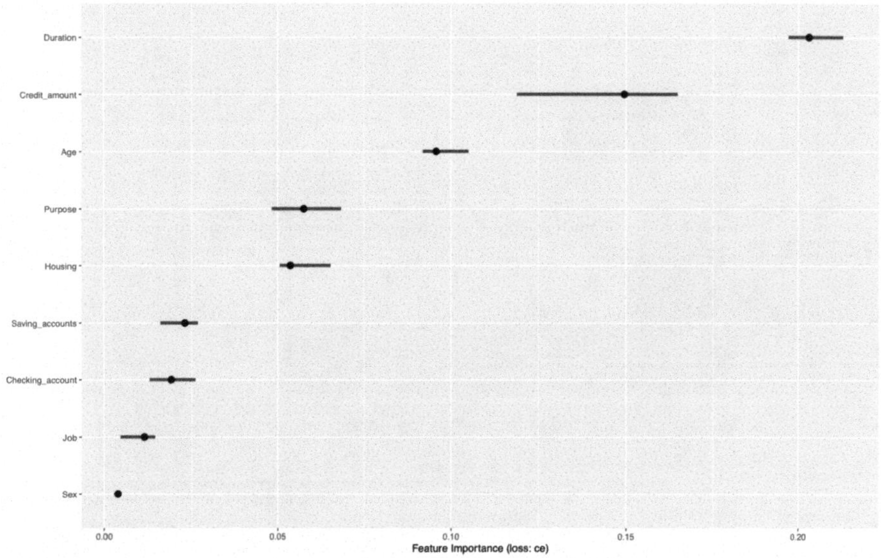

Fig. 3. Feature Importance with SHAP.

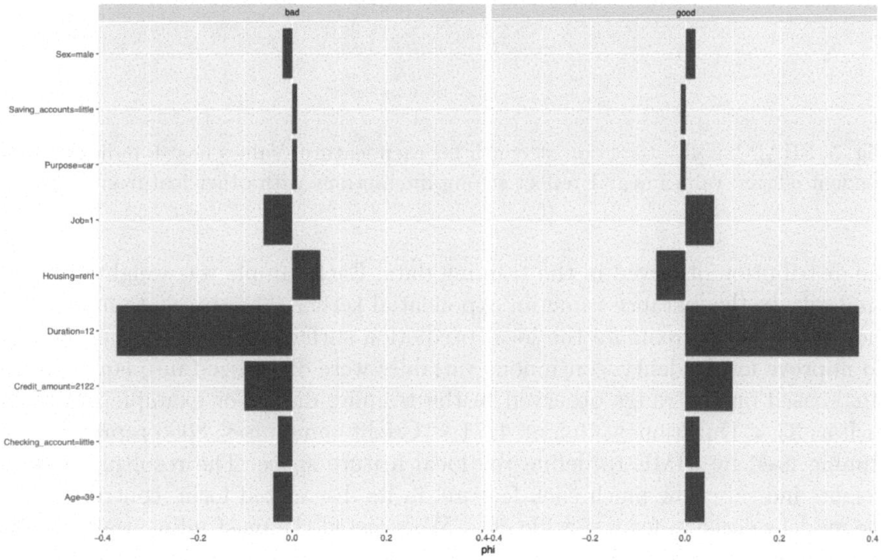

Fig. 4. Shapley values. The x-axis label "phi" corresponds to the individual SHAP values computed for each feature and observation.

be explained, a local surrogate model (linear regression) was trained on a set of perturbed samples generated in the neighborhood of the original instance. These perturbed samples were created by randomly modifying feature values based on

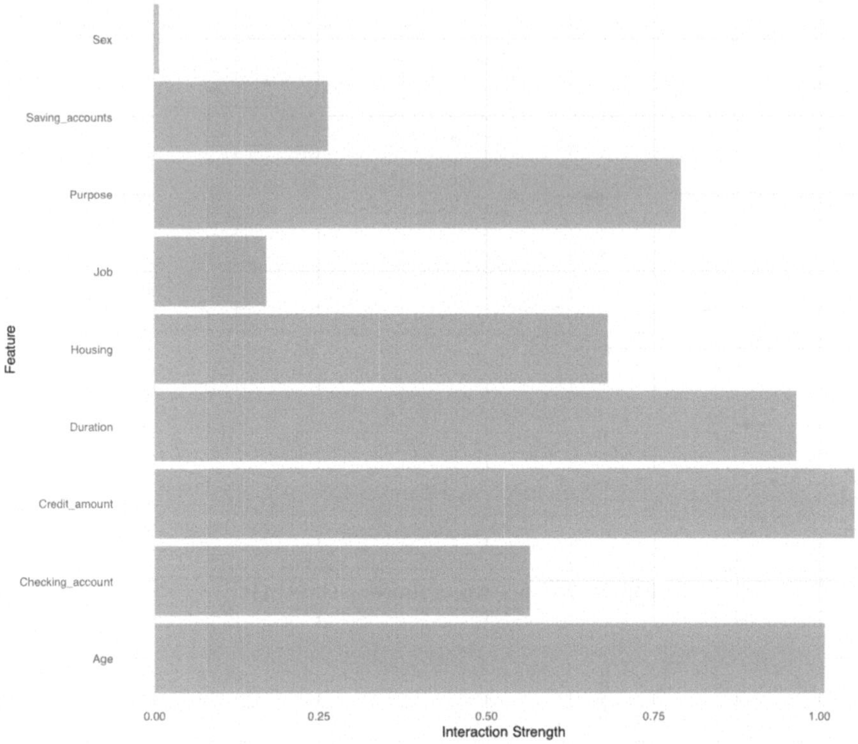

Fig. 5. SHAP-based interaction strength for each feature. Values near 0 indicate independent effects; values near 1 reflect strong interactions with other features.

the distribution observed in the training data. Each sample was weighted by its similarity to the instance using an exponential kernel. The surrogate model was then fitted to approximate the local prediction surface of the black-box model. To improve local fidelity, continuous variables were discretized into equal-width bins, based on the range observed in the training data. For example, intervals such as $33 < \text{Duration} \leq 46.5$ or $4171 < \text{Credit\_amount} \leq 8066$ represent the binning used by LIME to define the local feature space. The resulting feature weights indicate how much each feature, in its discretized form, contributed to the model's decision for a specific case. We selected the most influential features per instance (typically five to six) for interpretation. The model was explicitly treated as a classifier, allowing probabilities to be used in the explanation process. Default parameters from the `lime` package were used unless otherwise stated. LIME explanation plots (Fig. 6) show the weight of each feature in supporting or opposing the model's prediction. Features with positive weights push the prediction toward the predicted class, while negative weights indicate opposing influence. Notably, both cases displayed in the figure correspond to different predicted class labels (assigned based on the 0.5 probability threshold), even though the predicted probabilities themselves (0.73 and 0.52) are relatively close. The label "opposite classifications" refers to the predicted, not observed, class labels. The plots also report the Explanation Fit, which is the R^2 of the local surrogate model with respect to the black-box model's predicted probabilities in the perturbed neighborhood. While values of 0.1 or 0.2 may seem low, this is common when explaining complex models such as random forests. These values still offer meaningful insights, particularly when interpreted in context with known risk factors and domain knowledge. Across multiple cases, long loan durations (e.g., $33 < \text{Duration} \leq 46.5$ months) consistently drive predictions toward bad credit, reinforcing their role as a dominant risk factor. Similarly, renting or living for free increases the probability of bad credit, suggesting that homeownership may serve as a signal of financial stability. Low savings account balances frequently emerge as a negative factor, confirming their role in financial vulnerability. A key insight from LIME is that feature effects vary based on the applicant's overall profile. For example, older age ($47 < \text{Age} \leq 61$) tends to support good credit classification, while younger applicants may require additional stabilizing factors (e.g., lower credit amounts, stable housing) to reach a similar outcome. This demonstrates that the model adapts its decision-making based on contextual dependencies, rather than applying uniform rules across all individuals.

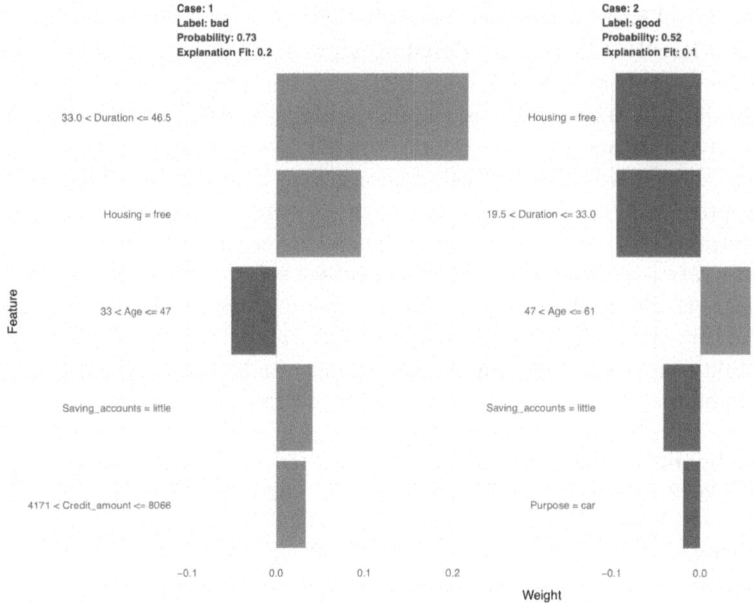

Fig. 6. LIME interpretation for two opposite classification.

4 Concluding Remarks

The combined application of E2Tree, SHAP, and LIME provides a comprehensive explainability framework for credit risk modeling, balancing global model structure with local instance-level explanations. E2Tree extracts and simplifies the decision structure of RF, revealing the hierarchical feature that drive predictions. SHAP confirms the dominant influence of loan duration and credit amount, while also identifying complex feature interactions that shape risk perception. LIME provides local insights, demonstrating how specific applicant characteristics affect individual predictions. Together, these methods enhance trust and transparency, allowing for interpretable decision-making without compromising predictive accuracy. The results consistently highlight loan duration, credit amount, housing status, and financial liquidity as key factors influencing creditworthiness. The hierarchical nature of E2Tree further reveals how different features, reinforcing that credit risk assessment is not determined by isolated variables but by a complex combination of financial behavior, loan conditions, and demographic factors. The ability to decompose these global and instance-level insights ensures that risk evaluation remains interpretable and actionable. Despite their strengths, each method has limitations. SHAP is computationally expensive and assumes feature independence potentially leading to misleading attributions in correlated datasets. LIME suffers from instability, as slight input variations can produce different explanations, and its reliance on random perturbations may generate unrealistic data points. Additionally, LIME lacks global insights, making it unsuitable for understanding the overall model behavior. E2Tree, while improving interpretability, is computationally demanding, as the dissimilarity matrix calculation can hinder scalability. Furthermore, its reliance on RF restricts flexibility in applications where alternative ensemble techniques, such as Gradient Boosting Machines, might be preferred. In datasets with sparse features or weak interactions, its added interpretability may offer limited benefits over simpler approaches. Overall, the integration of these methods offers a robust solution to RF interpretability, addressing both global structure and local decision processes. By leveraging their complementary strengths, this approach enhances model transparency, trust, and accountability, ensuring that complex ML models can be both accurate and interpretable in high-stakes applications such as credit risk assessment.

Acknowledgments. This research has been financed by the following research projects: PRIN-2022 "SCIK-HEALTH" (Project Code: 2022825Y5E; CUP: E53D2300611 006); PRIN-2022 PNRR "The value of scientific production for patient care in Academic Health Science Centres" (Project Code: P2022RF38Y; CUP: E53D23016650001).

References

1. Adadi, A., Berrada, M.: Peeking inside the black-box: a survey on explainable artificial intelligence (XAI). IEEE Access **6**, 52138–52160 (2018). https://doi.org/10.1109/ACCESS.2018.2870052

2. Aria, M., Cuccurullo, C., Gnasso, A.: A comparison among interpretative proposals for Random Forests. Mach. Learn. Appl. **6**, 100094 (2021). https://doi.org/10.1016/j.mlwa.2021.100094
3. Aria, M., Gnasso, A., Iorio, C., Pandolfo, G.: Explainable ensemble trees. Comput. Stat. **39**(1), 3–19 (2024). https://doi.org/10.1007/s00180-023-01387-4
4. Breiman, L.: Random forests. Mach. Learn. **45**, 5–32 (2001). https://doi.org/10.1023/A:1010933404324
5. Doshi-Velez, F., Kim, B.: Towards a rigorous science of interpretable machine learning. arXiv preprint arXiv:1702.08608 (2017)
6. Genuer, R., Poggi, J.-M., Tuleau-Malot, C.: Variable selection using Random Forests. Pattern Recogn. Lett. **31**(14), 2225–2236 (2010). https://doi.org/10.1016/j.patrec.2010.03.014
7. Guyon, I.: A scaling law for the validation-set training-set size ratio. AT&T Bell Laboratories **1**(11) (1997)
8. The Elements of Statistical Learning. SSS, Springer, New York (2009). https://doi.org/10.1007/978-0-387-84858-7
9. Linardatos, P., Papastefanopoulos, V., Kotsiantis, S.: Explainable AI: a review of machine learning interpretability methods. Entropy **23**(1), 18 (2020). https://doi.org/10.3390/e23010018
10. Lipton, P.: Contrastive explanation. Roy. Inst. Philos. Suppl. **27**, 247–266 (1990). https://doi.org/10.1017/S1358246100005130
11. Lipton, Z.C.: The mythos of model interpretability: in machine learning, the concept of interpretability is both important and slippery. Queue **16**(3), 31–57 (2018). https://doi.org/10.1145/3236386.3241340
12. Lundberg, S.M., et al.: From local explanations to global understanding with explainable AI for trees. Nat. Mach. Intell. **2**(1), 56–67 (2020). https://doi.org/10.1038/s42256-019-0138-9
13. Mantel, N.: The detection of disease clustering and a generalized regression approach. Can. Res. **27**, 209–220 (1967). https://doi.org/10.1007/978-1-4612-4380-9_22
14. Molnar, C.: Interpretable Machine Learning. Lulu.com (2020)
15. Molnar, C., Casalicchio, G., Bischl, B.: Interpretable machine learning – a brief history, state-of-the-art and challenges. In: Joint European Conference on Machine Learning and Knowledge Discovery in Databases, pp. 417–431. Springer (2020)
16. Ribeiro, M.T., Singh, S., Guestrin, C.: "Why should I trust you?" Explaining the predictions of any classifier. In: Proceedings of the 22nd ACM SIGKDD International Conference on Knowledge Discovery and Data Mining, pp. 1135–1144 (2016)
17. Rudin, C.: Stop explaining black box machine learning models for high stakes decisions and use interpretable models instead. Nat. Mach. Intell. **1**(5), 206–215 (2019). https://doi.org/10.1038/s42256-019-0048-x
18. Vishwarupe, V., Joshi, P.M., Mathias, N., Maheshwari, S., Mhaisalkar, S., Pawar, V.: Explainable AI and interpretable machine learning: a case study in perspective. Procedia Comput. Sci. **204**, 869–876 (2022). https://doi.org/10.1016/j.procs.2022.08.105

From Prediction to Explanation: Interpreting Risk Factors in Health Survey Analytics

Agostino Gnasso[1(✉)], Massimo Aria[1], and Roberta Siciliano[2]

[1] Department of Economics and Statistics, University of Naples Federico II, Naples, Italy
agostino.gnasso@unina.it
[2] Department of Electrical Engineering and Information Technologies, University of Naples Federico II, Naples, Italy

Abstract. Machine Learning methods have gained significant attention for their ability to deliver high predictive accuracy and reveal complex, non-obvious patterns in data. Among these, Random Forest stands out as a popular ensemble technique, appreciated for its robustness and practical applicability, particularly in contexts where minimizing prediction errors is crucial. Nevertheless, the inherent complexity and lack of transparency in such models often limit their interpretability and can hinder user trust. In this study, we employ the Random Forest algorithm to analyze data from the European Health Interview Survey (EHIS), conducted by the Italian National Institute of Statistics (ISTAT), with the aim of identifying key risk factors associated with depression in Italy. To improve the interpretability of the model's output, we apply the Explainable Ensemble Trees approach, which allows for a deeper understanding of the internal decision-making mechanisms of Random Forest. The aim is to shed light on the determinants of mental health conditions, providing valuable insights to better understand the factors contributing to the risk of depression.

Keywords: E2Tree · Machine Learning · Classification · Depression · EHIS

1 Introduction

Machine Learning (ML), a branch of Artificial Intelligence, enables the identification of complex patterns in large datasets with minimal human input [16]. Supervised learning methods, particularly ensemble approaches, have shown strong predictive performance by combining multiple weak learners into a robust model [10]. Among these, Random Forest (RF) is widely used due to its resistance to overfitting and high accuracy [6]. However, RF models are often considered "black boxes" due to their low interpretability, which limits their use in sensitive

A. D'Ambrosio et al. (Eds.): CLADAG-VOC 2025, SCDAKO, pp. 238–248, 2025.
https://doi.org/10.1007/978-3-032-03042-9_21

domains like healthcare. To address this, the field of Explainable Artificial Intelligence has developed tools to make ML models more transparent and trustworthy [12]. Interpretability focuses on understanding relationships between variables, while explainability clarifies how predictions are made [4]. In this study, we apply RF to data from the European Health Interview Survey (EHIS), conducted by the Italian National Institute of Statistics (ISTAT), to identify individual risk factors for depression in Italy. To improve interpretability, we use the recently proposed explainable ensemble-based methodology, called Explainable Ensemble Trees (E2Tree) [5], a method that reconstructs RF's decision logic into a simplified tree structure. The core idea of E2Tree is to identify the most informative classification or regression paths by analyzing the co-occurrence of observations in terminal nodes across the ensemble. The algorithm starts from a dissimilarity matrix, which quantifies how often pairs of observations fall into the same leaf across individual trees. Based on this matrix, E2Tree builds a new tree by applying classical tree-growing rules and additional stopping criteria aimed at preventing non-informative splits and overfitting. The result is an intuitive and interpretable structure that reflects the key decision patterns of the original model. Rather than focusing solely on algorithmic performance, our objective is to shed light on the underlying determinants of mental health conditions, offering insights that may support targeted public health interventions and inform evidence-based strategies.

2 Context

Depression is widely recognized in the literature as one of the most prevalent and disabling mental health conditions worldwide, with significant social and economic consequences. The World Health Organization (2023) identifies depression as a leading cause of disability, affecting over 280 million people globally. This growing burden has stimulated extensive research aimed at identifying the key risk factors and developing effective tools for early detection and intervention. Numerous studies have emphasized the multidimensional nature of depression, shaped by interactions between biological, psychological, and social determinants. Among the most frequently cited risk factors are chronic physical illness, female gender, advanced age, low educational and income levels, poor housing conditions, and lack of social support [13]. Social isolation and family structure have also been linked to increased vulnerability to depressive symptoms, highlighting the relevance of socio-demographic context in mental health research [19]. Gender disparities in depression prevalence are well-documented, with women consistently showing higher rates than men. These differences are attributed to both biological factors and unequal exposure to stressors such as caregiving responsibilities, economic disadvantage, and gender-based violence [2]. A further critical issue in the literature is the underdiagnosis and undertreatment of depression. A significant proportion of individuals affected by depressive symptoms do not receive clinical attention, often due to stigma, lack of awareness, or limited access to care services. This gap underscores the need for data-driven early warning systems capable of identifying individuals at risk even before clinical manifestation [11]. In recent years, machine learning (ML) techniques have gained attention in mental health research for their potential to detect hidden patterns in large health datasets. However, the use of black-box models in this domain raises concerns about transparency and accountability, particularly in clinical and policy settings. To address this, several explainability methods, such as SHAP [14], LIME [18], and rule-based models, have been proposed to improve model interpretability and guide actionable interventions. This paper contributes to this body of work by applying an interpretable ensemble-based methodology (E2Tree) [5] to identify relevant depression predictors in the Italian population, bridging the gap between predictive performance and interpretability in mental health analytics.

3 E2Tree

The Explainable Ensemble Trees (E2Tree) methodology [5] is designed to reconstruct an interpretable and explainable tree for RF by leveraging the internal structure of its ensemble model. The method operates entirely on the training set, from which it builds a proximity matrix $O \in \mathbb{R}^{n \times n}$, where each entry O_{ij} measures how frequently two observations u_i and u_j fall into a terminal node across the ensemble, weighted by the node's classification accuracy. From this proximity structure, a dissimilarity matrix $D = 1 - O$ is computed. This matrix guides the construction of a single explanatory tree that captures the key relational patterns identified by the ensemble. At each node, E2Tree selects the splitting variable and threshold that minimize within-node dissimilarity, defined as the average pairwise distance among observations grouped by the split. In addition to standard stopping rules (e.g., node size and minimum impurity decrease), E2Tree introduces an accuracy-based criterion: the tree-growing process halts when child nodes reach a high classification purity (typically above 95%). Unlike RF, which builds a large collection of randomized trees and aggregates their predictions, E2Tree provides a single, human-interpretable tree that reflects the core decision patterns learned by the ensemble. It is important to note that E2Tree is not retrained or evaluated on the test set; rather, it is a post-hoc tool aimed at explaining the RF's logic, not improving predictive performance.

4 Case Study

ML models are increasingly adopted as part of early warning systems (EWS), aimed at anticipating and preventing risks across various policy domains, including healthcare, education, public safety, and finance [3]. In these contexts, interpretability plays a key role, allowing stakeholders to understand the drivers of risk and to implement timely interventions. In line with this approach, the present section introduces the dataset used to model depression risk and presents the results obtained from the RF algorithm, further enriched through the application of the E2Tree method for model explainability.

4.1 Data Overview

The dataset employed in this study originates from the European Health Interview Survey (EHIS), a harmonized data collection initiative implemented across all EU Member States. The primary objective of the survey is to enable cross-country comparisons of health conditions, behavioral determinants, and healthcare service usage. These domains are examined within a broader socio-demographic framework, offering valuable insights for national health planning and European policy development. In Italy, the EHIS sample includes approximately 30,000 households from around 840 municipalities, representing a wide range of geographic and demographic contexts. Households are defined in a de facto sense, encompassing cohabiting individuals bound by familial or affective relationships. The sampling strategy follows a two-stage stratified design, with municipalities serving as primary sampling units and households as secondary units. Households are selected through random sampling from population registers (for small municipalities with fewer than 1,000 inhabitants) or from the household list used in the 2018 Permanent Census (for larger municipalities), ensuring the statistical representativeness of the sample. Data collection combines two methods: face-to-face interviews using paper-based questionnaires (PAPI) and self-administered questionnaires filled in by respondents. The most recent public release of the Italian EHIS dataset, provided by Istat, was used in this study to predict the likelihood of depression. The outcome variable reflects whether an individual reported experiencing depression in the 12 months preceding the interview. Two main challenges characterize this application: (i) the imbalance in the distribution of the target variable, with depression being a relatively rare condition in the general population, and (ii) the difficulty in accurately capturing depressive symptoms using available survey-based predictors. To address these issues, a stratified sample was drawn from the original dataset, including all respondents classified as depressed, along with a random sample of 7,000 non-depressed individuals. It is important to note that the resulting dataset does not preserve the original prevalence of depression in the general population. This sampling strategy, retaining all depressed individuals and a random subset of 7000 non-depressed ones, was adopted to address class imbalance and ensure adequate representation of the minority class. While this choice improves model training, it introduces a shift in base rate that must be considered when interpreting results. In particular, posterior probabilities estimated by the model are conditional on the sample distribution and cannot be directly interpreted as population-level probabilities. For future applications to new individuals drawn from the general population, recalibration techniques or prior probability corrections should be applied to ensure valid classification decisions. The final analytical dataset comprises 9,463 individuals, of whom 26% reported symptoms of depression. The model relies on 22 explanatory variables, covering a broad spectrum of socio-demographic, educational, household, and health-related characteristics (see Table 2). To capture the impact of comorbidities, a variable indicating the presence of at least one chronic illness in the past year was included. Additional covariates reflect individual perceptions of economic well-being, housing conditions, and social support, as well as psychological indicators such as fatigue, apathy, difficulty concentrating, eating disorders, and low self-esteem. Ordinal variables were numerically coded to facilitate model processing and improve result visualization. Observations with missing values were excluded from the analysis. An important methodological note concerns the use of proxy respondents. According to Istat's data collection guidelines [9], in certain situations, particularly for sensitive topics such as emotional well-being, responses may be provided by another household member if the selected respondent is absent or unwilling to answer. Although such proxy data may slightly affect measurement reliability, they are often used to reduce non-response bias and minimize data collection costs. In the context of mental health, where self-reporting may be compromised by the nature of the condition itself, proxy information can still provide meaningful insights. For this reason, this study includes data collected from proxy interviews, assuming that close household members are reasonably aware of the respondent's psychological condition (Table 1).

Table 1. Description and encoding of variables.

Variable Name	Variable Encoding/Range	Variable Type
SEX	Female; Male	Categorical
AGE	1 = 15–17; 2 = 18–24; 3 = 25–34; 4 = 35–44; 5 = 45–49; 6 = 50–54; 7 = 55–59; 8 = 60–64; 9 = 65–69; 10 = 70–74; 11 = $\geq$ 75	Numerical
GEO	C = Center; I = Island; N-E = Northeast; N-O = Northwest; S = South	Categorical
JOB	Worker; Unemployed; Retired; Other	Categorical
MARITAL.STATUS	Single; Married; Widowed; Divorced	Categorical
INCOME	1 = I; 2 = II; 3 = III; 4 = IV; 5 = V	Numerical
EDU	1 = Elementary; 2 = Middle School; 3 = Diploma; 4 = Post-secondary and Tertiary	Numerical
FAM.MEMBERS	1–7	Numerical
FAMILY.TYPE	Single Person; Single Parent; Couple with Children; Couple without Children; Other	Categorical
APATHY	1–4	Numerical
SLEEPING.DIS	1–4	Numerical
TIRED	1–4	Numerical
EATING.DIS	1–4	Numerical
SELF.EST.DIS	1–4	Numerical
ATTENTION.DIS	1–4	Numerical
SPORT	0–7	Numerical
BODY.MASS	1 = Underweight; 2 = Normal Weight; 3 = Overweight; 4 = Obese	Numerical
DISEASE	Yes; No	Categorical
CARE	1–5	Numerical
SMALL.HOUSE	Yes; No	Categorical
BAD.HOUSE	Yes; No	Categorical
ECO.STATUS	1 = Very Insufficient; 2 = Limited; 3 = Good; 4 = Excellent	Numerical

Posterior Probability Correction Under Prior Shift

As mentioned above, the training dataset was constructed through a stratified sampling design in which all depressed individuals were retained and a random subset of non-depressed individuals was selected. While this strategy addresses the issue of class imbalance during model training, it alters the marginal distribution of the target variable in the sample, introducing a prior probability shift that affects the interpretation of posterior probabilities. In this context, the probabilities estimated by the model, denoted $\hat{P}(Y = 1|X)$, are conditional on the modified sample distribution and cannot be interpreted as valid population-level probabilities. To enable meaningful application of the model to new instances drawn from the general population, it is necessary to recalibrate the predictions to reflect the true base rate [7]. A straightforward correction can be obtained through a Bayes-based adjustment formula, as follows:

$$\tilde{P}(Y = 1|X) = \frac{\hat{P}(Y = 1|X) \cdot \pi}{\hat{P}(Y = 1|X) \cdot \pi + (1 - \hat{P}(Y = 1|X)) \cdot (1 - \pi)}$$

where $\pi = \frac{P_{\text{pop}}(Y=1)}{P_{\text{train}}(Y=1)}$ denotes the ratio between the true prevalence of depression in the population and its prevalence in the training set. This correction rescales the posterior estimates to match the real-world class distribution and is particularly useful when applying the model for early-warning screening or clinical decision support.

4.2 Analysis and Results

The entire analysis was conducted within the R environment, using the randomForest package for model construction and the e2tree package for post-hoc model interpretation. The RF algorithm was trained following a standard train-test split approach: 70% of the dataset was allocated for training, while the remaining 30% was reserved for testing [8]. The RF model was configured with 500 trees, and the number of variables randomly selected at each node split was set to $\sqrt{p} = 5$, where p denotes the total number of predictors. The model's classification performance was evaluated using multiple metrics that are particularly suitable for imbalanced data scenarios. In addition to the confusion matrix (Table 3), we employed Balanced Accuracy, F1-Score, and Cohen's Kappa, which offer more robust insights than simple accuracy when class distributions are skewed [1]. As shown in Table 3, the Balanced Accuracy, the average between sensitivity (true

Table 2. Confusion matrix.

Observed	Predicted	
	Yes	No
Yes	472	265
No	187	1915

Table 3. Evaluation metrics.

Metrics	Value
Accuracy	0.841
Sensitivity	0.640
Specificity	0.911
Precision	0.716
Balanced Accuracy	**0.776**
F1-Score	**0.676**
Kappa	**0.571**

positive rate for depressed individuals) and specificity (true negative rate for non-depressed individuals), reaches a value of 0.776, indicating a good compromise between correctly identifying both classes. The F1-Score, reflecting the harmonic mean of precision and sensitivity, stands at 0.676, highlighting a moderate level of model precision. Lastly, Cohen's Kappa, which adjusts for agreement by chance, yields 0.571, suggesting a fair level of agreement between predicted and observed outcomes. To enhance the interpretability of the RF model, we employed E2Tree, which reconstructs a simplified tree-based representation that captures the decision logic embedded in the ensemble. The process begins with the creation of the co-occurrence matrix, which records how frequently pairs of observations co-occur in the same terminal nodes across the RF's decision trees. A heatmap of this co-occurrence matrix (Fig. 1a) visually represents these co-occurrence patterns, offering insights into the proximity structure generated by the RF model. The core output of E2Tree is a single, interpretable decision tree (Fig. 2), which summarizes the RF's internal logic in an accessible format. While not intended for direct prediction tasks, this tree serves as a tool for interpreting how the RF model generates its classifications. Each node in the E2Tree reflects a decision rule based on a specific variable, while terminal nodes provide the classification outcome along with the associated probabilities. This structure allows users to easily trace the reasoning path behind predictions, revealing the most influential variables and their hierarchical structure. In Fig. 1 a visual comparison between the heatmaps of the two proximity matrices (the one obtained from RF and the one estimated from E2Tree) demonstrates that E2Tree effectively preserves the relational structure encoded by the original RF, while significantly improving interpretability and transparency. While Fig. 1a represents the proximity matrix obtained from the RF, which appears dense and affected by background noise due to the aggregation of co-occurrence patterns from hundreds of trees, Fig. 1b shows the matrix estimated by E2Tree. This matrix is much cleaner and more structured, as E2Tree reconstructs a single tree that captures the dominant co-occurrence relationships without averaging across multiple weak learners. Despite their visual differences, the two matrices reflect similar proximity patterns. To statistically validate the similarity between the decision structures of RF and E2Tree, a Mantel test [15] was conducted to compare the pairwise sim-

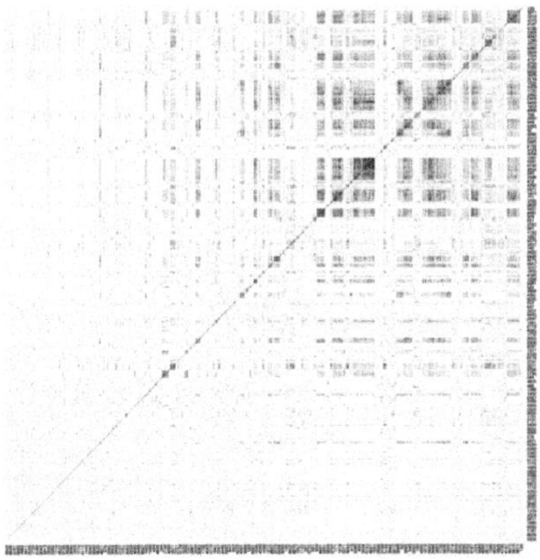

(a)

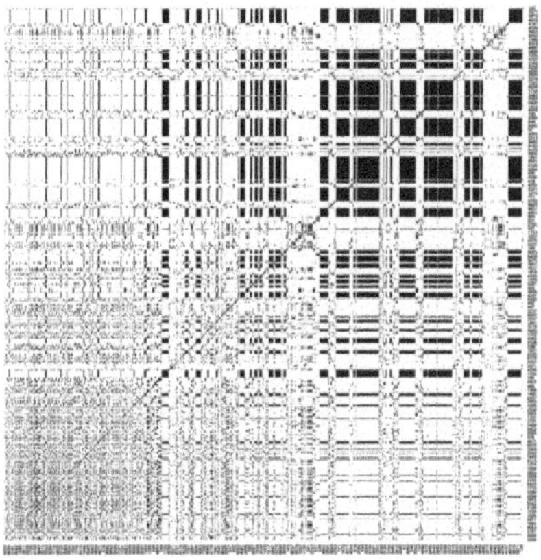

(b)

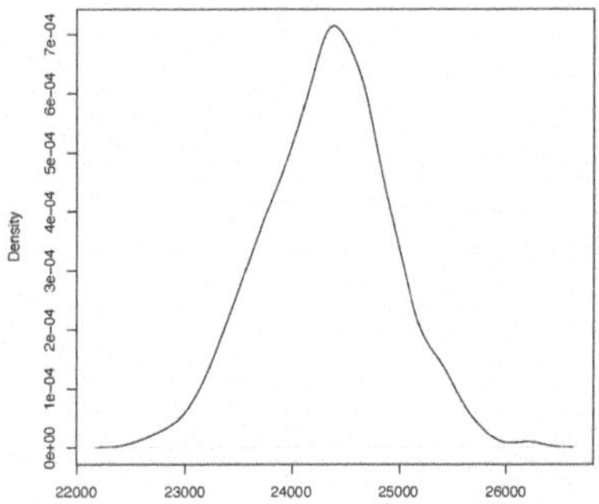

(c)

Fig. 1. Heatmap of the RF co-occurrence matrix (sub-plot a), heatmap of the estimated co-occurrence matrix by E2Tree (sub-plot b) and density plot of Mantel test permutations (sub-plot c).

ilarity matrices derived from both models. Figure 1c illustrates the permutation distribution of the Mantel test statistic under the null hypothesis that no association exists between the RF-based and E2Tree-based matrices. The Mantel test computes the correlation between the lower triangles of the two matrices and evaluates its significance through random permutations. Specifically, the rows and columns of one matrix are randomly permuted 999 times to generate a null distribution of correlation values. The observed Z-statistic (118,316) lies far in the right tail of this distribution, with a corresponding p-value of 0.001, indicating a strong and statistically significant structural correspondence between the original RF ensemble and its E2Tree approximation. The explanation tree generated by E2Tree (Fig. 2) provides a comprehensive overview of the most relevant decision paths used by the RF model. The initial split distinguishes individuals

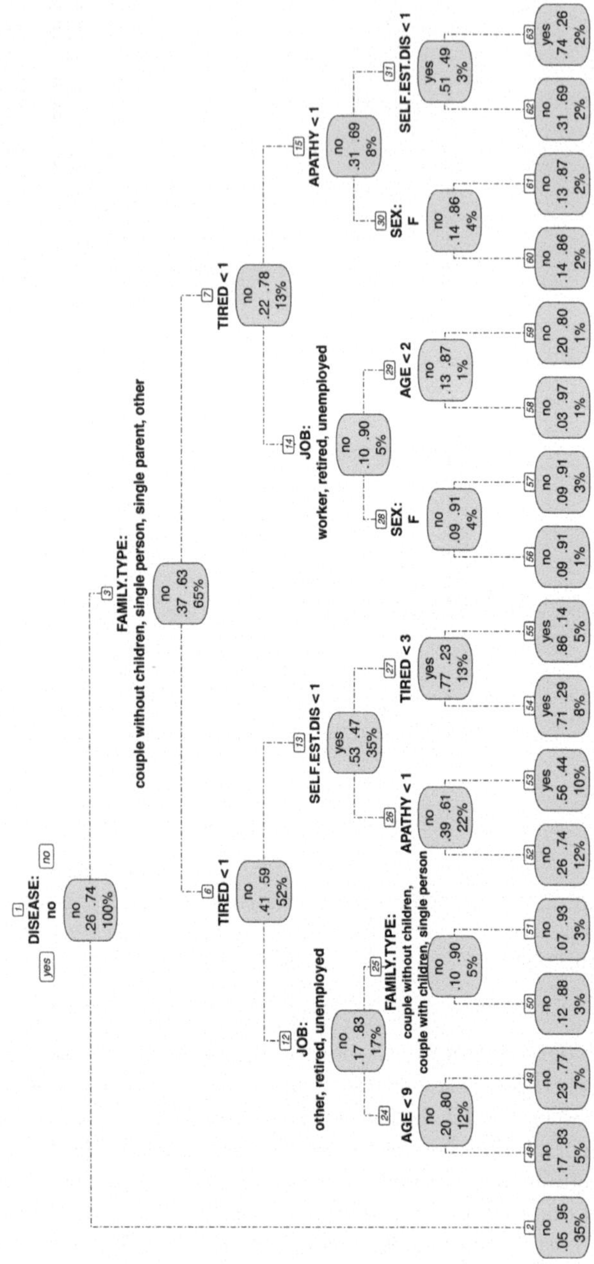

Fig. 2. E2Tree output: each node shows the distribution of observations across the different classes of the target variable, along with the percentage of the entire dataset represented by that node.

based on the presence of chronic illnesses other than depression. Those without long-term health conditions are directly classified as non-depressed. For individuals with chronic illnesses, the model explores additional splits, considering a combination of household structure and psychological symptoms. The tree then branches into two main subgroups based on family composition. One branch corresponds to "lonely families", including single individuals, couples without children, and single-parent households, while the other includes families with children under the age of 25. Among individuals in families with children, the presence of depression is strongly associated with symptoms such as fatigue, apathy, and low self-esteem. For those in other family types, even in the absence of low self-esteem, a lack of interest in daily activities is sufficient for the model to classify an individual as depressed. The majority of terminal nodes classify individuals as non-depressed, reflecting the natural class imbalance in the dataset. However, the paths leading to depressed classifications highlight consistent patterns: individuals with chronic illnesses, belonging to more isolated household types, and showing signs of emotional and motivational fatigue are more likely to be classified as depressed. This interpretative structure confirms that the RF model's predictive power is largely driven by a combination of clinical and psychosocial factors, such as chronic illness, apathy, tiredness, and diminished self-worth. E2Tree thus offers not only an interpretable summary of model behavior but also a practical tool for identifying actionable factors that may inform mental health policy and targeted interventions.

5 Conclusion

This study highlights key findings on the prediction of depression in the Italian population using data from the European Health Interview Survey (EHIS). The analysis confirms the strong predictive relevance of chronic illnesses, fatigue, low self-esteem, and apathy, factors that consistently emerge as central drivers of depressive outcomes. Additionally, social and demographic conditions such as family structure, employment status, and economic hardship contribute significantly to individual vulnerability. These results emphasize the potential of interpretable predictive models to uncover non-trivial relationships among clinical, psychological, and socio-demographic variables. While not designed for real-time diagnostic use, the findings contribute to a better understanding of depression risk profiles and may support the development of hypothesis-driven early warning strategies in future studies.

Acknowledgments. This research has been financed by the following research projects: PRIN-2022 "SCIK-HEALTH" (Project Code: 2022825Y5E; CUP: E53D23 00611006); PRIN-2022 PNRR "The value of scientific production for patient care in Academic Health Science Centres" (Project Code: P2022RF38Y; CUP: E53D23016650001). Roberta Siciliano is funded by "Fit for Medical Robotics" FIT4 MEDROB PNC0000007.

References

1. Akosa, J.: Predictive accuracy: a misleading performance measure for highly imbalanced data. In: Proceedings of the SAS Global Forum, vol. 12, pp. 1–4 (2017)
2. Albert, P.R.: Why is depression more prevalent in women? J. Psychiatry Neurosci. **40**(4), 219–221 (2015)
3. Amarasinghe, K., Rodolfa, K.T., Lamba, H., Ghani, R.: Explainable machine learning for public policy: use cases, gaps, and research directions. Data Policy **5**, e5 (2023). https://doi.org/10.1017/dap.2023.2
4. Aria, M., Cuccurullo, C., Gnasso, A.: A comparison among interpretative proposals for Random Forests. Mach. Learn. Appl. **6**, 100094 (2021). https://doi.org/10.1016/j.mlwa.2021.100094
5. Aria, M., Gnasso, A., Iorio, C., Pandolfo, G.: Explainable ensemble trees. Comput. Statistics **39**, 3–19 (2024). https://doi.org/10.1007/s00180-022-01312-6
6. Breiman, L.: Random forests. Mach. Learn. **45**, 5–32 (2001). https://doi.org/10.1023/A:1010933404324
7. Elkan, C.: The foundations of cost-sensitive learning. In: Proceedings of the 17th International Joint Conference on Artificial Intelligence (IJCAI), vol. 1, pp. 973–978. Lawrence Erlbaum Associates Ltd (2001)
8. Guyon, I.: A scaling law for the validation-set training-set size ratio. AT&T Bell Labs. **1**(11) (1997)
9. ISTAT – Istituto Nazionale di Statistica: Manuale di Tecniche di Indagine 6 – Il sistema di controllo della qualità dei dati. ISTAT, Roma (1989)
10. James, G., Witten, D., Hastie, T., Tibshirani, R.: An Introduction to Statistical Learning. STS, Springer, New York (2021). https://doi.org/10.1007/978-1-0716-1418-1_9
11. Kohn, R., Saxena, S., Levav, I., Saraceno, B.: The treatment gap in mental health care. Bull. World Health Organ. **82**(11), 858–866 (2004)
12. Linardatos, P., Papastefanopoulos, V., Kotsiantis, S.: Explainable AI: a review of machine learning interpretability methods. Entropy **23**(1), 18 (2020). https://doi.org/10.3390/e23010018
13. Lorant, V., Deliège, D., Eaton, W., Robert, A., Philippot, P., Ansseau, M.: Socioeconomic inequalities in depression: a meta-analysis. Am. J. Epidemiol. **157**(2), 98–112 (2003). https://doi.org/10.1093/aje/kwf182
14. Lundberg, S.M., Erion, G.G., Lee, S.-I.: Consistent individualized feature attribution for tree ensembles. arXiv preprint arXiv:1802.03888 (2018)
15. Mantel, N.: The detection of disease clustering and a generalized regression approach. Cancer Res. **27**(2_Part_1), 209–220 (1967). https://doi.org/10.1007/978-1-4612-4380-9_22
16. Mitchell, T.M.: Machine Learning. McGraw-Hill, New York (1997)
17. Moussavi, S., Chatterji, S., Verdes, E., Tandon, A., Patel, V., Ustun, B.: Depression, chronic diseases, and decrements in health: results from the World Health Surveys. The Lancet **370**(9590), 851–858 (2007). https://doi.org/10.1016/S0140-6736(07)61415-9
18. Ribeiro, M.T., Singh, S., Guestrin, C.: "Why should I trust you?" explaining the predictions of any classifier. In: Proceedings of the 22nd ACM SIGKDD International Conference on Knowledge Discovery and Data Mining, pp. 1135–1144 (2016)
19. Umberson, D., Karas Montez, J.: Social relationships and health: a flashpoint for health policy. J. Health Soc. Behav. **51**(1_suppl), S54–S66 (2010). https://doi.org/10.1177/0022146510383501

Conditional von Mises Bayesian Networks

Anna Gottard(✉) and Agnese Panzera

Department of Statistics, Computer Science, Applications University of Florence, Florence, Italy
{anna.gottard,agnese.panzera}@unifi.it

Abstract. Directed acyclic graphical models, or Bayesian networks, use a directed acyclic graph to represent the conditional independence relationships between a set of random variables. We introduce a novel class of Bayesian networks specifically designed for circular or angular variables, utilizing the properties of the von Mises distribution. We illustrate our proposal by applying these models to study the conditional independencies within a sequence of angles that characterize the structure of a glycopeptide.

Keywords: Circular data · Conditional independence · Dihedral angles

1 Introduction

Graphical models are a powerful probabilistic tool for analyzing conditional independence among random variables. This class of multivariate models represents the conditional independence structure of p variables using a graph $\mathcal{G} = (V, E)$, where $V = \{1, \ldots, p\}$ denotes the set of nodes and $E \subset V \times V$ represents the set of edges. Each node in V corresponds to a random variable, and the absence of an edge in E signifies a conditional independence relationship between the corresponding variables. The joint probability distribution associated with the model factorizes according to the graph structure. A graphical model is called *undirected graph model* or *Markov network* whenever $\mathcal{G}$ is undirected, meaning that $E \subseteq \{\{i, j\} : i, j \in V, i \neq j\}$ is a set of unordered pairs. This model treats all variables on equal footing. On the other hand, a graphical model is referred to as a *directed acyclic graphical model* or *Bayesian network* when $\mathcal{G}$ is a directed acyclic graph with only directed edges, i.e. $E \subseteq \{(i, j) : i, j \in V, i \neq j\}$ is a set of ordered pairs, with no cycles allowed. Bayesian networks assume a topological ordering of the variables and enable a convenient factorization of the joint distribution into univariate conditional distributions. Despite recent advancements, graphical models remain under-explored for dealing with circular variables.

Circular data can be viewed as points on the unit circle $\mathbb{T}$ and measured as angles which, in radians, range from $-\pi$ to π or, equivalently, from 0 to 2π. Examples of circular data include the flight directions of birds, wind directions,

A. D'Ambrosio et al. (Eds.): CLADAG-VOC 2025, SCDAKO, pp. 249–259, 2025.
https://doi.org/10.1007/978-3-032-03042-9_22

marine current directions, and angles in the polypeptide chains of protein structures. The key characteristic of circular data is its periodicity, which requires ad hoc statistical methods. *Circular statistics* encompasses statistical methods for circular data, presented, for example, by [7,11,14] and [12].

In some applications, data consists of p angles, which can be viewed as points on the surface of a p-dimensional torus, $\mathbb{T}^p$, formed by the p-fold Cartesian product of unit circles. A common example of toroidal data is found in three-dimensional protein structures, which are represented by a sequence of angles. In such applications, it can be of interest to describe the conditional independence structure of the set of angles, which can be effectively done using graphical models.

The first notable contribution on graphical models for circular variables, to the best of our knowledge, comes from [1], who explored the sequence and structure of proteins using models which combine Hidden Markov Models, treated as Bayesian networks, with distributions defined on the sphere and the bidimensional torus. Moreover, [16] investigated conditional independence among angles following a $p-$variate von Mises distribution, representing this structure through a factor graph. In addition, [10] consider the bivariate wrapped Cauchy distribution for tree-structured directed acyclic graphs, while [8] introduced the class of *torus graphs*, which are not based on a specific distribution but belong to a full exponential family with pairwise interactions. Recently, [5] introduced some Gaussian-related undirected graphical models for angular variables, focusing on the Wrapped Normal and Inverse Stereographic Normal distributions. The Inverse Stereographic Normal distribution has also been considered by [6] for defining a class of Bayesian networks.

In this work, we introduce the class of Bayesian networks for circular variables, based on the von Mises distribution. Specifically, after highlighting certain aspects of this distribution as an undirected graph model, we leverage the properties of the resulting univariate conditional distributions to define von Mises Bayesian networks. These models are useful when a natural ordering of the angles is known *a priori*, such as when studying the sequence of dihedral angles, where the primary structure of a protein can provide the node ordering.

The paper is organized as follows. In Sect. 2, we review some aspects of the multivariate von Mises distribution, also discussing some of its properties in terms of conditional independence and undirected graphical models. Section 3 introduces the new class of Bayesian networks, with Sect. 3.1 and Sect. 3.2 respectively focusing on inference and simulations for these models. Finally, in Sect. 4, we present a real-data application that studies the conditional independence relationships among dihedral angles in the structure of a glycopeptide. Finally, Sect. 5 collects some concluding remarks.

2 Multivariate von Mises Distribution

A circular random variable has a von Mises distribution with *mean direction* $\mu \in \mathbb{T}$, and *concentration parameter* $\kappa \geq 0$ if its probability density function, at $\theta \in \mathbb{T}$, is

$$f_{\Theta}(\theta) = \frac{1}{2\pi \mathcal{I}_0(\kappa)} \exp\left(\kappa \cos(\theta - \mu)\right),$$

where $\mathcal{I}_0(\kappa)$ is the modified Bessel function of the first kind and order 0.

The multivariate von Mises distribution on $\mathbb{T}^p$, as first introduced in [13], is also known as the multivariate sine distribution. Let $\boldsymbol{\mu} \in \mathbb{T}^p$, $\boldsymbol{\kappa} \in \mathbb{R}^p$, with entries $\kappa_i \geq 0$, and $\boldsymbol{\Lambda} \in \mathsf{Sym}^{p \times p}$, the set of symmetric matrices of order p, with entries $\lambda_{ii} = 0$ for $i \in \{1, \ldots, p\}$. A vector of random angles $\boldsymbol{\Theta} = (\Theta_1 \ldots \Theta_p)'$ has a $p-$variate sine distribution, $\boldsymbol{\Theta} \sim vM_p(\boldsymbol{\mu}, \boldsymbol{\kappa}, \boldsymbol{\Lambda})$, if its probability density function, at $\boldsymbol{\theta} \in \mathbb{T}^p$, is

$$f_{\boldsymbol{\Theta}}(\boldsymbol{\theta}) = \frac{1}{C(\boldsymbol{\kappa}, \boldsymbol{\Lambda})} \exp\left(\boldsymbol{\kappa}' c(\boldsymbol{\theta}, \boldsymbol{\mu}) + \frac{1}{2} s(\boldsymbol{\theta}, \boldsymbol{\mu})' \boldsymbol{\Lambda}\, s(\boldsymbol{\theta}, \boldsymbol{\mu})\right), \tag{1}$$

where $C(\boldsymbol{\kappa}, \boldsymbol{\Lambda})$ is a normalization factor and $c(\boldsymbol{\theta}, \boldsymbol{\mu})$ and $s(\boldsymbol{\theta}, \boldsymbol{\mu})$ are $p-$dimensional vectors having $\cos(\theta_i - \mu_i)$ and $\sin(\theta_i - \mu_i)$ as their respective i-th entry. The factor $C(\boldsymbol{\kappa}, \boldsymbol{\Lambda})$ has no closed form when $p > 2$.

Except for the case of independent random angles, if $\boldsymbol{\Theta} \sim vM_p(\boldsymbol{\mu}, \boldsymbol{\kappa}, \boldsymbol{\Lambda})$, the marginal distributions are not von Mises, while the univariate conditional distributions are von Mises. In particular, the distribution of Θ_j given $\boldsymbol{\Theta}_{\text{rest}}$, with rest $= \{1, \ldots, j-1, j+1, \ldots, p\}$ is

$$f_{\Theta_j | \boldsymbol{\Theta}_{\text{rest}}}(\theta_j \mid \boldsymbol{\theta}_{\text{rest}}) = \frac{\exp(\kappa_{j.\text{rest}} \cos(\theta_j - \mu_{j.\text{rest}}))}{2\pi \mathcal{I}_0(\kappa_{j.\text{rest}})},$$

whose parameters are functions of the parameters of the joint distribution, as follows

$$\mu_{j.\text{rest}} = \mu_j + \arctan\left(\sum_{i \neq j} \lambda_{ij} \sin(\theta_i - \mu_i)/\kappa_j\right), \tag{2}$$

$$\kappa_{j.\text{rest}} = \left\{\kappa_j^2 + \left(\sum_{i \neq j} \lambda_{ij} \sin(\theta_i - \mu_i)\right)^2\right\}^{1/2}. \tag{3}$$

Notice that the parameters $\mu_{j.\text{rest}}$ and $\kappa_{j.\text{rest}}$ depend on the off-diagonal elements λ_{ij} of the matrix $\boldsymbol{\Lambda}$. Specifically, if $\lambda_{ij} = 0$, then the conditional distribution of Θ_j given $\boldsymbol{\Theta}_{\text{rest}}$ does not depend on Θ_i, implying that $\Theta_j \perp\!\!\!\perp \Theta_i \mid \boldsymbol{\Theta}_{\text{rest}\setminus i}$. Indeed, the above conditional independence statement holds if and only if $\lambda_{ij} = 0$. This is because a zero value of λ_{ij} is the only way to ensure that the distribution of Θ_j given the rest does not depend on the value of Θ_i. Analogously, when the j-th row (and column for symmetry) of $\boldsymbol{\Lambda}$ is zero, $\lambda_{ij} = 0$ for each $i \neq j$, then $\Theta_j \perp\!\!\!\perp \boldsymbol{\Theta}_{\text{rest}}$, marginally.

When Θ_i and Θ_j are conditionally independent for all possible pairs i, j, $i \neq j$, then all off-diagonal elements of $\boldsymbol{\Lambda}$ are zero, and, as its diagonal elements are also zero, this would imply that $\boldsymbol{\Lambda}$ is the null matrix. For the *intersection* property of conditional independence [9], as the von Mises distribution is strictly positive, these pairwise conditional independence statements also imply the pairwise marginal independence of Θ_i and Θ_j for all $i \neq j$. In addition, when $\boldsymbol{\Lambda}$ is the null matrix, the multivariate von Mises distribution (1) factorizes into the p-fold product of univariate von Mises distributions. However, the conditional distributions of $\boldsymbol{\Theta}_{\boldsymbol{r}}$, for a subset $\boldsymbol{r}$ of $\{1, \ldots, p\}$ with cardinality greater than 1, are not von Mises. In summary, the sine distribution is not closed either under marginalisation or under conditioning, and λ_{ij} plays the role of the pairwise conditional independence parameter.

There are two main contributions where the sine distribution has been contemplated in the framework of graphical models. [16], thanks to the property of the sine distribution to involve only pairwise interactions, define a *factor graph* representation of this distribution. A factor graph adds to the classical graph an additional set of nodes, called factor nodes, used to represent the factorization of a probability distribution function. For the sine distribution, the factor nodes on the edges between two variable nodes are associated with the non-zero elements of the $\boldsymbol{\Lambda}$ matrix. The resulting graph obeys the pairwise Markov property, but, as the sine distribution is strictly positive, the global Markov property is implied by the pairwise for the Hammersley-Clifford Theorem ([9]). Therefore, the sine distribution is Markov with respect to the graph $\mathcal{G}$.

Undirected graphical models are typically also closed both under marginalisation and conditioning (see, for example, [3,17]), so that one can derive from the graph $\mathcal{G}$ of the joint distribution, the graph representing conditional independences holding in any marginal and conditional margin. However, [18] showed that for a joint distribution that is Markov with respect to a graph $\mathcal{G}$, all the marginal and conditional distributions, whatever their form, are also Markov to $\mathcal{G}$. This is also true for the sine distribution, even if it is not closed under marginalisation and conditioning. Nevertheless, for this lack of closedness, some desirable properties of undirected graphical models, such as decomposability and collapsibility, involving marginal distributions, cannot be used in this class of graphical models. As regards inference for the von Mises undirected graphical models, because of the lack of closed-form for the normalizing constant, $C(\boldsymbol{\kappa}, \boldsymbol{\Lambda})$, [16] utilized the full pseudo-likelihood approach as proposed by [13]. As an alternative, one can adopt a score–matching approach as recently proposed by [15] for directional distributions and by [19] and [20] for general manifolds. This approach does not require the computation of the normalizing constant.

The second contribution to undirected graphical models for the von Mises distribution is due to [8]. Here, the sine model is derived as a restriction of the natural parameter space of the torus graph density. While the torus graph model is a regular full exponential family, closed under marginalization and conditioning, the sine distribution does not belong to the regular exponential family. Therefore, it is not a torus graph model.

3 von Mises Bayesian Networks

In some applications, it could be of interest to consider a directed acyclic graph instead of an undirected graph. In some cases, an inherent variable ordering is available, and undirected graphs are inadequate. Here, we introduce a class of directed acyclic graphical models tailored for circular random variables. We define these models as a sequence of random angles having univariate conditional von Mises distributions. Letting $\mathrm{pa}(j)$ be the set of parents of the node $j \in \{1, \dots, p\}$, that is, the set of vertices such that there is an arrow from them to j, we provide the following definition.

Definition 1. (Conditional von Mises directed acyclic graphical model). *Let $\mathcal{G} = (V, E)$ be a directed acyclic graph with ordered set of vertices $V = (1, \dots, p)$ and set of directed edges $E \subset V \times V$. We say that $\boldsymbol{\Theta} = (\Theta_1 \dots \Theta_p)'$ is a Conditional von Mises directed acyclic graphical model with respect to $\mathcal{G}$ if the probability distribution of $\boldsymbol{\Theta}$ satisfies*

$$f_{\boldsymbol{\Theta}}(\boldsymbol{\theta}) = \prod_{j=1}^{p} f_{\Theta_j|\boldsymbol{\Theta}_{\mathrm{pa}(j)}}(\theta_j|\boldsymbol{\theta}_{\mathrm{pa}(j)}),$$

where

$$f_{\Theta_j|\boldsymbol{\Theta}_{\mathrm{pa}(j)}}(\theta_j|\boldsymbol{\theta}_{\mathrm{pa}(j)}) = \frac{\exp\{\kappa_{j.\mathrm{pa}(j)} \cos(\theta_j - \mu_{j.\mathrm{pa}(j)})\}}{2\pi \mathcal{I}_0(\kappa_{j.\mathrm{pa}(j)})},$$

with

$$\mu_{j.\mathrm{pa}(j)} = \mu_j + \arctan\left(\sum_{i \in \mathrm{pa}(j)} \lambda_{ij} \sin(\theta_i - \mu_i)/\kappa_j \right),$$

$$\kappa_{j.\mathrm{pa}(j)} = \left\{ \kappa_j^2 + \left(\sum_{i \in \mathrm{pa}(j)} \lambda_{ij} \sin(\theta_i - \mu_i) \right)^2 \right\}^{1/2}.$$

For each $i < j$,

$$i \notin \mathrm{pa}(j) \text{ and } (i, j) \notin E \iff \lambda_{ij} = 0.$$

Notice that for a node j with $\mathrm{pa}(j) = \emptyset$, the density $f_{\Theta_j|\boldsymbol{\Theta}_{\mathrm{pa}(j)}}$ is a marginal von Mises with parameters μ_j and κ_j. Assuming univariate conditional von Mises allows a model where both the mean direction and the concentration depend on the set of parents of each node. Conversely, one can assume only the mean direction to be dependent on the parents, adopting, for instance, the specification of conditional von Mises proposed by [4].

The joint distribution of $\boldsymbol{\Theta}$ is not von Mises, but it is the product of univariate von Mises. The Bayesian network factorization property ensures that this is a proper density, even if not closed under marginalization and conditioning. These properties are typically not required for Bayesian networks, as a marginalized/conditional distribution of a joint distribution that obeys a directed acyclic graph model will not generally be faithfully represented by any directed acyclic graph.

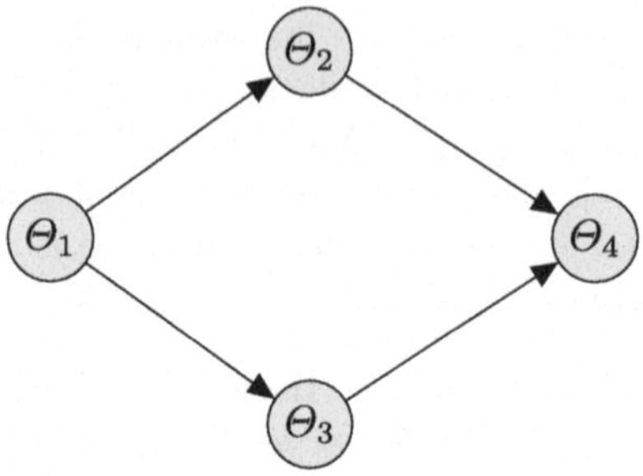

Fig. 1. Assumed DAG for both Scenario 1 and 2 in the simulation study

3.1 Inference and Graph Learning

The parameter estimates can be obtained, for instance, by maximizing the sequence of conditional likelihood functions according to the factorization property

$$\mathcal{L}(\boldsymbol{\mu}, \boldsymbol{\kappa}, \boldsymbol{\Lambda}) = \prod_{j=1}^{p} \frac{\exp\{\kappa_{j.\mathrm{pa}(j)} \cos(\theta_j - \mu_{j.\mathrm{pa}(j)})\}}{2\pi \mathcal{I}_0(\kappa_{j.\mathrm{pa}(j)})},$$

where $\mu_{j.\mathrm{pa}(j)}$ and $\kappa_{j.\mathrm{pa}(j)}$ are functions of the parameters in $\boldsymbol{\mu}$, $\boldsymbol{\kappa}$ and $\boldsymbol{\Lambda}$ as specified in (2) and (3).

Therefore, the parameters involved in a Conditional von Mises directed acyclic graphical model are μ_j, κ_j and λ_{ij} with $i \neq j \in \{1, \dots, p\}$. Notice that the μ_js are not of main interest in the graphical model selection. As argued by [13], the estimation of the μ_js does not affect the estimates of the conditional dependence parameters λ_{ij}. Therefore, the estimates of λ_{ij} can be obtained by maximizing the profile likelihood after plugging-in the sample circular means as estimates of the μ_js.

For learning the graph structure, we can start by including, for each node j, all the nodes in $\{1, \dots, j-1\}$ in the parent set pa(j). Then, edge selection can be performed as variable selection, for instance, by using the likelihood ratio test, as the absence of an edge corresponds to setting a zero the corresponding λ_{ij}, $i \in \{1, \dots, j-1\}$.

3.2 Simulation Study

We present some results of a simulation study to evaluate the performance of the proposed learning procedure and investigate the graph recovery rate. We report here four different scenarios with various sample sizes, all considering a Conditional von Mises directed acyclic graphical model on a random vector $\boldsymbol{\Theta} = (\Theta_1, \Theta_2, \Theta_3, \Theta_4)'$ whose density factorizes as follows

$$f_{\boldsymbol{\Theta}}(\boldsymbol{\theta}) = f_{\Theta_1}(\theta_1) \cdot f_{\Theta_2|\Theta_1}(\theta_2 \mid \theta_1) \cdot f_{\Theta_3|\Theta_1}(\theta_3 \mid \theta_1) \cdot f_{\Theta_4|\Theta_2\Theta_3}(\theta_4 \mid \theta_2, \theta_3).$$

The above factorization corresponds to the graph in Fig. 1 with local independencies $\Theta_2 \perp\!\!\!\perp \Theta_3 \mid \Theta_1$ and $\Theta_4 \perp\!\!\!\perp \Theta_1 \mid \Theta_2, \Theta_3$. Scenarios 1 and 2 both consider the case of high concentration, with $\kappa_j = 2$, while scenarios 3 and 4 consider lower concentration, $\kappa_j = 0.5$, $j \in \{1, \ldots, 4\}$. Scenarios 2 and 4 assume a weak dependence with the non-zero parameters $\lambda_{12} = \lambda_{13} = \lambda_{24} = \lambda_{34} = 0.5$, while scenarios 1 and 3 assume a stronger dependence with the non-zero parameters $\lambda_{12} = \lambda_{13} = \lambda_{24} = \lambda_{34} = 2$.

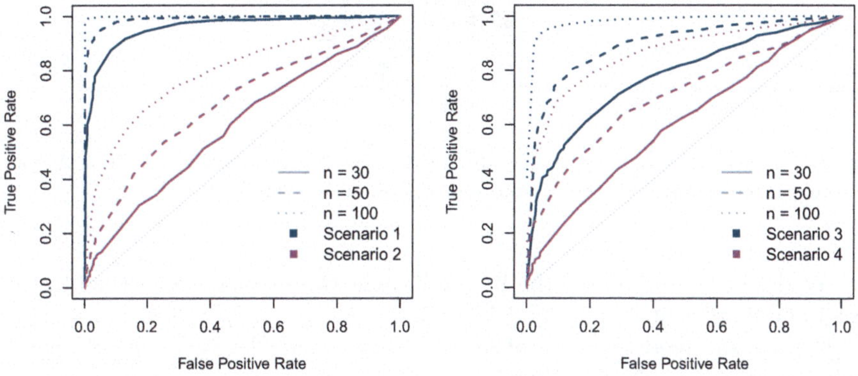

Fig. 2. Monte Carlo ROC curves over 250 simulation runs for data simulated from a Conditional von Mises directed acyclic graphical model with different sample sizes for scenarios 1 and 2 (left panel) and scenarios 3 and 4 (right panel).

Figure 2 reports Monte Carlo ROC curves for all four scenarios and sample sizes equal to 30, 50, and 100. As expected, the procedure performs better in the stronger dependence case and with larger sample sizes. In particular, in the case of higher concentration and stronger dependence, the procedure also performs well for moderate sample sizes.

4 The Structural Characteristics of a Mucin Glycopeptide Motif

We now examine a study on the structural characteristics of a mucin glycopeptide motif. The data were collected by [2], who investigated the secondary structure of a mucin glycopeptide derived from the N-terminal fragment of the cell surface glycoprotein CD43 using NMR spectroscopy.

Mucin glycoproteins constitute one of the most significant classes of cell surface molecules, playing a crucial role in various biological functions. Notably, [2] highlighted how mucin expression levels and structural organization can serve as markers for disease onset, particularly in carcinogenic transformations. Given this, a deeper understanding of the structural properties of mucin glycoproteins could be highly relevant for applications such as cancer vaccine development.

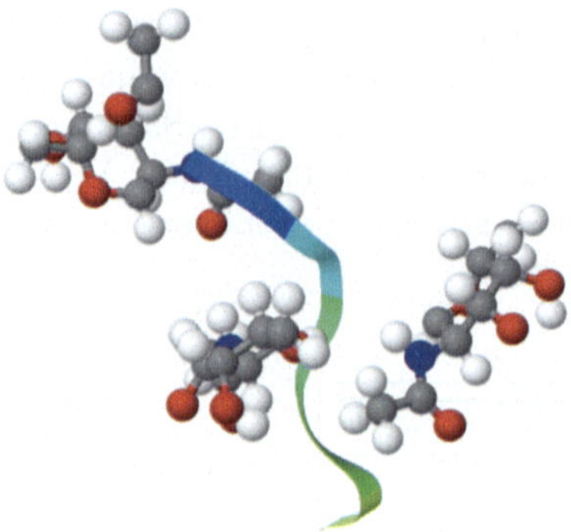

Fig. 3. The 3-D view biological assembly of the mucin glycopeptide 1KYJ, styled as a rainbow ribbon (*source*: https://www.rcsb.org/3d-view/jsmol/1kyj/1)

The data are available in the RCSB Protein Data Bank, under the identifier 1KYJ and consist of $n = 59$ models, each one measuring $p = 9$ dihedral angles: (Φ_1, Ψ_2), (ϕ, ψ) dihedral angles of serine, (Φ_3, Ψ_4) and (Φ_5, Ψ_6), pairs of dihedral angles of threonine, (Φ_7, Ψ_8), dihedral angles of alanine, and Φ_9, ϕ-angle of valine. Here, the subscripts refer to the angle ordering in the primary structure of the glycopeptide.

Typically, an entire protein structure is represented on the surface of a torus, and its dihedral-angle distribution is visualized using the Ramachandran plot, which displays all ϕ and Ψ angles on the same plane. Figure 4 shows the Ramachandran plots for the two threonine (THR) residues in 1KYJ. In Fig. 4(a), both residues are overlaid in the standard representation. In Fig. 4(b), the angles for the first residue (THR1) and the second residue (THR2) are distinguished.

A comparison of these plots suggests that the dependence between ϕ and Ψ varies between the two residues. Consequently, modeling all dihedral-angle pairs with a single bivariate distribution is overly simplistic, and it would be valuable to investigate the full dependence structure among all angles in the protein. A similar example of this issue can be found in [6]. Therefore, we distinguish the different residues in 1KYJ and we assume a Conditional von Mises directed acyclic graphical model, where the natural ordering of the nine nodes is provided by the primary structure of the peptide. Parameter estimation and edge selection are obtained using the likelihood-based approach described in Sect. 3.1. Figure 5 reports the selected graph. The dependence structure seems to follow the ribbon shape, also noted by [2] from the 3-D view of the glycopeptide, shown in Fig. 3. Interestingly, as quite common in our experience for ribbon-shaped peptides, the first and last dihedral angles are independent of the others.

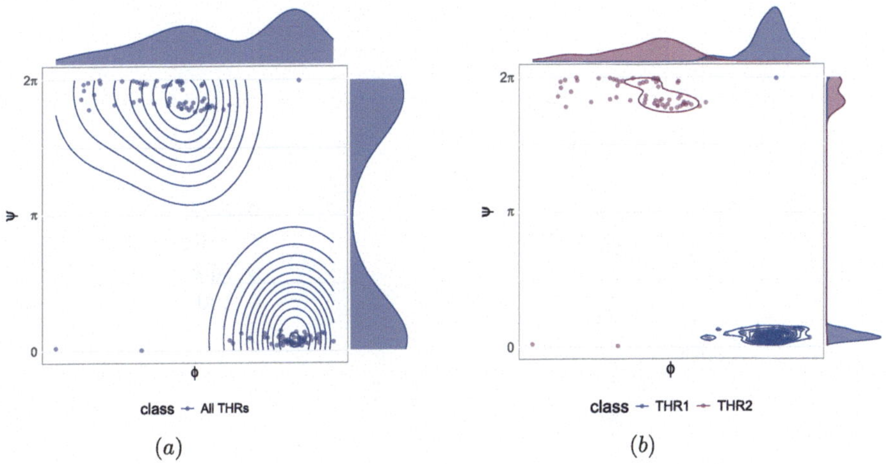

Fig. 4. Ramachandran plot of two amino acids (THR1 and THR2), in 1KYJ protein structures, (a) as a single bivariate distribution and (b) as two bivariate distributions.

5 Concluding Remarks

We propose the Conditional von Mises directed acyclic graphical model based on the von Mises distribution, where the pairwise dependence between two angles Θ_i and Θ_j given a known set of angular variables is encoded by the nonzero pattern of the parameters λ_{ij}.

By definition, the Conditional von Mises graphical model requires a known complete ordering of the involved variables. The knowledge of this ordering, even if it appears to be a stringent condition, is, in fact, common in many fields of application. For example, in studying the dependence among dihedral angles forming a protein, the ordering can be provided by its primary structure. Future research could be oriented to treat the node ordering as unknown and address its choice using suitably modified versions of algorithms for directed acyclic graph structural learning.

An interesting aspect of the proposed class of models is that it can be easily extended to the case of mixed-type variables, where some variables are circular, and others are not. In such a situation, the conditional distribution of the non-circular variable has to be replaced in the factorization with another conditional distribution suggested by the nature of the variable itself. These models could be suitable, for instance, to study a sequence of animal movements, together with the time of the day, the weather and the temperature. The Conditional von Mises graphical models, relying on a factorization of univariate distributions, make the estimation task and the possible extension to a high-dimensional setting easy.

Acknowledgments. This work was supported by the Italian Ministry of University and Research (MUR), Department of Excellence project 2023–2027 ReDS *Rethinking Data Science* - Department of Statistics, Computer Science, Applications - Univer-

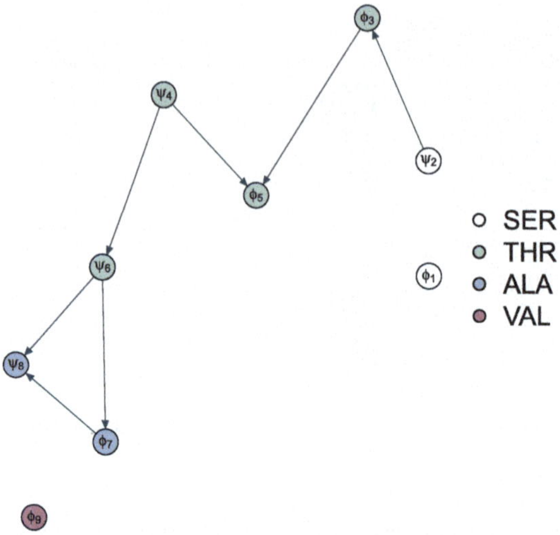

Fig. 5. Estimated directed acyclic graph, assuming the Conditional von Mises graphical model, for the angles of mucin glycopeptide 1KYJ, where SER stands for serine, THR stands for threonine, ALA stands for alanine, and VAL stands for valine.

sity of Florence. Financial support was also provided by the MUR-PRIN grant 2022 SMNNKY, CUP B53D23009470006 Unione europea- Next Generation EU, Missione 4 Componente 2 - CUP B53D23009470006.

References

1. Boomsma, W., Kent, J.T., Mardia, K.V., Taylor, C.C., Hamelryck, T.: Graphical models and directional statistics capture protein structure. Interdisc. Stat. Bioinf. **25**, 91–94 (2006)
2. Coltart, D.M., et al.: Principles of mucin architecture: structural studies on synthetic glycopeptides bearing clustered mono-, di-, tri-, and hexasaccharide glycodomains. J. Am. Chem. Soc. **124**(33), 9833–9844 (2002)
3. Cox, D.R., Wermuth, N.: Multivariate Dependencies: Models, Analysis and Interpretation. Chapman & Hall /CRC, Boca Raton (1996)
4. Fisher, N.I., Lee, A.J.: Regression models for an angular response. Biometrics **48**(3), 665–677 (1992)
5. Gottard, A., Panzera, A.: Gaussian-related undirected graphical models for circular variables. Statistics **58**(5), 1224–1247 (2024)
6. Gottard, A., Panzera, A.: Inverse stereographic Gaussian DAG models. In: Scientific Meeting of the Italian Statistical Society, pp. 136–141. Springer, Cham (2025)
7. Jammalamadaka, S.R., Sengupta, A.: Topics in Circular Statistics, vol. 5. World Scientific (2001)
8. Klein, N., Orellana, J., Brincat, S.L., Miller, E.K., Kass, R.E.: Torus graphs for multivariate phase coupling analysis. Ann. Appl. Stat. **14**(2), 635–660 (2020)
9. Lauritzen, S.L.: Graphical Models, vol. 17. Clarendon Press (1996)

10. Leguey, I., Bielza, C., Larrañaga, P.: Tree-structured bayesian networks for wrapped cauchy directional distributions. In: Luaces, O., Gámez, J.A., Barrenechea, E., Troncoso, A., Galar, M., Quintián, H., Corchado, E. (eds.) CAEPIA 2016. LNCS (LNAI), vol. 9868, pp. 207–216. Springer, Cham (2016). https://doi.org/10.1007/978-3-319-44636-3_19
11. Ley, C., Verdebout, T.: Modern Directional Statistics. CRC Press, Boca Raton (2017)
12. Ley, C., Verdebout, T.: Applied Directional Statistics: Modern Methods and Case Studies. CRC Press, Boca Raton (2018)
13. Mardia, K.V., Hughes, G., Taylor, C.C., Singh, H.: A multivariate von Mises distribution with applications to bioinformatics. Can. J. Stat. **36**(1), 99–109 (2008)
14. Mardia, K.V., Jupp, P.E.: Directional Statistics, vol. 494. John Wiley & Sons, Hoboken (2009)
15. Mardia, K.V., Kent, J.T., Laha, A.K.: Score matching estimators for directional distributions. arXiv preprint arXiv:1604.08470 (2016)
16. Razavian, N., Kamisetty, H., Langmead, C.J.: The von Mises graphical model: Regularized structure and parameter learning. Technical Report CMU-CS-11-129, Carnegie Mellon University (2011)
17. Richardson, T., Spirtes, P., et al.: Ancestral graph Markov models. Ann. Stat. **30**(4), 962–1030 (2002)
18. Sadeghi, K.: Stable mixed graphs. Bernoulli **19**(5B), 2330–2358 (2013)
19. Scealy, J.L., Wood, A.T.: Score matching for compositional distributions. J. Am. Stat. Assoc. 1–32 (2022)
20. Yu, S., Drton, M., Shojai, A.: Generalized score matching for general domains. Inf. Inference J. IMA (2021)

A Comparison of Estimation Methods in Latent Variable Models for Binary Panel Data

Lucia Guastadisegni(✉), Silvia Bianconcini, and Silvia Cagnone

Department of Statistical Sciences "Paolo Fortunati", University of Bologna, Bologna, Italy
{lucia.guastadisegni2,silvia.bianconcini,silvia.cagnone}@unibo.it

Abstract. This paper examines and compares various estimation methods for generalized linear latent variable models for multidimensional longitudinal binary data. In such cases, likelihood-based methods are problematic due to the high dimensional integrals involved, which lack analytical solution. Among the methods proposed in the literature to address this issue, we focus on approximate likelihood and composite likelihood methods. Specifically, within the first class, we examine the dimension-wise quadrature, while within the second class, we consider composite likelihood methods based on bivariate densities. The properties of these methods are evaluated through a comprehensive simulation study.

Keywords: Approximate likelihood methods · pairwise likelihood methods · high dimensional integrals

1 Introduction

Multivariate longitudinal data are common in many fields of research, such as biostatistics and psychology, and can be used to measure changes in outcomes over time or to identify the determinants of such changes in a set of individuals. Among the different models proposed in the literature to analyze multivariate longitudinal data (see [15] for a review), we focus on those that includes random effects and latent variables. We adopt the generalized linear latent variable model (GLLVM) described by [6], to handle various response types in a multivariate longitudinal setting. This model is convenient due to its high flexibility, as it incorporates time-specific latent variables to capture associations among different response variables at each time point, random effects to account for the association of each response over time, and an autoregressive structure to represent the dynamic nature of the time-dependent latent variables, allowing for the inclusion of covariates. When likelihood-based methods are applied to GLLVM for multidimensional discrete data, a challenge arises because the high dimensional integrals involved in the likelihood lack analytical solutions. Among

A. D'Ambrosio et al. (Eds.): CLADAG-VOC 2025, SCDAKO, pp. 260–271, 2025.
https://doi.org/10.1007/978-3-032-03042-9_23

the discrete outcomes, binary data are often regarded as the most complex case, as the limited information provided by binary responses further complicates the process of likelihood maximization.

Classic solutions to this problem include approximate likelihood methods, which are based on Gauss-Hermite quadrature approximations [5]. A popular method within this class is the Adaptive Gauss-Hermite (AGH) quadrature, originally introduced in a Bayesian framework to efficiently approximate posterior densities [12]. However, it becomes infeasible as the number of latent variables and random effects increases [3]. A simpler way to estimate the model is to use the first-order Laplace approximation of the likelihood [8], which is feasible even with a large number of latent variables and random effects, but it yields inaccurate results with discrete data [9]. A recent method for approximating the likelihood, is the dimension-wise (DW) quadrature, proposed by [4]. This method reduces the dimension of the multidimensional integrals by truncating the Taylor series expansion of the integrand as the Laplace approximation, but without requiring any derivative computation. The corresponding estimators are asymptotically as accurate as the AGH estimators when using the same number of quadrature points [4]. Moreover, DW has lower computational complexity compared to AGH, making it feasible even when the latter is not [4].

An alternative way to address the problem of the integrals involved in the likelihood of GLLVMs for multidimensional longitudinal discrete data is to use composite likelihood methods [10], that combine likelihood components, associated with small subsets of data. Within this class, we consider the pairwise likelihood (PL) approach [11], which is based on marginal densities for pairs of observations. The estimator resulting from the maximization of the PL is asymptotically unbiased but less efficient than the full maximum likelihood (ML) estimator [13]. The PL method can also be derived using the method of separate maximizations [7]. According to this approach, the composite likelihood is defined as the sum of all log pairwise likelihoods across pairs of items or times, with each pairwise likelihood maximized separately. The obtained parameter estimates are then combined to produce a single parameter estimate through different weighting matrices. We consider the average (AVE) method, where the weight matrix is designed to compute the arithmetic mean of the estimates [7] and the weighted mean (WM) method [14], where the weight matrix produces a combined estimator that is asymptotically equivalent to the classic PL estimator.

In the context of GLLVM for multidimensional longitudinal data, some comparisons have been made between the classical PL method and the DW quadrature method [2,3]. Overall, the DW quadrature method outperforms the PL method in terms of efficiency, but it has a higher computational burden.

In this work, we extend the comparisons by also considering PL methods based on separate maximizations. Specifically, we evaluate the performance of the DW quadrature method, the classic PL method, and the AVE and WM methods, for a GLLVM for multidimensional longitudinal binary data which, as outlined before, is the most challenging case. The performance of these methods is assessed through a simulation study, along with a comparison of different weighting approaches used in the separate maximizations methods.

The paper is organized as follows. Section 2 introduces the GLLVM for multidimensional longitudinal binary data. Section 3 describes the DW method. Section 4 reviews the classic PL method, while Sect. 5 discusses the PL methods based on separate maximizations. Section 6 presents a simulation study. Finally, Sect. 7 offers discussion.

2 The GLLVM for Longitudinal Binary Data

Let $\boldsymbol{y}_i = (\boldsymbol{y}_{i1}, \cdots, \boldsymbol{y}_{iT})$ be a $p \times T$ matrix, where $\boldsymbol{y}_{it} = (y_{i1t}, y_{i2t}, \cdots, y_{ipt})'$ is a vector of p binary variables measured at time t for individual i, with $i = 1, \ldots, n$. For each individual i, let $\boldsymbol{z}_i = (z_{i1}, z_{i2}, \ldots, z_{iT})^\top$ be a vector representing a latent variable measured at T time points, that accounts for the association among the p items at each time point, and $\boldsymbol{u}_i = (u_{i1}, u_{i2}, \ldots, u_{ip})'$ be a p-dimensional vector of random effects that accounts for the association of each item over time.

According to the GLLVM framework [1], we can express the joint probability of the individual response pattern $f(\boldsymbol{y}_i)$ as

$$f(\mathbf{y}_i) = \int_{\mathbb{R}^K} g(\mathbf{y}_i|\mathbf{u}_i, \mathbf{z}_i)h(\mathbf{u}_i, \mathbf{z}_i)d\mathbf{u}_i d\mathbf{z}_i, \tag{1}$$

where $g(\boldsymbol{y}_i|\boldsymbol{u}_i, \boldsymbol{z}_i)$ is referred to as the measurement part of the model and $h(\boldsymbol{u}_i, \boldsymbol{z}_i)$ as the structural part of the model. The dimension of the integral K is equal to the total number of latent variables and random effects.

For the measurement part of the model, the variables $\boldsymbol{y}_i$ are assumed to be conditionally independent given $\boldsymbol{u}_i$ and $\boldsymbol{z}_i$. This implies that

$$g(\mathbf{y}_i|\mathbf{u}_i, \mathbf{z}_i) = \prod_{j=1}^{p}\prod_{t=1}^{T} g(y_{ijt}|u_{ij}, z_{it}) = \pi_{ijt}(u_{ij}, z_{it})^{y_{ijt}}(1 - \pi_{ijt}(u_{ij}, z_{it}))^{1-y_{ijt}} \tag{2}$$

being $g(y_{ijt}|u_{ij}, z_{it})$ the random component and $\pi_{ijt}(u_{ij}, z_{it}) = P(y_{ijt} = 1|u_{ij}, z_{it})$. The latent variable z_{it}, $t = 1, \cdots, T$ and the random effects u_{ij}, $j = 1, \cdots, p$, determine the linear predictor that represents the systematic component of the model, that is $\eta_{ijt} = \alpha_{0jt} + \alpha_{1jt}z_{it} + u_{ij}$. The parameters α_{0jt} and α_{1jt} are the item- and time-specific intercept and loading, respectively. Under the assumption of measurement invariance, which must be tested on real data, the intercepts and factor loadings do not vary over time. This implies that $\alpha_{0jt} = \alpha_{0j}$ and $\alpha_{1jt} = \alpha_{1j}$ $\forall t$. The link between the systematic and the random component is the logit link function, $\eta_{ijt} = \ln\left(\frac{\pi_{ijt}(u_{ij}, z_{it})}{1-\pi_{ijt}(u_{ij}, z_{it})}\right)$ such that $\pi_{ijt} = \frac{\exp \eta_{ijt}}{1+\exp \eta_{ijt}}$.

The structural part of the model describes the dynamic nature of the model determined by the time-dependent latent variables z_{it}, $t = 1, \cdots, T$, and their relationship with the item-specific random effects u_{ij}, $j = 1, \cdots, p$. To account for the serial correlation in the latent variables, a time-homogeneous nonstationary autoregressive process of order one is specified. That is,

$$z_{it} = \phi z_{i(t-1)} + \delta_{it}, \qquad t = 2, \cdots, T, \tag{3}$$

with $z_{i1} \sim N(0, \sigma_1^2)$ and $\delta_{it} \sim N(0, 1)$ for identification purposes. The latent variables and the random effects are assumed to be independent and normally distributed with $\mathbf{0}$ mean vector and a block diagonal covariance matrix $\boldsymbol{\Sigma} = \begin{bmatrix} \boldsymbol{\Xi} & \mathbf{0} \\ \mathbf{0} & \boldsymbol{\Gamma} \end{bmatrix}$, where $\boldsymbol{\Xi} = \text{diag}_{j=1,\dots,p}(\sigma_{uj}^2)$, and $\boldsymbol{\Gamma}$ is the autocovariance matrix of the latent variables.

Full maximum likelihood is performed by maximizing the following log-likelihood function

$$l(\boldsymbol{\theta}) = \sum_{i=1}^{n} \log f(\boldsymbol{y}_i; \boldsymbol{\theta}) = \sum_{i=1}^{n} \log \int_{\mathbb{R}^K} g(\boldsymbol{y}_i | \boldsymbol{u}_i, \boldsymbol{z}_i) h(\boldsymbol{u}_i, \boldsymbol{z}_i) d\boldsymbol{u}_i d\boldsymbol{z}_i, \tag{4}$$

where $\boldsymbol{\theta}$ is the q-dimensional vector of model parameters. For binary data, the integrals cannot be solved analytically and K is equal to the number of random effects p plus the number of time-dependent latent variables T, that is $K = p+T$.

3 Dimension-Wise Quadrature Method

Let $\boldsymbol{b}_i = (\boldsymbol{u}_i, \boldsymbol{z}_i)'$ denotes the vector of latent variables and random effects. Similar to the AGH approximation, the DW quadrature method starts from the following representation of the marginal density function of the observed variables where, for simplicity, the individual subscript i has been omitted

$$f(\boldsymbol{y}; \boldsymbol{\theta}) = \int_{R^K} \frac{\prod_{j=1}^{p} \prod_{t=1}^{T} g(y_{jt} | \boldsymbol{b}) h(\boldsymbol{b})}{\phi(\boldsymbol{b}; \boldsymbol{b}_{mo}, \boldsymbol{\Sigma}_{mo})} \phi(\boldsymbol{b}; \boldsymbol{b}_{mo}, \boldsymbol{\Sigma}_{mo}) d\boldsymbol{b}, \tag{5}$$

where $\boldsymbol{b}_{mo} = \max_b g(\boldsymbol{y}|\boldsymbol{b}) h(\boldsymbol{b})$ and $\boldsymbol{\Sigma}_{mo}$ is minus the inverse of the corresponding Hessian matrix evaluated in the mode $\boldsymbol{b}_{mo}$. $\phi(\cdot)$ is the multivariate normal distribution, that is the posterior distribution of the latent vector $\boldsymbol{b}$. We can rewrite function (5) in terms of standardized variables

$$\begin{aligned} f(\boldsymbol{y}; \boldsymbol{\theta}) &= |\boldsymbol{C}_{mo}| \int_{R^K} \frac{\prod_{j=1}^{p} \prod_{t=1}^{T} g(y_{jt} | \boldsymbol{C}_{mo} \boldsymbol{b}^* + \boldsymbol{b}_{mo}) h(\boldsymbol{C}_{mo} \boldsymbol{b}^* + \boldsymbol{b}_{mo})}{\phi(\boldsymbol{b}^*; \mathbf{0}, \boldsymbol{I})} \phi(\boldsymbol{b}^*; \mathbf{0}, \boldsymbol{I}) d\boldsymbol{b}^* = \\ &= |\boldsymbol{C}_{mo}| \int_{R^K} m(\boldsymbol{b}^*) \phi(\boldsymbol{b}^*; \mathbf{0}, \boldsymbol{I}) d\boldsymbol{b}^* = |\boldsymbol{C}_{mo}| E_{\phi}[m(\boldsymbol{b}^*)], \end{aligned} \tag{6}$$

where $m(\boldsymbol{b}^*) = \frac{\prod_{j=1}^{p} \prod_{t=1}^{T} g(y_{jt} | \boldsymbol{C}_{mo} \boldsymbol{b}^* + \boldsymbol{b}_{mo}) h(\boldsymbol{C}_{mo} \boldsymbol{b}^* + \boldsymbol{b}_{mo})}{\phi(\boldsymbol{b}^*; \mathbf{0}, \boldsymbol{I})}$ is a positive function and $\boldsymbol{C}_{mo}$ is derived by the Cholesky decomposition of $\boldsymbol{\Sigma}_{mo} = \boldsymbol{C}_{mo} \boldsymbol{C}'_{mo}$. The DW method is applied to $E_{\phi}[m(\boldsymbol{b}^*)]$, and it considers the Taylor expansion of $m(\mathbf{b}^*)$ around 0 up to the s term. The function $m(\mathbf{b}^*)$ can be written as [4]

$$\hat{m}(\mathbf{b}^*) = \sum_{l=0}^{s} (-1)^l \binom{K - s + l - 1}{l} m_{s-l}(\mathbf{b}^*), \tag{7}$$

where $m_{s-l}(\mathbf{b}^*) = m(0, \cdots, b^*_{k_1}, 0, \cdots, 0, b^*_{k_{s-l}}, 0, \cdots, 0)$. Thus, $m_{s-l}(\mathbf{b}^*)$ is a function of just $s - l$ variables, with all the remaining variables fixed to 0. Substituting (7) to (6), we obtain the approximated marginal density function

$$f_a(\mathbf{y};\boldsymbol{\theta}) = f_L + |C_{mo}| \left[\sum_{l=0}^{s-1}(-1)^l \binom{K-s+l-1}{l} \int_{R^{s-l}} \sum_{k_1<\ldots<k_{s-l}} m_{s-l}(b^*)\phi(b^*_{k_1})\cdots\phi(b^*_{k_{s-l}})db^*_{k_1}..db^*_{k_{s-l}}\right], \tag{8}$$

where f_L is the Laplace approximation of the integral. The dimension of the integrals in (8) is determined by the choice of s. When s is small, these integrals can be easily approximated using the Gauss-Hermite quadrature.

When $s = 0$, DW is equivalent to the classical Laplace approximation. As s increases, the DW method consists of approximating the K-dimensional integral with the sum of one-, two-, up to s-dimensional integrals. When $s = K$, no dimension reduction is performed, and the method reduces to the AGH method. The DW estimators are consistent, as long as the number of observed variables and sample size increase, and asymptotically normally distributed, with a sandwich-type covariance matrix. The DW quadrature method is more accurate than the Laplace approximation and, for small values of s, at most 3, it can achieve the same accuracy as the AGH quadrature [4].

4 The Pairwise Likelihood Approach

The PL replaces the likelihood in (4) with bivariate marginal densities as follows [10]:

$$pl(\boldsymbol{\theta};\boldsymbol{y}) = \sum_{i=1}^{n} pl(\boldsymbol{\theta};\boldsymbol{y}_i) = \sum_{i=1}^{n}\sum_{j=1}^{p}\sum_{j'=j}^{p}\sum_{t,t'} \log f(y_{ijt}, y_{ij't'};\boldsymbol{\theta}), \tag{9}$$

where $t, t' = 1, \cdots, T$. Moreover, $t = t' \Rightarrow j < j'$ and $j = j' \Rightarrow t < t'$. Formula (9) can be rewritten as the sum of bivariate densities over all possible pairs of items or time points. We consider the one for pairs of items. The number of pairs of items in this model is given by $N_p = \binom{p}{2} + p$, where $\binom{p}{2}$ accounts for the bivariate densities formed by different items over time, and p accounts for the bivariate densities of the same item at different time points. Thus, formula (9) can be rewritten as

$$pl(\boldsymbol{\theta};\boldsymbol{y}) = \sum_{r=1}^{N_p} f_r(\boldsymbol{\theta}) \tag{10}$$

where $\sum_{r=1}^{N_p} = \sum_{j=1}^{p}\sum_{j'=j}^{p}$ and

$$f_r(\boldsymbol{\theta}) = \sum_{i=1}^{n} f_{ri}(\boldsymbol{\theta}) = \sum_{i=1}^{n}\sum_{t,t'} \log f(y_{ijt}, y_{ij't'};\boldsymbol{\theta}), \tag{11}$$

where $t, t' = 1, \cdots, T$. Moreover, $t = t' \Rightarrow j < j'$ and $j = j' \Rightarrow t < t'$. As clarified in Sect. 5, rewriting formula (9) as (10) is convenient for the formulation of the PL method based on separate maximization.

The PL estimator is obtained as

$$\hat{\boldsymbol{\theta}}_{PL} = \arg\max_{\boldsymbol{\theta}} pl(\boldsymbol{\theta}; \boldsymbol{y}). \tag{12}$$

Under standard regularity conditions [11], $\hat{\boldsymbol{\theta}}_{PL}$ is asymptotically [13]

$$\sqrt{n}(\hat{\boldsymbol{\theta}}_{PL} - \boldsymbol{\theta}) \xrightarrow{d} N(0, H(\boldsymbol{\theta})^{-1} J(\boldsymbol{\theta}) H(\boldsymbol{\theta})^{-1}), \tag{13}$$

where the matrices $H(\boldsymbol{\theta})$ and $J(\boldsymbol{\theta})$[1] can be estimated by their observed versions evaluated at $\hat{\boldsymbol{\theta}}_{PL}$, denoted as $\mathcal{H}(\hat{\boldsymbol{\theta}}_{PL})$ and $\mathcal{J}(\hat{\boldsymbol{\theta}}_{PL})$[2].

5 Pairwise Likelihood Methods Based on Separate Maximizations

[7] proposed an alternative formulation of the PL method, based on separate maximizations. Consider formula (10), where $pl(\boldsymbol{y}; \boldsymbol{\theta})$ is expressed as the sum of $f_r(\boldsymbol{\theta})$ across all possible pairs of items N_p. Instead of maximizing the entire $pl(\boldsymbol{y}; \boldsymbol{\theta})$, each $f_r(\boldsymbol{\theta})$ is maximized separately. Let $\boldsymbol{\theta}_r$ represent the vector of parameters of $f_r(\boldsymbol{\theta}_r)$. The ML estimator $\hat{\boldsymbol{\theta}}_r$ is defined as

$$\hat{\boldsymbol{\theta}}_r = \arg\max_{\boldsymbol{\theta}_r} \ f_r(\boldsymbol{\theta}_r). \tag{14}$$

Let us denote $\hat{\boldsymbol{\Theta}}' = (\hat{\boldsymbol{\theta}}_1, \ldots, \hat{\boldsymbol{\theta}}_{N_p})'$, and let $\boldsymbol{\Theta}$ represent the corresponding true parameter vector. Both $\hat{\boldsymbol{\Theta}}$ and $\boldsymbol{\Theta}$ are Q_p-dimensional vectors, where $Q_p = N_p q$, and q is the dimension of $\boldsymbol{\theta}$. It is important to note that the parameter vectors $\boldsymbol{\theta}$ and $\boldsymbol{\Theta}$ are not equivalent. Specifically, certain parameters in $\boldsymbol{\theta}$ have more counterparts in $\boldsymbol{\Theta}$ than others [7]. For instance, the parameters ϕ and σ^2 can be estimated from all $f_r(\boldsymbol{\theta}_r)$, meaning they have N_p counterparts in $\boldsymbol{\Theta}$. In contrast, the item-specific intercepts and slopes can only be estimated from the functions $f_r(\boldsymbol{\theta}_r)$ that involve the respective items and therefore have fewer counterparts in $\boldsymbol{\Theta}$.

The estimator $\hat{\boldsymbol{\Theta}}$ shares the same asymptotic properties as its constituting elements, which are ML estimators. For this reason, asymptotically [7]

$$\sqrt{n}(\hat{\boldsymbol{\Theta}} - \boldsymbol{\Theta}) \xrightarrow{d} N(\mathbf{0}, H(\boldsymbol{\Theta})^{-1} J(\boldsymbol{\Theta}) H(\boldsymbol{\Theta})^{-1}). \tag{15}$$

The asymptotic covariance matrix of $\hat{\boldsymbol{\Theta}}$, of dimension $Q_p \times Q_p$, has the form

$$V(\boldsymbol{\Theta}) = H(\boldsymbol{\Theta})^{-1} J(\boldsymbol{\Theta}) H(\boldsymbol{\Theta})^{-1}, \tag{16}$$

where $H(\boldsymbol{\Theta})$ and $J(\boldsymbol{\Theta})$ are $Q_p \times Q_p$ block matrices, where $H(\boldsymbol{\Theta})$ is block diagonal. Each block is denoted as H_{rr} and J_{ru}[3], for $r, u = 1, \ldots, N_p$. These matrices

---

[1] $H(\boldsymbol{\theta}) = -\frac{1}{n} E\left[\frac{\partial^2 pl(\boldsymbol{\theta}; y)}{\partial \boldsymbol{\theta} \partial \boldsymbol{\theta}'}\right]$, $J(\boldsymbol{\theta}) = \frac{1}{n} E\left[\left(\frac{\partial pl(\boldsymbol{\theta}; y)}{\partial \boldsymbol{\theta}}\right)\left(\frac{\partial pl(\boldsymbol{\theta}; y}{\partial \boldsymbol{\theta}}\right)'\right]$.

[2] $\mathcal{H}(\hat{\boldsymbol{\theta}}_{PL}) = -\frac{1}{n} \sum_{i=1}^{n} \frac{\partial^2 pl(\mathbf{y}_i, \hat{\boldsymbol{\theta}}_{PL})}{\partial \boldsymbol{\theta} \partial \boldsymbol{\theta}'}$, $\mathcal{J}(\hat{\boldsymbol{\theta}}_{PL}) = \frac{1}{n} \sum_{i=1}^{n} \left(\frac{\partial pl(\boldsymbol{y}_i, \hat{\boldsymbol{\theta}}_{PL})}{\partial \boldsymbol{\theta}}\right)\left(\frac{\partial pl(\boldsymbol{y}_i, \hat{\boldsymbol{\theta}}_{PL})}{\partial \boldsymbol{\theta}}\right)'$.

[3] $H_{rr} = -\frac{1}{n} \sum_{i=1}^{n} E\left[\frac{\partial^2 f_{ri}(\boldsymbol{\theta}_r)}{\partial \boldsymbol{\theta}_r \partial \boldsymbol{\theta}_r'}\right]$, $J_{ru} = \frac{1}{n} \sum_{i=1}^{n} E\left[\left(\frac{\partial f_{ri}(\boldsymbol{\theta}_r)}{\partial \boldsymbol{\theta}_r}\right)\left(\frac{\partial f_{ui}(\boldsymbol{\theta}_u)}{\partial \boldsymbol{\theta}_u}\right)'\right]$.

can be estimated by their observed versions, evaluated at $\hat{\boldsymbol{\Theta}}$, denoted as $\mathcal{H}(\hat{\boldsymbol{\Theta}})$ and $\mathcal{J}(\hat{\boldsymbol{\Theta}})$. The matrix $\mathcal{H}(\hat{\boldsymbol{\Theta}})$ is block diagonal, and $\mathcal{J}(\hat{\boldsymbol{\Theta}})$ is a block matrix, both of size $Q_p \times Q_p$, with each block denoted as $\mathcal{H}_{rr}(\hat{\boldsymbol{\theta}}_r)$ and $\mathcal{J}_{ru}(\hat{\boldsymbol{\theta}}_r, \hat{\boldsymbol{\theta}}_u)$[4] for $r, u = 1, \ldots, N_p$.

The estimators in $\hat{\boldsymbol{\Theta}}$ should be combined to obtain a single estimator for the model parameter vector $\boldsymbol{\theta}$. This can be achieved by pre-multiplying $\hat{\boldsymbol{\Theta}}$ by a matrix of weights $\boldsymbol{A}$ [7]. Consequently,

$$\sqrt{n}(\boldsymbol{A}\hat{\boldsymbol{\Theta}} - \boldsymbol{A}\boldsymbol{\Theta}) \xrightarrow{d} N(\mathbf{0}, \boldsymbol{A}V(\boldsymbol{\Theta})\boldsymbol{A}'). \tag{17}$$

The asymptotic covariance matrix of the combined estimator takes the form $\boldsymbol{A}V(\boldsymbol{\Theta})\boldsymbol{A}'$ and can be estimated as $\boldsymbol{A}\mathcal{V}(\hat{\boldsymbol{\Theta}})\boldsymbol{A}'$, where $\mathcal{V}(\hat{\boldsymbol{\Theta}}) = \mathcal{H}(\hat{\boldsymbol{\Theta}})^{-1}\mathcal{J}(\hat{\boldsymbol{\Theta}})\mathcal{H}(\hat{\boldsymbol{\Theta}})^{-1}$. Among the various weighting matrices proposed in the literature, we consider those based on the AVE and WM methods.

5.1 The AVE Method

The estimators in $\hat{\boldsymbol{\Theta}}$ are averaged to obtain one single estimator for the model parameter in $\boldsymbol{\theta}$. The estimator $\hat{\boldsymbol{\theta}}_{AVE}$ is obtained as

$$\hat{\boldsymbol{\theta}}_{AVE} = \boldsymbol{A}\hat{\boldsymbol{\Theta}}. \tag{18}$$

In this case, $\boldsymbol{A} = [A_1 \cdots A_{N_p}]$ has dimension $q \times qN_p$. Each matrix A_r is diagonal, with the weights on the main diagonal denoted as $\alpha_{hh}^{(r)}$, for $r = 1, \ldots, N_p$ and $h = 1, \ldots, q$. If θ_h is estimated in the r-th pair, the corresponding weight in A_r can be computed as

$$\alpha_{hh}^{(r)} = \frac{1}{n_h}, \tag{19}$$

where n_h is the number of times θ_h is estimated in the N_p separate maximizations. Moreover, it holds that $\sum_{r=1}^{N_p}\sum_{h=1}^{q}\alpha_{hh}^{(r)} = 1$.

5.2 The Weighted Mean Method

[14] demonstrated that $\hat{\boldsymbol{\theta}}_{PL}$ is asymptotically a weighted estimator of estimates obtained from separate maximizations. The detailed proof is provided in [14]. We refer to this approach as the weighted mean (WM) method. The weight matrix A_r, for $r = 1, \ldots, N_p$, that ensures the asymptotic equivalence between the WM and the PL methods is defined as

$$A_r = \left(\sum_{g=1}^{N_p} \frac{\partial^2 f_g(\hat{\boldsymbol{\theta}}_g)}{\partial\boldsymbol{\theta}\partial\boldsymbol{\theta}'}\right)^{-1} \frac{\partial^2 f_r(\hat{\boldsymbol{\theta}}_r)}{\partial\boldsymbol{\theta}\partial\boldsymbol{\theta}'}, \tag{20}$$

where the index g denotes the summation across item pairs, with r held fixed.

[4] $\mathcal{H}_{rr}(\hat{\boldsymbol{\theta}}_r) = -\frac{1}{n}\sum_{i=1}^{n}\frac{\partial^2 f_{ri}(\hat{\boldsymbol{\theta}}_r)}{\partial\boldsymbol{\theta}_r\partial\boldsymbol{\theta}_r'}$, $\mathcal{J}_{ru}(\hat{\boldsymbol{\theta}}_r, \hat{\boldsymbol{\theta}}_u) = \frac{1}{n}\sum_{i=1}^{n}\left(\frac{\partial f_{ri}(\hat{\boldsymbol{\theta}}_r)}{\partial\boldsymbol{\theta}_r}\right)\left(\frac{\partial f_{ui}(\hat{\boldsymbol{\theta}}_u)}{\partial\boldsymbol{\theta}_u}\right)'$.

The final estimator $\hat{\boldsymbol{\theta}}_{WM}$ is obtained as

$$\hat{\boldsymbol{\theta}}_{WM} = \boldsymbol{A}\hat{\boldsymbol{\Theta}}, \tag{21}$$

where $\boldsymbol{A} = [A_1 \cdots A_{N_p}]$ is a matrix of dimension $q \times N_p q$.

6 Simulation Study

We conduct a simulation study to evaluate the performance of DW, PL, WM, and AVE methods for a GLLVM applied to multidimensional longitudinal binary data. Some preliminary results on the performance of the classical PL and DW methods for this type of model have been presented in [2], for different values of T and p fixed to 3. That study shows that the performance of DW and PL remains consistent for $T = 6, 15$. In this work, we also explore the WM and AVE methods and compare their respective weighting schemes. We focus on the case $T = 6$ time points, which represents a plausible setting for panel data, and consider sample sizes $n = 200$ and $n = 1000$. For consistency with the study of [2], we set the number of items to $p = 3$.

The same data-generating values used in the study of [2] are considered: $\phi = 0.5$, $\sigma_1^2 = 2$, and $\sigma_{u_j}^2$ (for $j = 1, 2, 3$) ranging from 1 to 2. The item intercepts are generated from a log-normal distribution with a mean of 0 and a standard deviation (SD) of 0.1, while the item slopes are generated from a log-normal distribution with a mean of 0 and an SD of 0.5. We consider the classic PL method and the AVE and WM methods. For DW, we set $s = 0, 1, 2, 3$. The values of s are not increased further because, already for $s = 2$ or 3, a high level of accuracy can be obtained [2,3]. Moreover, increasing s would also lead to higher computational time.

The optimization for all methods is performed in R using the *optim* function, which employs a Quasi-Newton algorithm, combined with Fortran. For all methods, the number of quadrature points is fixed at eight, as this represents a good trade-off between computation time and estimator accuracy. We consider 500 replications for each condition of the study.

6.1 Results

The following figures display the boxplots of parameter estimates for ϕ, σ_1^2, $\sigma_{u_3}^2$, and α_{1_2}. The remaining parameter estimates exhibit similar behavior and are omitted for brevity.

The boxplots of parameter estimates for $p = 3$, $T = 6$, $n = 200$ are reported in Fig. 1.

Starting from the composite likelihood methods we observe several differences between the classic PL and WM, with the latter producing less accurate results, particularly for the parameters ϕ and σ_1^2. The AVE method exhibits noticeable bias for $\sigma_{u_3}^2$ and α_{1_2}. DW demonstrates accurate results for $s = 2$ and $s = 3$, and shows lower variability than the composite likelihood methods for most parameters.

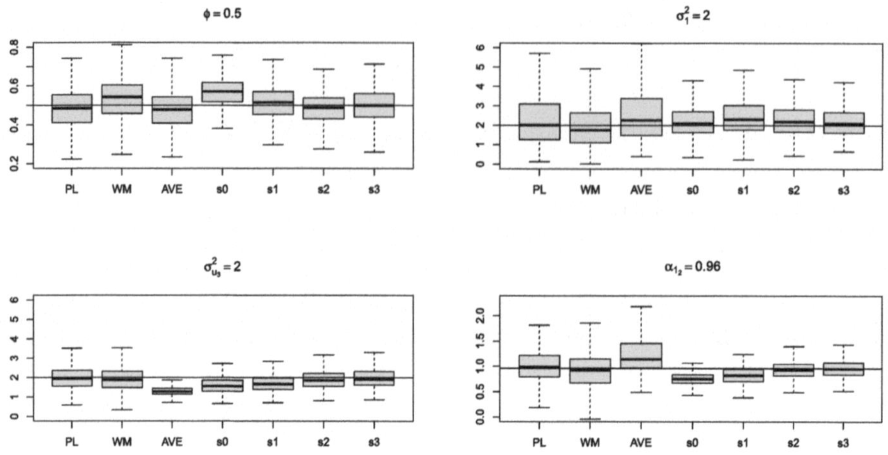

Fig. 1. Boxplots of parameter estimates, $n = 200$.

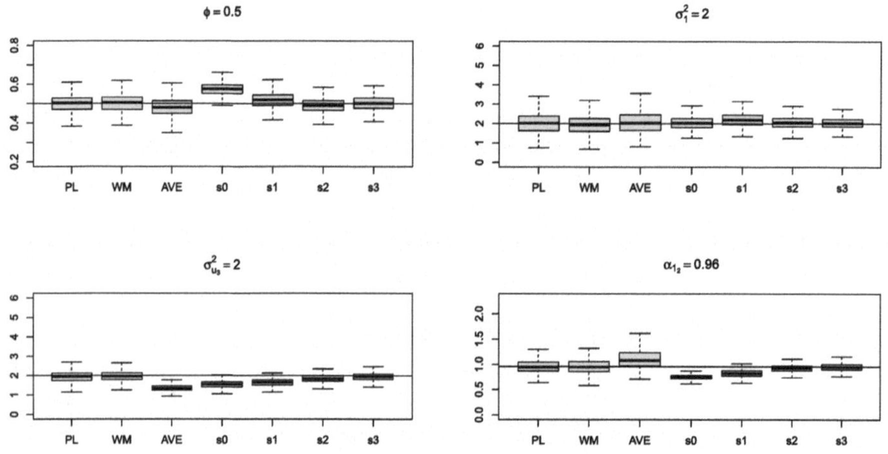

Fig. 2. Boxplots of parameter estimates, $n = 1000$.

For $p = 3$, $T = 6$, $n = 1000$ (Fig. 2), all methods show reduced variability compared to $n = 200$. Notably, also for large sample sizes, DW demonstrates lower variability than the composite likelihood methods for most parameters and produces accurate results for $s = 2$ and $s = 3$. Concerning the composite likelihood methods, as the sample size increases, WM and PL exhibit a similar performances. However, with AVE, the bias in the estimates of $\sigma^2_{u_3}$ and α_{1_2} does not reduce as the sample size increases.

Further insights into the results can be obtained by examining the behaviour of the weights.

Figure 3 presents scatter plots of the WM and AVE weights corresponding to the estimates of ϕ, σ^2_1, $\sigma^2_{u_3}$ and α_{1_2} for $p = 3$, $T = 6$, and $n = 1000$. On the x-

axis, the WM weights are plotted, while the corresponding parameter estimates are shown on the y-axis. The red vertical line identifies the AVE weights, which are the same for all estimates, while the black horizontal line represents the true parameter value.

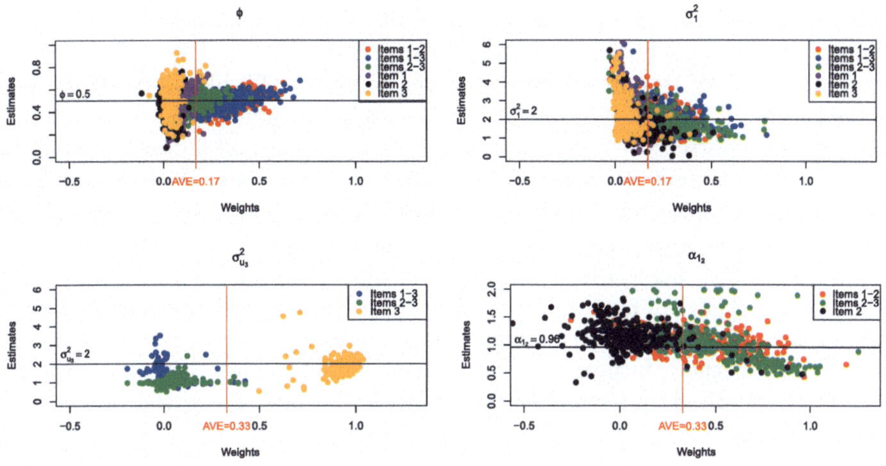

Fig. 3. Scatterplots for parameter estimates and weights for $n = 1000$.

We can notice that the weights of the WM method are very different from those provided by the AVE method for all model parameters. Interestingly, we observe a clear trend for $\sigma^2_{u_3}$: the WM weights concentrate around 1 for the estimates obtained from $f_r(\boldsymbol{\theta}_r)$ formed by bivariate densities of the same item over time, and around 0 for the estimates obtained from $f_r(\boldsymbol{\theta}_r)$ formed by bivariate densities of different items over time. A similar behavior is also observed for the variances of the remaining random effects. This reflects the fact that the random effects account for the variability of the same item over time. In contrast, for α_{1_2}, higher WM weights are observed for the estimates obtained from $f_r(\boldsymbol{\theta}_r)$ formed by bivariate densities of different items over time. This is also true for the loading of the third item, even though it is not reported in the Figure. Indeed, the loadings account for the correlations between different items. The AVE weights are equal to 0.33 for all variances of the random effect and loading estimates, without accounting for any differences between them. Interestingly, for ϕ and σ^2_1, we observe that the WM weights, which concentrate around 0, correspond to the less accurate estimates, which are those farther from the black horizontal line. The AVE weights are equal to 0.17 for all estimates, regardless of their distance from the true values.

Even though the AVE method is simpler and faster, it provides rough weights for the estimates and does not capture the specific characteristics of the parameters as effectively as the WM method.

7 Conclusion

In this work, we compared several composite likelihood methods and the DW method to address the high-dimensional integrals involved in the likelihood of GLLVM models for multidimensional longitudinal binary data, which do not have analytical solutions. Among the composite likelihood methods, we evaluated the classic PL method as well as some PL methods based on separate maximizations, namely the WM and AVE methods.

Generally, DW is more efficient than the composite likelihood methods considered for $s = 2$ and $s = 3$, but it involves a high computational cost. Similar findings are reported by [2] for longitudinal binary data and [3] for longitudinal ordinal data. Among the separate maximizations methods, WM outperforms AVE, which, despite its simplicity, produces inconsistent results, particularly for the variances of the random effects and the loadings. WM is asymptotically equivalent to the classic PL, but it does not offer a significant practical advantage over the latter. Therefore, it is valuable to explore weights that minimize the variance of the combined estimator, such as those proposed by [14]. Regarding the DW method, future work should focus on improving the computational time of the method and studying its asymptotic properties in the presence of longitudinal data.

Acknowledgements. This study was funded by the European Union - NextGenerationEU, in the framework of the "GRINS -Growing Resilient, INclusive and Sustainable project" (PNRR - M4C2 - I1.3 - PE00000018 – CUP J33C22002910001). The views and opinions expressed are solely those of the authors and do not necessarily reflect those of the European Union, nor can the European Union be held responsible for them.

References

1. Bartholomew, D.J., Knott, M., Moustaki, I.: Latent Variable Models and Factor Analysis: A Unified Approach. Wiley (2011). https://doi.org/10.1002/9781119970583
2. Bianconcini, S., Cagnone, S.: Estimation issues in multivariate panel data. In: Studies in Classification, Data Analysis, and Knowledge Organization, Springer (Forthcoming)
3. Bianconcini, S., Cagnone, S.: Comparison between different estimation methods of factor models for longitudinal ordinal data. Wiberg M.; Molenaar D.; González J.; Böckenholt U.; Kim J.S., Quantitative Psychology: The 85th Annual Meeting of the Psychometric Society, Virtual, Springer Proceedings in Mathematics & Statistics, 9–21 (2021). https://doi.org/10.1007/978-3-030-74772-5_2
4. Bianconcini, S., Cagnone, S., Rizopoulos, D.: Approximate likelihood inference in generalized linear latent variable models based on the dimension-wise quadrature. Electron. J. Stat. **11**(2), 4404–4423 (2017). https://doi.org/10.1214/17-EJS1360
5. Bock, R.D., Aitkin, M.: Marginal maximum likelihood estimation of item parameters: application of an EM algorithm. Psychometrika **46**(4), 443–459 (1981). https://doi.org/10.1007/BF02293801

6. Dunson, D.B.: Dynamic latent trait models for multidimensional longitudinal data. J. Am. Stat. Assoc. **98**(463), 555–563 (2003). https://doi.org/10.1198/016214503000000387
7. Fieuws, S., Verbeke, G.: Pairwise fitting of mixed models for the joint modeling of multivariate longitudinal profiles. Biometrics **62**(2), 424–431 (2006). https://doi.org/10.1111/j.1541-0420.2006.00507.x
8. Huber, P., Ronchetti, E., Victoria-Feser, M.-P.: Estimation of generalized linear latent variable models. J. R. Stat. Soc. Ser. B Stat. Methodol. **66**(4), 893–908 (2004). https://doi.org/10.1111/j.1467-9868.2004.05627.x
9. Joe, H.: Accuracy of Laplace approximation for discrete response mixed models. Comput. Stat. Data Anal. **52**(12), 5066–5074 (2008). https://doi.org/10.1016/j.csda.2008.05.002
10. Lindsay, B.G.: Composite likelihood methods. Contemp. Math. **80**, 221–239 (1988). https://doi.org/10.1090/conm/080/999014
11. Molenberghs, G., Verbeke, G.: Models for Discrete Longitudinal Data. Springer, Heidelberg (2005). https://doi.org/10.1007/0-387-28980-1
12. Naylor, J.C., Smith, A.F.M.: Applications of a method for the efficient computation of posterior distributions. J. R. Stat. Soc.: Ser. C: Appl. Stat. **31**(3), 214–225 (1982). https://doi.org/10.2307/2347995
13. Varin, C., Vidoni, P.: A note on composite likelihood inference and model selection. Biometrika **92**(3), 519–528 (2005). https://doi.org/10.1093/biomet/92.3.519
14. Vasdekis, V., Rizopoulos, D., Moustaki, I.: Weighted pairwise likelihood estimation for a general class of random effects models. Biostatistics **15**(4), 677–689 (2014). https://doi.org/10.1093/biostatistics/kxu018
15. Verbeke, G., Fieuws, S., Molenberghs, G., Davidian, M.: The analysis of multivariate longitudinal data: a review. Stat. Methods Med. Res. **23**(1), 42–59 (2014). https://doi.org/10.1177/0962280212445834

Mixture of Experts Latent Trait Analyzers

Dalila Failli[1], Maria Francesca Marino[2(✉)], and Francesca Martella[3]

[1] Dipartimento di Scienze Politiche, Università degli Studi di Perugia, Via Pascoli, 20, 06123 Perugia, Italy
dalila.failli@unipg.it

[2] Dipartimento di Statistica, Informatica, Applicazioni, Università degli Studi di Firenze, Viale Morgagni 59, 50134 Firenze, Italy
mariafrancesca.marino@unifi.it

[3] Dipartimento di Scienze Statistiche, Sapienza Università di Roma, Piazzale Aldo Moro 5, 00185 Roma, Italy
francesca.martella@uniroma1.it

Abstract. The Mixture of Latent Trait Analyzers (MLTA) represents a model-based clustering approach specifically tailored to multivariate categorical data. It accommodates clustering of units through a finite mixture specification, while also accounting for the residual latent variability of units within each cluster through a set of multidimensional latent variables (traits). The original formulation is extended to account for the effect of concomitant variables (covariates). These are allowed to affect cluster formation, the conditional outcome distribution, both (as in standard mixtures of experts models), or neither. Overall, the proposal improves the model's flexibility and its capacity to reflect the complexity of the data.

Keywords: Model-based clustering · Finite mixtures · Concomitant variables · EM algorithm · Variational inference

1 Introduction

Complex data are increasingly common across various fields due to advancements in technology that allow for the collection of large and diverse datasets. Examples include healthcare, bioinformatics, finance, as well as social and environmental sciences, which often involve high-dimensional data and complex dependence structures. All these features emphasize the need for advanced analytical methods to effectively extract insights from such data. A key aspect of complex data analysis often involves the identification of homogeneous groups (clusters) of units. To this aim, a finite mixture approach can be used.

In the context of multivariate categorical data, [6] introduced a Mixture of Latent Trait Analyzers (MLTA), which combines features proper of latent class and latent trait analysis [1]. Specifically, the MLTA assumes that units can be

A. D'Ambrosio et al. (Eds.): CLADAG-VOC 2025, SCDAKO, pp. 272–281, 2025.
https://doi.org/10.1007/978-3-032-03042-9_24

grouped into homogeneous clusters identified via a finite mixture specification, as in the latent class model. In addition, the propensity of each unit toward specific values for the categorical outcomes is assumed to depend on a multi-dimensional continuous latent variable (trait), as in the latent trait framework. Therefore, compared to the latent class model, the MLTA also allows to capture the residual latent variability within each cluster, thus permitting to overcome problems related to the local independence assumption upon which the latent class model is based. Such a model specification is also linked to the mixture of Rash models [11] and, more in general, to the mixture of item response theory models [10], as well as to the factor mixture model [9]. [4] and [5] extended the original MLTA specification to account for the effect that concomitant variables (covariates) may have on clustering formation. In this regard, a multinomial logit specification for the prior probabilities of the finite mixture is considered.

In this paper, we further extend the MLTA with concomitant variables by allowing these latter to influence cluster formation, the (conditional) outcome distribution, both models, or neither, as in standard mixtures of experts models [7]. This approach offers several advantages. First, including covariates into the latent layer of the model allows to understand how the observed characteristics of units influence their clustering, better reflecting the underlying structures of the data. Second, allowing the distribution of response variables to also depend on covariates allows to capture the relationship between them. Overall, the inclusion of covariates in both the observed and latent layers of the model improves its flexibility and ability to reflect the complexity of the data. The simplest alternatives in which concomitant variables influence one layer only or neither are obtained as special cases, giving great flexibility to the proposal.

A practical context in which covariates may impact both clustering and response variables is the analysis of biomedical data. For example, when studying the effectiveness of a treatment on patients with different clinical characteristics, covariates can be used to identify subgroups of patients (clusters) with similar clinical features. The same covariates can also directly influence the response to a specific treatment. In addition, the multidimensional latent trait in the MLTA specification would allow to account for the influence of residual unobserved factors on the outcomes of interest.

The paper is organized as follows. In Sect. 2, we extend the original MLTA model, also describing model assumptions, parameter estimation, and model selection. Section 3 contains concluding remarks and further extensions of the approach.

2 Methodology

In this section, we extend the MLTA for multivariate categorical data in a mixture of experts perspective. Specifically, in Sect. 2.1, we introduce the model for binary variables for simplicity, while noting that the approach can be easily extended to categorical variables. Section 2.2 discusses parameter estimation and clustering, and Sect. 2.3 outlines the procedure for computing standard errors and model selection.

2.1 Model Assumptions

Let Y_{ik} denote the k-th binary random variable and y_{ik} the corresponding observed value for the i-th unit, with $i = 1, \ldots, N$, and $k = 1, \ldots, R$. The MLTA model assumes that each unit belongs to one out of G unobserved groups (clusters) identified by the latent variable $\boldsymbol{z}_i = (z_{i1}, \ldots, z_{iG})' \overset{iid}{\sim}$ Multinomial$(1, (\eta_1, \ldots, \eta_G)')$, whose generic element is given by

$$z_{ig} = \begin{cases} 1 \text{ if unit } i \text{ belongs to cluster } g, \\ 0 \text{ otherwise.} \end{cases}$$

The parameter η_g denotes the probability that a randomly selected unit belongs to cluster g, with $g = 1, \ldots, G$, under the constraints $\sum_{g=1}^{G} \eta_g = 1$ and $\eta_g > 0$, $\forall\, g = 1, \ldots, G$.

To account for the possible effect that concomitant variables may have on the clustering, we follow the approach suggested by [4] and relax the homogeneity assumption of the latent class prior probabilities η_g, $g = 1, \ldots, G$. Let $\boldsymbol{x}_i = (1, x_{i1}, \ldots, x_{iJ})'$ denote a $(J+1)$-dimensional design vector for unit i, and let $\boldsymbol{\beta}_g = (\beta_{0g}, \beta_{1g}, \ldots, \beta_{Jg})'$ denote the corresponding vector of coefficients for the g-th latent class. We assume that unit-specific prior probabilities are related to the observed $\boldsymbol{x}_i$'s via a multinomial logit link function as follows:

$$\eta(\boldsymbol{x}_i; \boldsymbol{\beta}_g) = \Pr(z_{ig} = 1 | \boldsymbol{x}_i; \boldsymbol{\beta}_g) = \frac{\exp\{\boldsymbol{x}_i' \boldsymbol{\beta}_g\}}{1 + \sum_{g'=2}^{G} \exp\{\boldsymbol{x}_i' \boldsymbol{\beta}_{g'}\}}, \qquad g = 2, \ldots, G. \tag{1}$$

Note that the first cluster is the reference, so that $\boldsymbol{\beta}_1 = \boldsymbol{0}$. Therefore, the β_{jg} coefficient in the vector $\boldsymbol{\beta}_g$ measures the effect of the j-th concomitant variable on the logit of belonging to the g-th cluster with respect to the reference.

The MLTA model also assumes the existence of a continuous D-dimensional latent trait $\boldsymbol{u}_i$, which is distributed according to a D-variate Gaussian density, with zero mean vector and identity covariance matrix, i.e., $\boldsymbol{u}_i \sim \mathcal{N}_D(\boldsymbol{0}, \boldsymbol{I})$. The model postulates that the conditional distribution of the R-dimensional vector $\boldsymbol{Y}_i = (Y_{i1}, \ldots, Y_{iR})'$, given that unit i belongs to the g-th cluster, is specified by a latent trait model with parameters b_{gk} and $\boldsymbol{w}_{gk}$, with $g = 1, \ldots, G$, and $k = 1, \ldots, R$. In this sense, conditional on $\boldsymbol{u}_i$ and $z_{ig} = 1$, variables in $\boldsymbol{Y}_i$ are assumed to be independent Bernoulli random variables with parameters $\pi_{gk}(\boldsymbol{u}_i)$. These are modeled via the following logistic function:

$$\pi_{gk}(\boldsymbol{u}_i) = \Pr(Y_{ik} = 1 \mid \boldsymbol{u}_i, z_{ig} = 1) = \frac{1}{1 + \exp\{-[b_{gk} + \boldsymbol{w}_{gk}' \boldsymbol{u}_i]\}}. \tag{2}$$

Here, b_{gk} is the model intercept and represents the baseline tendency of units in the g-th cluster to "positively" respond to the k-th variable. More specifically, a positive (respectively, negative) value for this parameter indicates a tendency toward a "positive" (respectively, "negative") response to the k-th variable (item) for units in the g-th cluster. Furthermore, parameters $\boldsymbol{w}_{gk}$ are the cluster- and variable-specific slopes associated with the latent trait $\boldsymbol{u}_i$. They are meant to describe unit-specific deviations from the baseline levels dictated by b_{gk}. A larger value of $\boldsymbol{w}_{gk}$ indicates greater heterogeneity of units within the g-th cluster in the way they respond to the k-th variable. A simpler and more parsimonious version of the model is obtained by assuming homogeneity of the slope parameters $\boldsymbol{w}_{gk}$ with respect to clusters (i.e., $\boldsymbol{w}_{gk} = \boldsymbol{w}_k \; \forall g = 1, \ldots, G$), meaning that the latent trait has the same effect in all clusters.

Here, coherently with mixtures of experts regression models, we further extend the MLTA specification by allowing the outcome variable distribution to also depend on the vector of concomitant variables $\boldsymbol{x}_i$. In this respect, the logistic function in Equation (2) is modified as

$$\pi_{gk}(\boldsymbol{u}_i) = \Pr(Y_{ik} = 1 \mid \boldsymbol{u}_i, z_{ig} = 1, \boldsymbol{x}_i) = \frac{1}{1 + \exp\{-[b_{gk} + \boldsymbol{x}_i' \boldsymbol{\gamma}_{gk} + \boldsymbol{w}_{gk}' \boldsymbol{u}_i]\}}, \tag{3}$$

where $\boldsymbol{\gamma}_{gk} = (\gamma_{0gk}, \gamma_{1gk}, \ldots, \gamma_{Jgk})'$ denotes a $(J+1)$-dimensional vector of cluster- and variable-specific coefficients and, for identifiability purposes, $\gamma_{0gk} = 0, \forall g = 1, \ldots, G$ and $k = 1, \ldots, R$.

The proposed extension allows concomitant variables to influence the cluster memberships in Equation (1), the probability distribution of the outcomes through Equation (3), both, or neither. Consequently, this results in four possible MLTA specifications within the mixture of experts framework, providing significant flexibility. See the path diagrams in Fig. 1 for a deeper understanding.

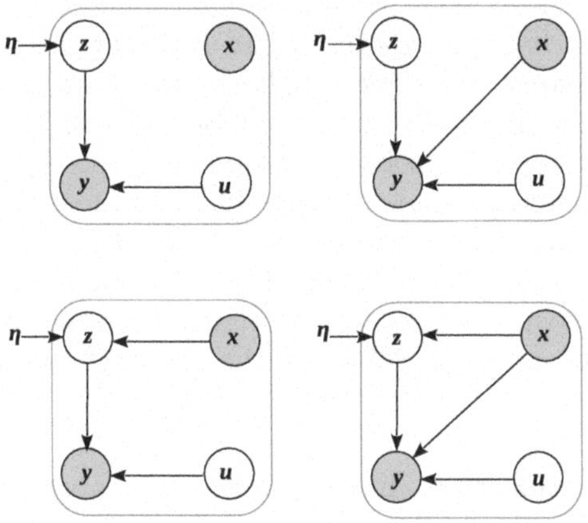

Fig. 1. Path diagrams representing the four possible model specifications: the standard MLTA (upper left), the MLTA with concomitant variables (bottom left), the Mixture of Latent Trait Regression model (upper right), and the Mixture of Experts Latent Trait Analyzers (bottom right).

Based on the adopted specification, parameter interpretation varies and so do the goals one wants to achieve. When concomitant variables are allowed to affect the conditional outcome distribution, the latent variables in the model are meant to account for unobserved heterogeneity, i.e., heterogeneity between units that cannot be explained through the observed covariates. On the other hand, when concomitant variables are allowed to affect cluster memberships, the main focus is on measuring the impact that covariates have on cluster formation.

2.2 Parameter Estimation

Let $\boldsymbol{\theta}$ represent the overall parameter vector. Starting from the assumptions detailed in the previous section, the model log-likelihood function can be written as:

$$\ell(\boldsymbol{\theta}) = \sum_{i=1}^{N} \log \left(\sum_{g=1}^{G} \eta(\boldsymbol{x}_i; \boldsymbol{\beta}_g) \int_{\mathbb{R}^D} f(\boldsymbol{y}_i \mid \boldsymbol{u}_i, z_{ig} = 1, \boldsymbol{x}_i) f(\boldsymbol{u}_i) d\boldsymbol{u}_i \right). \quad (4)$$

Due to the local independence assumption upon which the model is based, the following factorization holds:

$$f(\boldsymbol{y}_i \mid \boldsymbol{u}_i, z_{ig} = 1, \boldsymbol{x}_i) = \prod_{k=1}^{R} f(y_{ik} \mid \boldsymbol{u}_i, z_{ig} = 1, \boldsymbol{x}_i),$$

where

$$f(y_{ik} \mid \boldsymbol{u}_i, z_{ig} = 1, \boldsymbol{x}_i) = (\pi_{gk}(\boldsymbol{u}_i))^{y_{ik}} (1 - \pi_{gk}(\boldsymbol{u}_i))^{1-y_{ik}}.$$

A key aspect to notice is that the integral appearing in Equation (4) cannot be computed analytically, as it does not admit a closed-form expression. A standard solution when dealing with latent variables consists in relying on an indirect approach based on the use of an EM-type algorithm [2]. However, in the present framework, the direct application of such an algorithm is not a viable solution as it still requires the evaluation of multidimensional integrals unavailable in closed form. To solve the issue, [6] proposed to combine the EM algorithm with a variational approximation [8,13] of the integral in Eq. (4). The resulting estimation procedure corresponds to a double EM algorithm based on a variational approximation of the likelihood function. Here, clearly, the algorithm needs to be modified with respect to the original proposal by [6] to account for the inclusion of concomitant variables on the observed and latent layers of the model.

Variational Approximation. In the present framework, the objective of the variational approach is to approximate the joint conditional distribution of variables in $\boldsymbol{Y}_i$, i.e., $f(\boldsymbol{y}_i \mid \boldsymbol{u}_i, z_{ig} = 1, \boldsymbol{x}_i)$, with a lower bound that is conjugate to the Gaussian distribution assumed for $\boldsymbol{u}_i$ so as to allow for the solution of the integral in Eq. (4). This lower bound is expressed as a function of some auxiliary parameters, $\boldsymbol{\xi}_{ig} = (\xi_{ig1}, \ldots, \xi_{igR})'$, with $\xi_{igk} \neq 0$, $\forall\, k = 1, \ldots, R$, known as variational parameters. As highlighted in [8] and [13], the logarithm transform of the integral in Equation (4), i.e.,

$$\mathcal{L}_{ig} = \log(f(\boldsymbol{y}_i \mid z_{ig} = 1, \boldsymbol{x}_i)) = \log \left(\int_{\mathbb{R}^D} f(\boldsymbol{y}_i \mid \boldsymbol{u}_i, z_{ig} = 1, \boldsymbol{x}_i) f(\boldsymbol{u}_i) d\boldsymbol{u}_i \right),$$

may be approximated by the following lower bound:

$$\begin{aligned}\mathcal{L}(\boldsymbol{\xi}_{ig}) &= \log(\tilde{f}(\boldsymbol{y}_i \mid z_{ig}=1,,\boldsymbol{x}_i;\boldsymbol{\xi}_{ig}))\\ &= \log\left(\int_{\mathbb{R}^D} \tilde{f}(\boldsymbol{y}_i \mid \boldsymbol{u}_i, z_{ig}=1,,\boldsymbol{x}_i;\boldsymbol{\xi}_{ig}) f(\boldsymbol{u}_i) d\boldsymbol{u}_i\right),\end{aligned}$$

where

$$\begin{aligned}\tilde{f}(\boldsymbol{y}_i \mid \boldsymbol{u}_i, z_{ig}=1,\boldsymbol{x}_i;\boldsymbol{\xi}_{ig}) &= \prod_{k=1}^{R} \tilde{f}(y_{ik} \mid \boldsymbol{u}_i, z_{ig}=1,,\boldsymbol{x}_i;\xi_{igk}),\\ \tilde{f}(y_{ik} \mid \boldsymbol{u}_i, z_{ig}=1,\boldsymbol{x}_i;\xi_{igk}) &= \sigma(\xi_{igk})\exp\left\{\frac{A_{igk}-\xi_{igk}}{2} + \lambda(\xi_{igk})(A_{igk}^2 - \xi_{igk}^2)\right\},\\ \sigma(\xi_{igk}) &= (1+\exp\{-\xi_{igk}\})^{-1},\\ A_{igk} &= (2y_{ik}-1)(b_{gk} + \boldsymbol{x}_i'\gamma_{gk} + \boldsymbol{w}_{gk}'\boldsymbol{u}_i),\\ \lambda(\xi_{igk}) &= \frac{1/2 - \sigma(\xi_{igk})}{2\xi_{igk}}.\end{aligned}$$

After some little algebra, we have

$$\begin{aligned}\mathcal{L}(\boldsymbol{\xi}_{ig}) &= \log\left(\int_{\mathbb{R}^D} \tilde{f}(\boldsymbol{y}_i \mid \boldsymbol{u}_i, z_{ig}=1,\boldsymbol{x}_i;\boldsymbol{\xi}_{ig}) f(\boldsymbol{u}_i) d\boldsymbol{u}_i\right)\\ &= \log\Bigg(\int_{\mathbb{R}^D} \prod_{k=1}^{R} \sigma(\xi_{igk})\exp\Big\{\Big(y_{ik}-\frac{1}{2}\Big)b_{gk} + \Big(y_{ik}-\frac{1}{2}\Big)\boldsymbol{x}_i'\gamma_{gk} - \frac{\xi_{igk}}{2}\\ &+ \lambda(\xi_{igk})b_{gk}^2 + \lambda(\xi_{igk})(\boldsymbol{x}_i'\gamma_{gk})^2 - \lambda(\xi_{igk})\xi_{igk}^2 + 2\lambda(\xi_{igk})b_{gk}\boldsymbol{x}_i'\gamma_{gk}\\ &+ \frac{\boldsymbol{\mu}_{ig}'\boldsymbol{\Sigma}_{ig}^{-1}\boldsymbol{\mu}_{ig}}{2}\Big\} \mid \boldsymbol{\Sigma}_{ig} \mid^{1/2}\\ &\times (2\pi)^{-D/2} \mid \boldsymbol{\Sigma}_{ig} \mid^{-1/2} \times \exp\Big\{-\frac{1}{2}(\boldsymbol{u}_i - \boldsymbol{\mu}_{ig})\boldsymbol{\Sigma}_{ig}^{-1}(\boldsymbol{u}_i - \boldsymbol{\mu}_{ig})'\Big\} d\boldsymbol{u}_i\Bigg).\end{aligned}$$

In the last part of the above equation, it is possible to recognize the kernel of a multivariate Gaussian density with covariance matrix and mean vector given by

$$\boldsymbol{\Sigma}_{ig} = \left[\boldsymbol{I} - 2\sum_{k=1}^{R} \lambda(\xi_{igk})\boldsymbol{w}_{gk}\boldsymbol{w}_{gk}'\right]^{-1}, \tag{5}$$

$$\boldsymbol{\mu}_{ig} = \boldsymbol{\Sigma}_{ig}\left[\sum_{k=1}^{R} y_{ik} - \frac{1}{2} + 2\lambda(\xi_{igk})b_{gk} + 2\lambda(\xi_{igk})\boldsymbol{x}_i'\gamma_{gk}\right]\boldsymbol{w}_{gk}, \tag{6}$$

respectively. As we are now dealing with conjugate distributions, it is possible to derive an approximate closed form for $\mathcal{L}_{ig}$. That is,

$$\mathcal{L}_{ig} \approx \mathcal{L}(\boldsymbol{\xi}_{ig}) = \log\left(\prod_{k=1}^{R} \sigma(\xi_{igk}) \exp\left\{\left(y_{ik} - \frac{1}{2}\right)b_{gk} + \left(y_{ik} - \frac{1}{2}\right)\boldsymbol{x}_i'\boldsymbol{\gamma}_{gk} - \frac{\xi_{igk}}{2}\right.\right.$$
$$+ \lambda(\xi_{igk})b_{gk}^2 + \lambda(\xi_{igk})(\boldsymbol{x}_i'\boldsymbol{\gamma}_{gk})^2 - \lambda(\xi_{igk})\xi_{igk}^2 + 2\lambda(\xi_{igk})b_{gk}\boldsymbol{x}_i'\boldsymbol{\gamma}_{gk}$$
$$\left.\left. + \frac{\boldsymbol{\mu}_{ig}'\boldsymbol{\Sigma}_{ig}^{-1}\boldsymbol{\mu}_{ig}}{2} + \frac{\log \mid \boldsymbol{\Sigma}_{ig} \mid}{2}\right\}\right).$$

In turn, the log-likelihood function is approximated as

$$\ell(\boldsymbol{\theta}) \approx \ell(\boldsymbol{\theta}, \boldsymbol{\xi}_{ig}) = \sum_{i=1}^{N} \log\left(\sum_{g=1}^{G} \eta(\boldsymbol{x}_i; \boldsymbol{\beta}_g)\tilde{f}(\boldsymbol{y}_i \mid z_{ig} = 1, \boldsymbol{x}_i; \boldsymbol{\xi}_{ig})\right),$$

where $\tilde{f}(\boldsymbol{y}_i \mid z_{ig} = 1, \boldsymbol{x}_i; \boldsymbol{\xi}_{ig}) = \exp\{\mathcal{L}(\boldsymbol{\xi}_{ig})\}$.

EM Algorithm. Parameter estimates can be derived starting from the following complete-data log-likelihood function

$$\begin{aligned}\ell_c(\boldsymbol{\theta}) = &\sum_{i=1}^{N}\sum_{g=1}^{G} z_{ig} \log[f(\boldsymbol{y}_i \mid \boldsymbol{u}_i, z_{ig} = 1, \boldsymbol{x}_i)] \\ &+ \sum_{i=1}^{N} \log[f(\boldsymbol{u}_i)] \\ &+ \sum_{i=1}^{N}\sum_{g=1}^{G} z_{ig} \log[\eta(\boldsymbol{x}_i; \boldsymbol{\beta}_g)]. \end{aligned} \tag{7}$$

To estimate model parameters, we may proceed by iteratively alternating the steps described below.

At the generic iteration of the algorithm, the Expectation step (E-step) consists of computing the expected value of the complete-data log-likelihood function in Eq. (7), conditional on the observed data $\boldsymbol{y}_i = (y_{i1}, \dots, y_{iR})'$, and the current value of parameter estimates, $\boldsymbol{\theta}^{(t)}$. Removing the dependence on the iteration index, we need to compute

$$\begin{aligned}Q(\boldsymbol{\theta} \mid \boldsymbol{\theta}^{(t)}) = &\sum_{i=1}^{N}\sum_{g=1}^{G} \int_{\mathbb{R}^D} \hat{z}_{ig} f(\boldsymbol{u}_i \mid z_{ig} = 1, \boldsymbol{y}_i) \log[f(\boldsymbol{y}_i \mid \boldsymbol{u}_i, z_{ig} = 1, \boldsymbol{x}_i)] d\boldsymbol{u}_i \\ &+ \sum_{i=1}^{N} \int_{\mathbb{R}^D} f(\boldsymbol{u}_i \mid \boldsymbol{y}_i) \log[f(\boldsymbol{u}_i)] d\boldsymbol{u}_i \\ &+ \sum_{i=1}^{N}\sum_{g=1}^{G} \hat{z}_{ig} \log[\eta(\boldsymbol{x}_i; \boldsymbol{\beta}_g)]. \end{aligned}$$

Here, $\hat{z}_{ig}$ denotes the posterior expectation of the binary indicator variables z_{ig}'s and can be computed as

$$\hat{z}_{ig} = \frac{\hat{\eta}(\boldsymbol{x}_i; \hat{\boldsymbol{\beta}}_g) \exp\left\{\mathcal{L}(\hat{\boldsymbol{\xi}}_{ig})\right\}}{\sum_{g=1}^{G} \hat{\eta}(\boldsymbol{x}_i; \hat{\boldsymbol{\beta}}_g) \exp\left\{\mathcal{L}(\hat{\boldsymbol{\xi}}_{ig})\right\}}.$$

In the Maximization step (M-step), the following sub-steps are carried out.

1. The multinomial logit coefficients $\boldsymbol{\beta}_g$ are estimated via a Newton-Raphson algorithm by maximizing the likelihood of a multinomial logit model, with weights provided by $\hat{z}_{ig}$ derived at the E-step. Prior probabilities $\eta(\boldsymbol{x}_i; \boldsymbol{\beta}_g)$ are updated accordingly.
2. The integral in the complete-data log-likelihood is approximated by using the variational approach. To this aim, a second EM algorithm nested within the first is required, and this is described in the following.
   (a) In the second-level E-step, we compute the approximate posterior distribution of the latent trait $\boldsymbol{u}_i$, conditional on the observed variables $\boldsymbol{y}_i$, the posterior probabilities $\hat{z}_{ig}$, and the variational parameters $\hat{\boldsymbol{\xi}}_{ig}$; that is, we compute $\tilde{f}(\boldsymbol{u}_i \mid \boldsymbol{y}_i, \hat{z}_{ig} = 1; \hat{\boldsymbol{\xi}}_{ig})$, for $g = 1, \ldots, G$ and $i = 1, \ldots, N$. This approximate posterior distribution corresponds to a Gaussian distribution, with covariance matrix, $\boldsymbol{\Sigma}_{ig}$, and mean vector, $\boldsymbol{\mu}_{ig}$, given by Equations (5) and (6), respectively.
   (b) In the second-level M-step, we maximize the approximate expected value of the complete-data log-likelihood function with respect to the variational parameters, $\xi_{igk} \in \boldsymbol{\xi}_{ig}$, and the parameters defining cluster proportions, $b_{gk}, \boldsymbol{w}_{gk}$, and γ_{gk}, for $g = 1, \ldots, G$, and $k = 1, \ldots, R$.

Denoting with t and $\delta > 0$ the iteration index and the tolerance level, respectively, the two steps of the algorithm detailed so far are iterated until the following stopping rule is satisfied

$$|| \ \ell(\hat{\boldsymbol{\theta}}^{(t+1)}) - \ell(\hat{\boldsymbol{\theta}}^{(t)}) \ || < \delta.$$

2.3 Standard Errors and Model Selection

To evaluate uncertainty associated with the estimates obtained from the EM algorithm, we employ a non-parametric bootstrap approach [3], as suggested by [4]. Given the $(N \times R)$-dimensional observed data matrix, $\boldsymbol{y}$, we draw with repetition S samples having the same size of $\boldsymbol{y}$. In detail, for each bootstrap replicate, N rows of $\boldsymbol{y}$ are drawn with repetition, so that each unit can appear several times, and the proposed extended MLTA model is estimated for fixed G and D. Let $\hat{\boldsymbol{\theta}}_{(s)}$ denote the vector of estimates obtained from the s-th bootstrap sample $(s = 1, \ldots, S)$. By paying attention to the label switching issue via a sorting of model parameters at each iteration, bootstrap standard errors correspond to the square root of the diagonal elements of the following matrix:

$$\hat{V}(\hat{\boldsymbol{\theta}}) = \frac{1}{S} \sum_{s=1}^{S} (\hat{\boldsymbol{\theta}}_{(s)} - \hat{\boldsymbol{\theta}}_{(.)})(\hat{\boldsymbol{\theta}}_{(s)} - \hat{\boldsymbol{\theta}}_{(.)})',$$

with $\hat{\boldsymbol{\theta}}_{(.)}$ being the empirical mean vector, i.e., $\hat{\boldsymbol{\theta}}_{(.)} = \frac{1}{S} \sum_{s=1}^{S} \hat{\boldsymbol{\theta}}_{(s)}$.

Following standard practice in latent class and latent trait analysis, the number of clusters, G, as well as the size of the continuous latent trait, D, are not considered as model parameters, but rather as fixed quantities. To identify their optimal values, the model is estimated for increasing values of G and D. The specification corresponding to the smallest value of the chosen information criterion, such as the Bayesian Information Criterion (BIC; [12]), is selected as the optimal one. Note that when computing the BIC, a Gauss-Hermite quadrature approximation of the log-likelihood obtained at the convergence of the EM algorithm is used.

The same penalized likelihood criterion can be also used for selecting the concomitant variables to include in the model, as well as the optimal specification among the four possibilities offered by the proposed mixture of experts latent trait analyzers.

3 Conclusions

We extend the Mixture of Latent Trait Analyzers (MLTA), first introduced in the statistical literature by [6], adopting a mixture of experts perspective. This extension allows concomitant variables to influence different layers of the model: the latent layer (i.e., the cluster membership of units in the sample, as in [4] and [5]), the observed layer (i.e., the conditional outcomes' distribution), both layers simultaneously, or neither.

This formulation offers several important advantages. First, incorporating covariates into the latent layer may enhance our understanding of how observed characteristics drive cluster formation. By explicitly modeling the relationship between unit-specific features and the latent group membership, we introduce greater flexibility and interpretability with respect to the original MLTA specification, making the model better suited to capture the underlying structure of the data. Second, allowing the conditional distribution of the outcome variable to depend on covariates enables the model to explain how unit-specific characteristics shape the observed response patterns. This leads to a more refined and data-driven representation of heterogeneity, improving both predictive performance and inferential insights.

Furthermore, our approach naturally encompasses a hierarchy of models, where the effect of concomitant variables can be selectively introduced. The intermediate cases, in which covariates influence only one of the two layers (either the latent or observed structure), or neither, arise as special instances within this generalized framework. This flexibility makes the model adaptable to a broad range of applications, ensuring that it remains both parsimonious and interpretable, while still capturing essential data complexities.

References

1. Bartholomew, D., Knott, M., Moustaki, I.: Latent Variable Models and Factor Analysis: A Unified Approach 3rd ed. Wiley (2011)
2. Dempster, A.P., Laird, N.M., Rubin, D.B.: Maximum likelihood from incomplete data via the EM algorithm. J. R. Stat. B. **39**(1), 1–38 (1977)
3. Efron, B.: Bootstrap methods: another look at the jackknife. Ann. Stat. **7**(1), 1–26 (1979)
4. Failli, D., Marino, M.F., Martella, F.: Finite mixtures of latent trait analyzers with concomitant variables for bipartite networks: an analysis of COVID-19 data. Multivar. Behav. Res. **59**(4), 801–817 (2024)
5. Failli, D., Arpino, B., Marino, M.F.: A finite mixture approach for the analysis of digital skills in Bulgaria, Finland and Italy: the role of socio-economic factors Stat. Methods Appl. **33**, 1483–1511 (2024)
6. Gollini, I., Murphy, T.B.: Mixture of latent trait analyzers for model-based clustering of categorical data. Stat. Comput. **24**(4), 569–588 (2014)
7. Gormley, I.C., Fruhwirth-Schnatter, S.: Mixture of experts models. In: Fruhwirth-Schnatter, S., Celeux, G., Robert, C.P. (eds.) (1st ed.). Chapman and Hall/CRC (2019). https://doi.org/10.1201/9780429055911
8. Jaakkola, T.S., Jordan, M.I.: Bayesian logistic regression: a variational approach. In: Madigan, D., Smyth, P. (eds.) Proceedings of the 1997 Conference on Artificial Intelligence and Statistics (1997)
9. Lubke, G.H., Muthén, B.O.: Investigating population heterogeneity with factor mixture models. Psychol. Methods **10**(1), 21–39 (2005)
10. Maij-de Meij, A.M., Kelderman, H., van der Flier, H.: Fitting a mixture item response theory model to Personality Questionnaire Data: characterizing latent classes and investigating possibilities for improving prediction. Appl. Psychol. Meas. **32**(8), 611–631 (2008)
11. Rost, J.: Rasch models in latent classes: an integration of two approaches to item analysis. Appl. Psychol. Meas. **14**(3), 271–282 (1990)
12. Schwarz, G.: Estimating the dimension of a model. Ann. Stat. **6**, 461–464 (1978)
13. Tipping, M.E.: Probabilistic visualisation of high-dimensional binary data. In: Proceedings of the 11th International Conference on Neural Information Processing Systems (NIPS'98), pp. 592–598. MIT Press (1998)

An Hybrid Preference Learning Framework to Refine the Consensus Ranking

Maurizio Romano(✉) and Gianpaolo Zammarchi

Department of Business and Economic, University of Cagliari, Cagliari, Italy
{romano.maurizio,gianpaolo.zammarchi}@unica.it

Abstract. The study of preference rankings (p. learning) is becoming increasingly important in many scientific fields. Preferences are expressed when a group of judges (raters) evaluate a collection of elements (items), assigning an order to objects based on which ones are preferred over others. When there are many judges and also a large number of items to be evaluated, an aggregate measure is needed in order to solve the rank aggregation problem, providing an interpretable comparison of the ranked items and assessing the overall level of agreement among judges. The rank aggregation problem is NP-hard because it becomes more difficult as the number of items increases significantly. Approaches such as the branch-and-bound can be applied to problems with a limited number of items (i.e. < 200). When the number of items grows, heuristic techniques have been developed to provide approximate solutions. Many of these heuristics are based on Kemeny's axiomatic approach, which has shown to be valid with tied rankings. In this paper, we propose a framework that aims at providing more than just a choice between *a* slow but extremely accurate algorithm and *a faster but not so accurate one.* Thus, following a hybrid approach, a trade-off between the two options is permitted. A simulation study shows the performance of the proposal in a controlled environment. Furthermore, a real world data set with a large number of items is considered. Results show significant improvements in the solution found, with a reasonable additional amount of computational time. This improvement is mainly investigated whilst using the recently proposed PSOPR algorithm (as the faster one) and the state of the art QUICK (as the slowest one).

Keywords: Preference learning · Kemeny problem · tied rankings · heuristics · particle swarm optimization

MSC Classification: 62-08 · 90-08 · 68W25

1 Introduction

Preference rankings is an approach based on the analysis of the preferences expressed by a group of judges (or raters) who evaluate a collection of objects (or

G. Zammarchi—Contributing authors.

A. D'Ambrosio et al. (Eds.): CLADAG-VOC 2025, SCDAKO, pp. 282–293, 2025.
https://doi.org/10.1007/978-3-032-03042-9_25

items). According to Fürnkranz and Hüllermeier [1], preference learning is a more recent term that encompasses the traditional field of preference ranking analysis [2]. The concepts of judges and items can be very broad. Different authors have defined judges as e.g. individuals, agencies, online platform users, whereas the item may be a service, a university, genes, and so forth [see, for example, 3–5]. The problems that arise when a large number of items are evaluated by a large number of judges are the computational complexity (the rank aggregation problem is an NP-hard problem) and the definition of an aggregate ranking. In some fields, this problem is defined as social choice problem, or also rank aggregation problem, or also Kemeny problem, median ranking problem, central ranking problem [6,7].

The problem of aggregating preferences relies upon formal axiomatic statements that specify the geometric space in which the problem is to be characterized as well as the metrics of agreement between rankings [8]. In this paper, we adopt the Kemeny distance, which has been shown to satisfy all of the Kemeny and Snell's axioms [8], imposing the definition of a strict and generally recognized set of axioms that any distance defined on preference rankings should adhere to. The Kemeny distance is strictly connected to the well-known Kendall's distance and is fully defined by the permutation polytope, the convex hull of preference rankings with n items in a $\mathbb{R}^n$, which has become the geometric reference space for preference rankings [2,9–11]. Other distances can also be used, although they may present problems. One of the issues underlying the possible problems relies upon considering only the case of full ranking and not the case of ties. For example, Spearman's distance is influenced by the so-called "irrelevant alternatives" [6], whilst Kendall's distance violates the triangular inequality. Situations like this can lead to inconsistent results during the rank aggregation process. For further details, refer, for instance, to Marden [2]. From an interpretative perspective, when the number of items to be ranked is large, it is generally assumed that the universe of rankings contains all possible tied rankings. That is, tied rankings are not considered expressions of indifference but rather "positive statements of agreement" [6]. As previously noted, the rank aggregation problem is extremely complex and is addressed, in particular when the number of elements to be ranked is high, with heuristic solutions [see for instance, 12–18].

Based on a recently published method, the Particle Swarm Optimization for Preference Rankings (PSOPR) [19] we developed an approach that tackles the problem in two steps: the first step is to realize an approximate solution using existing algorithms, without sacrificing too much on quality. The second step is based on the result of the first one, significantly reducing the computational burden and producing a more accurate solution. Particle Swarm Optimization was originally designed to optimize continuous functions and, therefore, would not be suitable for preference rankings data, but PSOPR has been specifically developed to be applied in the discrete case. We provide an illustrative example where all types of rankings (full, constrained, partial) were addressed using our method, and we show the results.

2 The Kemeny Axiomatic Framework

Rank data can be represented as n-dimensional rank vectors where the i-th element of the vector indicates the position for that element relative to the rest, after the placement of the elements is established in any arbitrary order [2]. The best case scenario is when the ranking is complete and each element has an unique position (full rankings), but that's not always the case. There are two other possible cases: one is when the ranking is incomplete (partial ranking) and not all the elements have been evaluated so their position in the ranking is not known, whilst the other is when some elements have been evaluated in the same way (tied ranking) and therefore occupy the same position.

Consider two rankings of n items, X and Y. The two ranks can be expressed as the *score matrices* **X** and **Y**, which are skew-symmetric matrices of generic elements x_{ij} and y_{ij}, respectively.

$$x_{ij}\ (y_{ij}) = \begin{cases} 1 & \text{if object } i \text{ is ranked ahead of object } j \\ -1 & \text{if object } i \text{ is ranked behind object } j \\ 0 & \text{if the objects are tied, or if } i = j \end{cases} \tag{1}$$

The distance between two rankings, defined by Kemeny and Snell [8], can be expressed as $d(X,Y) = \frac{1}{2}\sum_{i,j=1}^{n}|x_{ij} - y_{ij}|$, which is called the Kemeny distance and has some important properties: (1) is a metric; (2) is invariant under addition or subtraction of equally ranked first and/or last objects; (3) is invariant under the same permutation of all objects; (4) its minimum positive value is equal to one.

For the set of all the possible rankings with n items, Good [20] showed that $|\mathcal{H}^n| = \sum_{r=0}^{n} r!\left\{{n \atop r}\right\}$, where $|\mathcal{H}^n|$ is the size of are all the possible ways to partition a set of n objects (i.e. $\mathcal{H}^n$) into r non-empty subsets (i.e. $\left\{{n \atop r}\right\}$), sometimes called 'buckets'. Moreover, we define $A_1, \ldots, A_m$ as a set of rankings of n items produced by m judges. An aggregate measure is the median ranking $\hat{S}$ which is defined as one (or more) ranking(s) that minimizes the sum of distances $\hat{S} = \underset{S\in\mathcal{H}^n}{\arg\min}\sum_{k=1}^{m} d(A_k, S)$.

Clearly, even when dealing with a limited number of items, the computational burden becomes unsustainable. For this reason, several scholars have begun to focus on an approximate solution. For example, Emond and Mason [6] proposed an algorithm based on the decomposition of the original problem into simpler subproblems capable of finding the exact solution in a short amount of time, even when there is more than one solution (i.e. branch-and-bound).

Additionally, Emond and Mason [6] proposed a different score matrix specific for tied rankings. Let x_{ij} be the generic elements of the score matrix **X** related to the ranking X: this element will be equal to 1 if either the i-th object is preferred to the j-th item or if they are in a tie. Thus, we can define $\tau_X(X,Y)$ as:

$$\tau_X(X,Y) = \sum_{i,j=1}^{n} \frac{x_{ij}y_{ij}}{n(n-1)}. \tag{2}$$

$\tau_X(X, Y)$ provides a clear interpretation of the rank aggregation solution, since

$$\hat{S} = \underset{S \in \mathcal{H}^n}{\arg\min} \sum_{k=1}^{m} d(A_k, S) = \underset{S \in \mathcal{H}^n}{\arg\max} \frac{1}{mn(n-1)} \sum_{i,j=1}^{n} c_{ij} s_{ij}, \quad (3)$$

where c_{ij} are the elements of the *combined input matrix* $\mathbf{C}$, which is the sum of the m score matrices of the sample of judges, namely $\mathbf{C} = \sum_{j=1}^{m} \mathbf{A}_j$, and s_{ij} are the elements of the score matrix of the consensus ranking $\hat{S}$. Basically, when comparing two solutions, it is possible to rely on a single value to evaluate which solution is better, since the one with the smallest τ_X value will be the one that minimizes the distances.

3 Particle Swarm Optimization for Preference Rankings (PSOPR)

Whilst trying to replicate bird flocks and fish schools behaviors in an algorithm, Particle Swarm Optimization (PSO) methodology has been introduced as an effective optimization heuristic that focuses on a loss function to iteratively improve a certain starting solution [21]. The philosophy of PSO is to consider a swarm of birds (i.e. particles) that are looking for the global optimum whilst moving into a multidimensional search space. Furthermore, each particle has both a cognitive (i.e. individual) memory of the best solution found, and a social (i.e. global) memory of the best solution that all swarm' particles have found. This implies that each particle can impact on each others' movements [22]. Thus, whilst computing the new position, each member of the swarm consider a combination of the following: a random component, the last movement vector, cognitive and social memories. Romano et al. [19] proposed the PSOPR algorithm (see Algorithm 1), a modified version of the original implementation of PSO [23] that works with discrete data. Thus allowing it to find the consensus ranking of some preference data. In the following, the previously mentioned heuristic is summarized. Let $\mathbf{A}$ be a $m \times n$ matrix composed of m judges expressing preferences for n items, and W_k as a frequency weight vector. According to the default PSOPR parameters, $n_p = \lfloor 10 + 2\sqrt{n} \rceil$ particles are defining the swarm (i.e. vectors of n positive integers each) $P \subset (\mathbb{Z}^+)^n \in [\mathbf{b_l}; \mathbf{b_u}]$ s.t.:

$$\mathbf{b_l} = (\underbrace{1, \quad 1, \quad \cdots, \quad 1}_{n \text{ elements}}) \qquad \mathbf{b_u} = (\underbrace{n, \quad n, \quad \cdots, \quad n}_{n \text{ elements}}),$$

where $\lfloor \cdot \rceil = \text{round}(\cdot)$, $\mathbb{Z}^+$ is the set of all positive integers, and $\mathbf{b_l}$ ($\mathbf{b_u}$) is the lower (upper) bound vector used to represent the range of values (i.e. rank) for each of the n items. Considering that PSOPR has to deal with rankings, differently from a classical PSO, each value of the bounds (i.e. each dimension) is always fixed. In fact, by definition of ranking with permitted ties, each item rank can vary in the following range $[1; n]$. Considering the Euclidean norm $||\cdot||$, the search space diameter (i.e. swarm' breadth) is $d = ||\mathbf{b_u} - \mathbf{b_l}||$.

Each particle $\mathbf{p}_i \in P$ with $i = 1, \ldots, n_p$ is composed by three components: position $\mathbf{p_i.pos}$, velocity $\mathbf{p}_i.v$, and the personal (i.e. individual) best solution found by the particle $\mathbf{p_i.pbest}$. The position $\mathbf{p_i.pos}$, represents a possible consensus ranking whilst each value of the n dimensions of the particle is the rank of the assigned item (ties are allowed). The velocity $\mathbf{p}_i.v$ is a parameter used to compute how far a certain particle should move from the previous $\mathbf{p_i.pos}_t$ to the next one $\mathbf{p_i.pos}_{t+1}$, whilst the inertia $\omega = \frac{1+2\ln(2)}{8}$ defines the amount of velocity to keep at each update. Considering that each particle keep the personal best solution found, the final solution estimated in the swarm is the global best $\mathbf{p_{best}} = \max(\mathbf{p_i.pbest}) \quad \forall \mathbf{p}_i \in P$. The goodness of the estimated solution is evaluated using the τ_X (Eq. 2) coefficient. Inizializing each particle is the first phase of PSOPR. Thus, position and velocity are randomly assigned by sampling with replacement from a uniform distribution of integers $U\{\min, \max\}$, s.t.:

$$\mathbf{p_i.pos} \sim U\{1, n\} \quad \mathbf{p}_i.v \sim U\{-|n-1|, |n-1|\}. \tag{4}$$

After that, there is an update phase. Each particle starts moving to a new position by first updating the velocity:

$$\mathbf{p}_i.v_{t+1} = \omega \cdot \mathbf{p}_i.v_t + \phi_p \cdot \mathbf{r_p} \cdot (\mathbf{p_i.pbest}_t - \mathbf{p_i.pos}_t) + \phi_g \cdot \mathbf{r_g} \cdot (\mathbf{p_{best}.pbest}_t - \mathbf{p_i.pos}_t) \tag{5}$$

where $\mathbf{r_p} \sim U(0,1)$ and $\mathbf{r_g} \sim U(0,1)$ are two random numbers sampled with replacement from a uniform distribution of reals $U(\min, \max)$. A fundamental stochastic component is considered for the particle new position by $\mathbf{r_p}$ and $\mathbf{r_g}$. They act as weights, making the particle randomly considering more the personal best ($\mathbf{p_i.pbest}$) position and/or the swarm best ($\mathbf{p_{best}.pbest}$) position respectively; ϕ_p and ϕ_g are two constants $\left(\phi_p = \phi_g = \frac{1+2\ln(2)}{2}\right)$, working as weights to a-priori force the particle to quantify how much it follows the personal best and/or the swarm best directions, respectively. Thus, the updated velocity provides the information to move the particle to a new place s.t. $\mathbf{p_i.pos}_{t+1} = \lfloor \mathbf{p_i.pos}_t + \mathbf{p_i.v_{t+1}} \rceil$. Therefore, the updated particle position is evaluated, and both $\mathbf{p_i.pbest}$ and $\mathbf{p_{best}}$ are also updated according to the new values of $\tau_X(\mathbf{p_i.pos}, \mathbf{A}, W_k)$ computed for each $\mathbf{p}_i \in P$.

The process is iterated until one of the following stopping criteria is met: (1) maximum number of iterations reached (max_j). This count does not reset if the algorithm restarts; (2) maximum number of "stagnate" (i.e. no improvement) iterations (max_s); (3) some particular cases (all-tie solution, $\tau_X(\cdot) = 1$).

Moreover, to avoid cases such as the so call *stagnate* (i.e. all particles stacked together) which could be a local maximum if not the global, a lower bound threshold to allows restarting PSOPR $\alpha_r = 0.0018$ is specified; α_r is the percentile of the swarm' breadth $d = \|\mathbf{b_u} - \mathbf{b_l}\|$, where $\mathbf{b_l}$ and $\mathbf{b_u}$ are updated at each iteration as follows:

$$\mathbf{b_l} = \min \mathbf{p_i.pos} \qquad \mathbf{b_u} = \max \mathbf{p_i.pos} \qquad \forall \mathbf{p_i} \in P \tag{6}$$

Algorithm 1. Particle Swarm Optimization for Preference Rankings (PSOPR)

Input:

$\mathbf{A}$; n; m; $W_k = \text{NULL}$; $n_p = \lfloor 10 + 2\sqrt{n} \rceil$;

$\omega = \frac{1 + 2\ln(2)}{8}$; $\phi_p = \frac{1 + 2\ln(2)}{2}$; $\phi_g = \frac{1 + 2\ln(2)}{2}$;

$max_j = 1000$; $max_s = \infty$; $\alpha_r = 0.0018$;

Phase 1: Initialization
$P = \{\mathbf{p}_1, \ldots, \mathbf{p}_{n_p}\}$
$\mathbf{b_l} = (\underbrace{1, \ 1, \ \cdots, \ 1}_{n \text{ elements}})$
$\mathbf{b_u} = (\underbrace{n, \ n, \ \cdots, \ n}_{n \text{ elements}})$
$d = \|\mathbf{b_u} - \mathbf{b_l}\|$
$\alpha_r = d \cdot \alpha_r$
$\mathbf{p_{best}} = \mathbf{p}_1$
for $i = 1$ **to** n_p **do**
    $\mathbf{p_i.pos} \sim U\{1, n\}$ **Eq. 4**
    $\mathbf{p_i}.v \sim U\{-|n-1|, |n-1|\}$ **Eq. 4**
    $\mathbf{p_i.pbest} = \mathbf{p_i.pos}$
    if $\tau_X(\mathbf{p_i.pos}, \mathbf{A}, W_k) > \tau_X(\mathbf{p_{best}.pos}, \mathbf{A}, W_k)$ **then**
        $\mathbf{p_{best}} = \mathbf{p}_i$
    end if
end for

Phase 2: Iterative update
for $j = 1$ **to** max_j **do**
    if A stopping criterion is met then
        STOP
    end if
    for $i = 1$ **to** n_p **do**
        $\mathbf{r_p}, \mathbf{r_g} \sim U(0, 1)$
        $\mathbf{p_i}.v = \omega \cdot \mathbf{p_i}.v + \phi_p \cdot \mathbf{r_p} \cdot (\mathbf{p_i.pbest} - \mathbf{p_i.pos}) + \phi_g \cdot \mathbf{r_g} \cdot (\mathbf{p_{best}.pbest} - \mathbf{p_i.pos})$
        $\mathbf{p_i.pos} = \lfloor \mathbf{p_i.pos} + \mathbf{p_i}.v \rceil$
        if $\tau_X(\mathbf{p_i.pos}, \mathbf{A}, W_k) > \tau_X(\mathbf{p_i.pbest}, \mathbf{A}, W_k)$ **then**
            $\mathbf{p_i.pbest} = \mathbf{p_i.pos}$
        end if
        if $\tau_X(\mathbf{p_i.pos}, \mathbf{A}, W_k) > \tau_X(\mathbf{p_{best}.pos}, \mathbf{A}, W_k)$ **then**
            $\mathbf{p_{best}} = \mathbf{p}_i$
        end if
    end for
    $\mathbf{b_l} = \min \mathbf{p_i.pos} \quad \forall \mathbf{p}_i \in P$
    $\mathbf{b_u} = \max \mathbf{p_i.pos} \quad \forall \mathbf{p}_i \in P$
    $d = \|\mathbf{b_u} - \mathbf{b_l}\|$
    if $d < \alpha_r$ **then**
        Restart
    end if
end for

Output: $\{\hat{\mathbf{S}} = \mathbf{p_{best}.pos}\}$; $\{\hat{S}_{\tau_X} = \tau_X(\mathbf{p_{best}.pos}, \mathbf{A}, W_k)\}$

Once one of the stopping criteria is met the algorithm terminates, producing in output the estimated solution $\hat{\mathbf{S}} = \mathbf{p_{best}.pos}$ that is the consensus ranking for the data matrix $\mathbf{A}$, and the associated measure of the adequacy of the solution $\hat{S}_{\tau_X} = \tau_X(\mathbf{p_{best}.pos}, \mathbf{A}, W_k)$.

4 The Hybrid-COnsensus Refinement Framework (h-CORF)

Considering the state of the art, PSOPR have been previously compared [19] whilst relying on the `ConsRank` and `RankAggSIgFUR` `R` packages [24,25], which implements the following algorithms for finding the consensus ranking: DECOR, QUICK, FAST, and SIGFUR [12,14,26,27]. PSOPR obtains τ_X values at the same levels of the other state of the art algorithms but it is much faster. In fact, computational time of PSOPR is almost invariant to n whilst depending only on the number of particles n_p. Therefore, despite still being feasible for $n \leq 100$, these results suggest using PSOPR for a data set with $n \geq 250$, especially if $n \gg 250$.

Nevertheless, there is still room for improvement. In this section, the hybrid-COnsensus Refinement Framework (h-CORF) is introduced as a methodology to improve the quality of the consensus ranking produced by a faster algorithm (i.e. PSOPR) by using more accurate and slower ones (i.e. QUICK) when $n \gg 250$. As the name suggests, the basic idea is to refine a portion of good previously produced consensus ranking. Whilst comparing two good solutions (S_1 and S_2) it is unlikely that a certain item is ranked as one of the top in S_1 and in one of the bottom in S_2. Thus, the idea is to run a slower (but more accurate) algorithm considering only a portion of the data that we call head and tail for top and bottom ranked items respectively. Whilst for a small dataset ($n \ll 250$), this approach might be viewed as a compromise of the goodness of the results and computational time, things change when most accurate algorithms are not feasible anymore ($n \gg 250$). In fact, such a framework permits selective running of those algorithms where, usually, it is not possible.

To assert the usefulness of h-CORF, a comparison with the results obtained in Romano et al. [19] is conducted in terms of two main aspects: goodness of the estimated solution (in terms of τ_X), and computational efficiency (in terms of running time expressed in seconds). In the following, the same simulation study is first compared to assess the performance of h-CORF in a controlled environment (Sect. 4.1). Finally, our empirical investigation is completed in the analysis of a real world and public data sets presenting a large number of items (Sect. 4.2).

4.1 Simulation Study

Following the same simulation study adopted in the first introduction of PSOPR [19], we used the `rgmm` function implemented in the `PerMallows` `R` package [28]. Considering a number of generated items (n), a central permutation (σ), and a dispersion (i.e. concentration) parameter θ, a sample of m permutations is generated from a Generalized Mallows Model $GMM(n, \sigma, \theta)$. Note that the greater the θ value, the greater the concentration of the rankings in the sample of judges around the consensus ranking. Moreover, a θ equals to zero implies that each ranking has an equal probability of being the consensus ranking.

Considering the two main aspects needed to compare the algorithms (goodness of the estimated solution and computational efficiency), different sets of data are sampled from GMM with all the combinations of the following parameters: $m = 20$, $n \in \{100, 250, 500, 1000\}$, $\sigma = \{1, \ldots, n\}$, $\theta \in \{0.0125, 0.1, 0.8\}$.

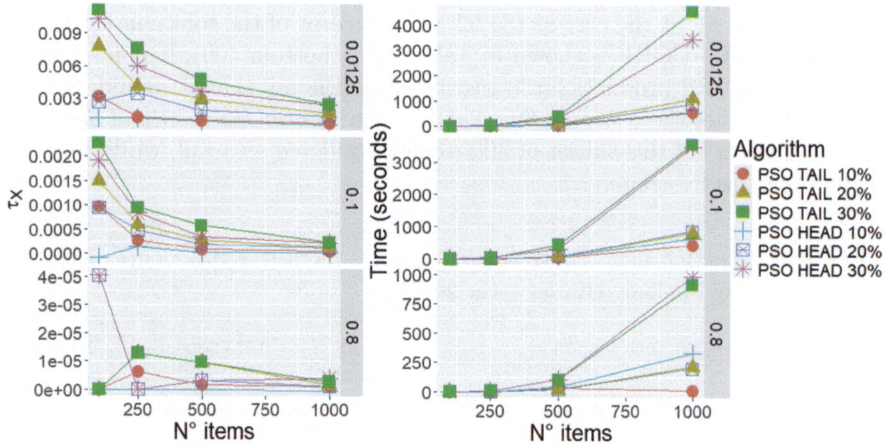

Fig. 1. Varying the dispersion parameter (θ) and the number of items (n): **[left panel]** algorithms improvement of the best solution found by PSOPR (the higher τ_X, the better); **[right panel]** algorithms additional computational time (in seconds) within respect to standard PSOPR

As it is possible to notice in Fig. 1 the hybrid approach improves the quality of the consensus ranking produced by PSOPR. This is especially true for more *noised* data (i.e. $\theta = 0.0125$) and, obviously, with a larger window (i.e. 30%). Notably, the improvement is more evident whilst refining the tail rather than the head. As we can expect, the larger the considered tail (head) the greater the amount of time required for the refinement. Thus, as a trade-off between improvement and additional needed time, simulations results suggest using the tail 30% approach for $n < 500$ and tail 20–10% for $n \gg 500$.

4.2 Global University Rankings

A huge ($n = 1,173$) public data set is now considered. Available at preflib.org [29], Boehmer and Schaar [30] provided the *Global University Ranking* data set, that contains rankings of universities according to different criteria for year 2015 (i.e. file *00046-00000004.soi*) but solutions are not known nor published by the authors. Therefore, considering that the Branch and Bound (BB) algorithm can't produce solutions in acceptable time with a huge n, with few obvious exceptions (i.e. all judges with the same rankings), there is no way to know the "real" solution.

Considering the above-mentioned simulation study and $n \gg 500$, h-CORF is executed with the tail 10% approach.

As shown in Romano et al. [19], the only algorithm (in the set of considered ones) able to estimate a solution with a huge number of items in an acceptable computational time is the PSOPR ($\tau_X = 0.4375569$, time = 6.579 s). Executing h-CORF further improves the consensus ranking by adding relevant +0.03770116 to the previous τ_X at the cost of 194.92 s. An excerpt of the solutions estimated by PSOPR and h-CORF is shown in Table 1 (i.e. bottom 400). It is possible to notice that PSOPR produces far less ex-aequo in the tail than it should. In fact, h-CORF identifies the *National Sun Yat-Sen University* (*University of Seoul*) as the top (bottom) of this subset of 400 items considered as a tail, whilst putting all the others as ex-aequo.

Table 1. Bottom 400 universities as estimated by PSOPR and h-CORF using the Global University Ranking data set (year 2015)

PSOPR ($\tau_X = 0.4375569$, time = 6.579 s)		h-CORF ($\tau_X = 0.47525806$, time = 201.499 s)	
Rank	University	Rank	University
772	{Indian Institute of Technology Kanpur	772	National Sun Yat-Sen Univ.
773	La Trobe Univ.	773	{Örebro Univ.
774	National Polytechnic Institute	774	École centrale de Lyon
775	Open Univ. (UK)	775	École Centrale Paris
776	State Univ. of New York Upstate Medical Univ.	776	École normale supérieure de Cachan
777	Univ. of Patras	777	École Polytechnique de Montréal
778	Univ. of Pretoria}	778	Adam Mickiewicz Univ. in Pozna
...	...	...	...
1163	{Alexandria Univ.	1163	Wrocław Univ. of Technology
1164	Babe-Bolyai Univ.	1164	Xiangtan Univ.
1165	China Pharmaceutical Univ.	1165	Xidian Univ.
1166	Federal Univ. of Ceará	1166	Yamagata Univ.
1167	Henan Normal Univ.	1167	Yamaguchi Univ.
1168	Hunan Normal Univ.	1168	Yangzhou Univ.
1169	Southwest Jiaotong Univ.	1169	Yeungnam Univ.
1170	Stockholm School of Economics	1170	Yuan Ze Univ.
1171	Univ. of A Coruña	1171	Zhejiang Univ. of Technology
1172	Univ. of the Algarve	1172	Zhengzhou Univ.}
1173	Yuan Ze Univ.}	1173	Univ. of Seoul

5 Concluding Remarks

In this paper, the hybrid-COnsensus Refinement Framework for the rank aggregation problem completely defined within the Kemeny's axiomatic approach has been introduced. Principally born to refine PSOPR solutions, it can be used with any algorithm that follows the same approach. For instance, it can be used to improve DECOR (QUICK) whilst using QUICK (Branch and Bound). A comparison with an already proposed differential evolution algorithm and PSOPR shows that the quality of the results can be improved by refining them with h-CORF. Thus, this framework allows the partial use of branch-and-bound-like proposals also in contexts that have a huge number of objects ($n \gg 250$). A simulation study and a practical application of the rank aggregation problem with real data were shown. Results prove the quality improvement, unbinding the user from the binary choice (*a* slow but extremely accurate algorithm or *a faster but not so accurate one*) whilst permitting a trade-off between them as such a hybrid approach is. The latest has been shown to be consistent not only in real data but also in a simulation study.

Declarations

Funding. Maurizio Romano acknowledge financial support under the National Recovery and Resilience Plan (NRRP), Mission 4 Component 2 Investment 1.5 - Call for tender No.3277 published on December 30, 2021 by the Italian Ministry of University and Research (MUR) funded by the European Union – Next Generation EU. Project Code ECS0000038 – Project Title eINS Ecosystem of Innovation for Next Generation Sardinia – CUP F53C22000430001- Grant Assignment Decree No. 1056 adopted on June 23, 2022 by the MUR.
This study was funded by the European Union - NextGenerationEU, Mission 4, Component 2, in the framework of the GRINS -Growing Resilient, INclusive and Sustainable project (GRINS PE00000018 – CUP F53C22000760007). The views and opinions expressed are solely those of the authors and do not necessarily reflect those of the European Union, nor can the European Union be held responsible for them.

References

1. Fürnkranz, J., Hüllermeier, E.: Preference learning and ranking by pairwise comparison. In: Fürnkranz, J., Hüllermeier, E. (eds.) Preference Learning: An Introduction, pp. 1–17. Springer, Heidelberg (2011). https://doi.org/10.1007/978-3-642-14125-6_1
2. Marden, J.I.: Analyzing and Modeling Rank Data. Chapman & Hall, London (1996)
3. Csató, L., Tóth, C.: University rankings from the revealed preferences of the applicants. Eur. J. Oper. Res. **286**(1), 309–320 (2020)
4. D'Ambrosio, A., Iorio, C., Staiano, M., Siciliano, R.: Median constrained bucket order rank aggregation. Comput. Stat. **34**(2), 787–802 (2019). https://doi.org/10.1007/s00180-018-0858-z
5. Vitelli, V., et al.: Transcriptomic pan-cancer analysis using rank-based Bayesian inference. Mol. Oncol. **17**(4), 548–563 (2023). https://doi.org/10.1002/1878-0261.13354
6. Emond, E.J., Mason, D.W.: A new rank correlation coefficient with application to the consensus ranking problem. J. Multi-Criteria Decis. Anal. **11**(1), 17–28 (2002)
7. Crispino, M., Antoniano-Villalobos, I.: Informative priors for the consensus ranking in the Bayesian mallows model. Bayesian Anal. **1**(1), 1–24 (2022)

8. Kemeny, J.G., Snell, J.L.: Preference rankings: an axiomatic approach. In: Kemeny, J.G., Snell, J.L. (eds.) Mathematical Models in the Social Sciences, pp. 9–23. The MIT Press Cambridge (1962)
9. Thompson, G.L.: Generalized permutation polytopes and exploratory graphical methods for ranked data. Ann. Stat. **21**(3), 1401–1430 (1993)
10. Heiser, W.J., D'Ambrosio, A.: Clustering and prediction of rankings within a kemeny distance framework. In: Lausen, B., Poel, D., Ultsch, A. (eds.) Algorithms from and for Nature and Life, pp. 19–31. Springer, Cham (2013). https://doi.org/10.1007/978-3-319-00035-0_2
11. Alvo, M., Yu, P.L.H.: Statistical Methods for Ranking Data, vol. 1341. Springer, New York (2014)
12. D'Ambrosio, A., Amodio, S., Iorio, C.: Two algorithms for finding optimal solutions of the Kemeny rank aggregation problem for full rankings. Electron. J. Appl. Stat. Anal. **8**(2), 198–213 (2015). https://doi.org/10.1016/j.ejor.2015.08.048
13. Amodio, S., D'Ambrosio, A., Siciliano, R.: Accurate algorithms for identifying the median ranking when dealing with weak and partial rankings under the kemeny axiomatic approach. Eur. J. Oper. Res. **249**(2), 667–676 (2016)
14. D'Ambrosio, A., Mazzeo, G., Iorio, C., Siciliano, R.: A differential evolution algorithm for finding the median ranking under the kemeny axiomatic approach. Comput. Oper. Res. **82**, 126–138 (2017)
15. Aledo, J.A., Gámez, J.A., Rosete, A.: Approaching rank aggregation problems by using evolution strategies: the case of the optimal bucket order problem. Eur. J. Oper. Res. **270**(3), 982–998 (2018)
16. Badal, P.S., Das, A.: Efficient algorithms using subiterative convergence for kemeny ranking problem. Comput. Oper. Res. **98**, 198–210 (2018)
17. Aledo, J.A., Gámez, J.A., Rosete, A.: A highly scalable algorithm for weak rankings aggregation. Inf. Sci. **570**, 144–171 (2021)
18. Acampora, G., Iorio, C., Pandolfo, G., Siciliano, R., Vitiello, A.: A memetic algorithm for solving the rank aggregation problem. In: Hošková-Mayerová, Š, Flaut, C., Maturo, F. (eds.) Algorithms as a Basis of Modern Applied Mathematics. SFSC, vol. 404, pp. 447–460. Springer, Cham (2021). https://doi.org/10.1007/978-3-030-61334-1_23
19. Romano, M., Conversano, C., Siciliano, R., D'Ambrosio, A.: Particle swarm optimization for preference rankings. Adv. Data Anal. Classif. (2025). https://doi.org/10.1007/s11634-025-00626-9
20. Good, I.: The number of orderings of n candidates when ties are permitted. Fibonacci Q. **13**, 11–18 (1975)
21. Kennedy, J., Eberhart, R.: Particle swarm optimization. In: Proceedings of ICNN'95 - International Conference on Neural Networks, vol. 4, pp. 1942–1948 (1995). https://doi.org/10.1109/ICNN.1995.488968
22. Meissner, M., Schmuker, M., Schneider, G.: Optimized particle swarm optimization (OPSO) and its application to artificial neural network training. BMC Bioinform. **7**(1), 125 (2006). https://doi.org/10.1186/1471-2105-7-125
23. Claus, B.: PSO: Particle Swarm Optimization. (2022). R package version 1.0.4. https://CRAN.R-project.org/package=pso
24. D'Ambrosio, A.: ConsRank: Compute the Median Ranking(s) According to the Kemeny's Axiomatic Approach. (2023). R package version 2.1.4. https://CRAN.R-project.org/package=ConsRank
25. Parker, H., Singh, R., Badal, P.S.: RankAggSIgFUR: Polynomially Bounded Rank Aggregation Under Kemeny's Axiomatic Approach (2023). R package version 1.0.0. https://CRAN.R-project.org/package=RankAggSIgFUR

26. Albano, A., Plaia, A.: Element weighted Kemeny distance for ranking data. Electron. J. Appl. Stat. Anal. **14**(1), 117–145 (2021). https://doi.org/10.1285/i20705948v14n1p117
27. Badal, P.S., Das, A.: Efficient algorithms using subiterative convergence for kemeny ranking problem. Comput. Oper. Res. **98**, 198–210 (2018). https://doi.org/10.1016/j.cor.2018.06.007
28. Irurozki, E., Calvo, B., Lozano, J.A.: PerMallows: an R package for mallows and generalized mallows models. J. Stat. Softw. **71**(12), 1–30 (2016). https://doi.org/10.18637/jss.v071.i12
29. Mattei, N., Walsh, T.: Preflib: a library of preference data. In: Proceedings of the 3rd International Conference on Algorithmic Decision Theory (ADT 2013). Lecture Notes in Artificial Intelligence. Springer, ADT 2013, Bruxelles, Belgium (2013). http://preflib.org
30. Boehmer, N., Schaar, N.: Collecting, classifying, analyzing, and using real-world elections. arXiv preprint arXiv:2204.03589 (2022)

Topic Homogeneity Test-Based Fuzzy Document Clustering

Gian Mario Sangiovanni[1,3](✉), Louisa Kontoghiorghes[2,3], Ana Colubi[2,3], and Maria Brigida Ferraro[1,3]

[1] Department of Statistical Sciences, Sapienza University of Rome, P.le Aldo Moro 5, 00185 Rome, Italy
gianmario.sangiovanni@uniroma1.it
[2] King's Business School of London, Bush House, 30 Aldwych, London, WC2B 4BG, UK
[3] Faculty of Economics and Business Studies, Justus Liebig University of Giessen, Licher Str. 64, 35394 Giessen, Germany

Abstract. A new fuzzy document clustering algorithm based on topic homogeneity is introduced. In detail, a novel dissimilarity measure is proposed, derived from the p-value of a hypothesis test that assesses the homogeneity of topic distributions between two documents. First, the topic distributions are derived through Latent Dirichlet Allocation, and then a bootstrap procedure is applied to obtain the p-value. Finally, the resulting dissimilarity matrix is integrated into the fuzzy relational clustering procedure. The performance of the proposal is evaluated using a benchmark dataset.

Keywords: document clustering · topic homogeneity test · bootstrap · latent Dirichlet allocation

1 Introduction

Document clustering involves partitioning text documents into similar, homogeneous clusters based on their content. It finds extensive applicability across various domains, from large-scale web searches [7] to hierarchical models that classify web services using concise descriptors [15]. Additionally, it plays a pivotal role in the automatic organization of large corpora [3].

To extract meaningful information that can be utilized in clustering methods that rely on numerical data, a corpus must be transformed into a numerical representation suitable for clustering algorithms. This process is known as the Vector Space Model (VSM) [23], which translates a collection of documents into numerical features. In detail, it is the statistical model used to determine the relevance between the documents in the collection and the words within each document. In the VSM, the documents are encoded by a set of multidimensional features spanned by the term vectors representing the vocabulary (obtained as the remaining list of unique words after performing the pre-processing). Thus,

A. D'Ambrosio et al. (Eds.): CLADAG-VOC 2025, SCDAKO, pp. 294–303, 2025.
https://doi.org/10.1007/978-3-032-03042-9_26

in the VSM, a corpus of M documents with $|V|$ unique terms is converted into an $M \times |V|$ matrix, known as Document Term Matrix (DTM). The main issue, depending on the VSM representation, is that the relatively large number of words within a document results in high dimensionality.

This complexity can significantly hinder the effectiveness of many clustering algorithms [2]. This poses several challenges, such as the curse of dimensionality [22] and difficulties in visualizing and interpreting individual components. Therefore, before applying clustering techniques, it is advisable to transform the DTM to reduce its sparsity. The two most commonly used transformations for this purpose are Principal Component Analysis (PCA) and Singular Value Decomposition (SVD). However, the main issue remains an interpretation problem, as the newly created variables are no longer interpretable. The same reasoning can be generalized to embeddings, where dimensionality reduction techniques are commonly employed. Without any loss of generality, the issue of high dimensionality in text clustering remains an open problem, particularly in the context of word embeddings [18].

Document clustering techniques range from prototype-based [9,14] and graph-based methods [10] to density-based [5,21] and model-based methods [12,24] (see, for a review, [6]). Unfortunately, traditional distance metrics, such as the Euclidean distance, become inefficient for document clustering [1]. Consequently, identifying an alternative measure to determine the similarity between feature vectors is crucial in this context. In the literature, the most commonly used similarity and distance measures include the Jaccard correlation coefficient, Kullback-Leibler divergence, and cosine similarity. However, conventional approaches often struggle to effectively capture the semantic nuances and relationships between documents. This suggests that introducing a novel similarity measure based on the homogeneity of topic distributions emerges as a promising alternative. The foundational idea is inspired by [16]. This measure leverages the topic distribution within documents, which also enhances the interpretability of the underlying processes.

This paper introduces a fuzzy relational clustering procedure based on the above described dissimilarity measure. The method aims to reduce dimensionality and enhance result interpretability. It also leverages the flexibility of the fuzzy approach, allowing each document to belong to multiple clusters with membership grades ranging from 0 to 1. The two-stage process starts with a Latent Dirichlet Allocation (LDA) model [4] to tackle DTM sparsity, identify main topics, and improve interpretability. Following the LDA output, topic homogeneity is tested, and p-values are estimated using a bootstrap approach [16]. The second stage uses the dissimilarity matrix, based on the p-values, as input for the relational data algorithm.

The rest of this paper is organized as follows: The test-based similarity measure is defined in Sect. 2. Section 3 introduces the topic homogeneity test-based fuzzy document clustering method. An application is provided in Sect. 4. Finally, Sect. 5 offers comments and concluding remarks.

2 Test-Based Similarity Measure

Given a corpus $\mathcal{C} = \{\mathbf{d}_1, \dots, \mathbf{d}_M\}$ of M documents, where each document $\mathbf{d}_i = \{w_1^i, \dots, w_{n_{\mathbf{d}_i}}^i\}$ is defined as an ensemble of words, with $n_{\mathbf{d}_i}$ being the total number of words in the i-th document, and a vocabulary $V = \{w_1, \dots, w_{|V|}\}$[1], a new measure of similarity between documents is defined. The heuristic idea of comparing pairs of documents, as outlined in [16], suggests that two documents can be considered "similar" if they share the same topics. Formally, given two documents $\mathbf{d}_i$ and $\mathbf{d}_j$ $(i \neq j)$, it is possible to build this hypothesis test:

$$\begin{cases} H_0 : (\theta_{i1}, \dots, \theta_{iG}) = (\theta_{i^*1}, \dots, \theta_{i^*G}) \\ H_1 : (\theta_{i1}, \dots, \theta_{iG}) \neq (\theta_{i^*1}, \dots, \theta_{i^*G}) \end{cases} \tag{1}$$

where $\boldsymbol{\theta}_i = (\theta_{i1}, \dots, \theta_{iG})$ is the probability distribution for the topics within the i-th document and G is the number of topics. After fitting an LDA model, it is possible to define $\widehat{\boldsymbol{\theta}}_i$ as the estimated posterior distribution of topics associated to the i-th document and $\widehat{\boldsymbol{\beta}}_j$ as the estimated posterior distribution of words within the j-th topic. Therefore, in the hypothesis test, the posterior distributions for the two chosen documents are compared. To quantify the similarity between these two distributions, the Kullback-Leibler divergence can be used as a test statistic[2]. However, since the probability distribution of this test statistic is not known a priori, the associated p-value must be estimated. To accomplish this, a bootstrap method is applied in conjunction with sampling from the posterior predictive distribution of the words in the two selected documents. More in detail:

1. For each pair of distinct documents $\mathbf{d}_i$ and $\mathbf{d}_j$, where $i \neq j$, calculate the test statistic considering $\widehat{\boldsymbol{\theta}}_i$ and $\widehat{\boldsymbol{\theta}}_j$ as reference probability distributions.
2. Under the null hypothesis of topic distribution homogeneity, combine the two documents into a synthetic one, denoted as $\mathbf{d}_{i-j}$. Practically, since the data for the model is the DTM, simply add the corresponding rows together to create a new $(M-1) \times |V|$ matrix and replace it in the position corresponding to $\mathbf{d}_i$. Then, delete the row in the DTM corresponding to $\mathbf{d}_{i^*}$, obtaining DTM^{merged}, fit again the LDA model on $\mathcal{C} \bigcup \{\mathbf{d}_{i-j}\}/\{\mathbf{d}_i, \mathbf{d}_j\}$ and obtain the new estimates $\widehat{\boldsymbol{\Theta}}_{merged} \in (M-1) \times G$ and $\widehat{\mathrm{B}}_{merged} \in G \times |V|$.
3. Sample with replacement B times from the pseudo-population $\mathbf{d}_{i-j}$ to generate $\mathbf{d}_i^{new}$ and $\mathbf{d}_j^{new}$. To accomplish this task, compute the dot product between the topic distribution within the pseudo-population and the generic column of the estimated Topic-Word matrix, obtaining to weight each word with the probability of observing a specific topic. Then, draw from a Multinomial distribution $n_{\mathbf{d}_i}$ times for $\mathbf{d}_i$ and $n_{\mathbf{d}_j}$ times for $\mathbf{d}_j$ (Dirichlet-Multinomial distribution).

[1] This representation aligns with the Bag-Of-Words (BOW) paradigm, wherein the sequence and semantic relationships between words are disregarded.

[2] In principle, other test statistics can be used to accomplish this task, such as the Bhattacharyya Distance.

Since higher values of the Kullback-Leibler divergence indicate that two distributions are dissimilar, the p-value is estimated by calculating the proportion of times the estimated test statistic in the bootstrap samples exceeds the observed value. If the bootstrap estimate of the p-value falls below the chosen significance level α, the null hypothesis of document homogeneity can be rejected. Conversely, if the bootstrap estimate exceeds the significance level, there is insufficient evidence to reject the null hypothesis. Therefore, this procedure can be interpreted as a form of similarity measure.

In principle, this similarity can be computed for all pairs of documents. At the end of the process, an $M \times M$ symmetric matrix of p-values is obtained. It is important to note that, although symmetry is not guaranteed in the initial estimates, after a sufficient number of bootstrap iterations, the values tend to stabilize and become nearly symmetric. Assuming symmetry is reasonable, as an asymmetric similarity measure would be counterintuitive. Furthermore, under the assumption of symmetry, only $(M \times (M-1))/2$ computations are required, since the diagonal elements are assumed to be equal to $1-\alpha$, reflecting that each document is perfectly similar to itself. Determining the necessary number of bootstrap replicates to obtain reliable p-values is an important issue. According to [11], a sufficiently large number of replicates is required, usually at least 1000 for precise estimation of p-values. It is worth noting that, although using 1000 bootstrap replicates yields stable p-value estimates, the computational burden may become prohibitive for very large document collections. To mitigate this, one could explore accelerated strategies such as early-stopping rules based on monitoring p-value convergence. These optimizations will be investigated in future work.

3 Test-Based Fuzzy Document Clustering Proposal

Document clustering aims to partition documents into clusters where documents within the same cluster exhibit high similarity (cohesion), while those in different clusters show lower similarity (separation).

The proposed method follows a well-established approach known as Tandem Analysis, a two-stage process. The first stage addresses the inherent sparsity of the DTM, which makes the direct application of clustering techniques highly inefficient, as reported in Sect. 1. To mitigate this issue, the proposed strategy leverages LDA to identify the main topics within the corpus under analysis. This transformation reduces the dimensionality of the DTM from an $M \times |V|$ matrix to an $M \times G$ matrix, $\widehat{\Theta} = (\widehat{\boldsymbol{\theta}}_1, \cdots, \widehat{\boldsymbol{\theta}}_M)$, with $G \ll |V|$, and a $G \times |V|$ matrix, $\widehat{\mathrm{B}} = (\widehat{\boldsymbol{\beta}}_1, \cdots, \widehat{\boldsymbol{\beta}}_G)$. Compared to other dimensionality reduction methods, this approach offers the advantage of improved interpretability, even though the principal aim of LDA is not to reduce the dimensionality of the data matrix, but rather to find common topics inside a corpus and define a set of words linked to each topic.

Starting from $\widehat{\Theta}$, the p-value associated with the homogeneity test can be estimated (see Sect. 2). Essentially, it can be interpreted as a measure of similarity between any two documents, where lower values indicate greater dissimilarity and higher values suggest stronger similarity. This interpretation is directly linked to the definition of the p-value: a low value suggests rejecting the null hypothesis (indicating dissimilarity), whereas a high value implies insufficient evidence to reject the null hypothesis (suggesting similarity). A straightforward approach to quantifying dissimilarity is to define:

$$D_p(\mathbf{d}_i, \mathbf{d}_j) = 1 - \widehat{p}_{ij} \quad i \neq j \tag{2}$$

where $D_p(\mathbf{d}_i, \mathbf{d}_j)$ takes values in the range $[0, 1]$. It should be noted that this is not a true distance metric, as it does not necessarily satisfy all the properties required for a distance function[3]. By applying this calculation to every pair of documents, a dissimilarity matrix is constructed.

This dissimilarity matrix is used as the input argument of a fuzzy relational clustering algorithm. The term "relational" indicates that clustering is performed based solely on pairwise similarity measures among units. One of the most prominent algorithms in this context is FANNY [13]. However, this algorithm assumes that similarities are derived from an L_1 norm, which may limit its applicability. To overcome this limitation, a generalization involving the use of any measure of dissimilarity, known as fuzzy relational clustering (FRC), is introduced in [8]. FRC is capable of operating directly on arbitrary dissimilarity matrices, without requiring full metric properties, making it well-suited to our setting.

Starting with FRC, a variant for document clustering, involving $D_p(\cdot, \cdot)$ as the dissimilarity measure, is proposed (FRC-p). To partition M documents into K clusters, the constrained minimization problem can be formulated as follows:

$$\begin{aligned} \min_{\mathbf{U}} J_{\text{FRC}-p} &= \sum_{k=1}^{K} \frac{\displaystyle\sum_{i=1}^{M}\sum_{j=1}^{M} u_{ik}^m u_{jk}^m D_p(\mathbf{d}_i, \mathbf{d}_j)}{\displaystyle 2\sum_{i=1}^{M} u_{ik}^m} \\ \text{s.t.} \quad & u_{ik} \in [0, 1], \quad \forall i = 1, \ldots, M, \quad \forall k = 1, \ldots, K, \\ & \sum_{k=1}^{K} u_{ik} = 1, \quad \forall i = 1, \ldots, M, \end{aligned} \tag{3}$$

where m is the general fuzzifier taking values greater than 1 and $\mathbf{U}$ is defined as the membership degree matrix. Each element of $\mathbf{U}$, u_{ik}, indicates the membership degree of the i-th document to the k-th cluster. The information contained in the membership degree matrix is very useful for discovering documents that may have intermediate characteristics between two or more clusters. Note that the precision of each p-value depends on the number of bootstrap replicates used in the estimation phase. Consequently, the computational complexity of the proposed algorithm is influenced by the construction of the dissimilarity matrix, which can be computationally expensive. Thus, an optimal balance

[3] It may not satisfy the triangle inequality.

between accuracy and efficiency must be considered when selecting the number of bootstrap replicates. As suggested in [11], 1000 bootstrap replicates generally provide a good balance. Lastly, although the significance level α is used in defining the null hypothesis for homogeneity, it plays only a marginal role in the clustering process itself. In the context of this method, α influences only the diagonal entries of the dissimilarity matrix (i.e., self-comparisons), which are not directly used in clustering and therefore have little practical effect on the partitioning outcome.

Lastly, unlike classical algorithms such as K-means or hierarchical clustering, which operate directly on vector-space representations, our approach leverages a two-stage, test-driven approach. By framing pairwise comparison as a homogeneity test on posterior topic distributions, we capture subtle shifts in thematic composition that point-wise distances would miss. Whereas K-means implicitly assumes spherical clusters in the original space and hierarchical methods rely on predefined linkage criteria, our p-value–based dissimilarity adapts to the empirical variability of topics and naturally accommodates documents with overlapping themes. This makes it particularly suitable for corpora where topics blend or evolve gradually, which are scenarios in which rigid geometric distances often fail to reflect true semantic relatedness.

4 Application

To verify the adequacy of the proposed method, the widely recognized 20 Newsgroups dataset is used. It consists of roughly 20,000 documents classified into 20 different groups, covering topics ranging from science to politics and motorcycles. Following the methodology outlined in [19], a subset of the original dataset is selected to reduce computational complexity. Specifically, the science-related macro-area is chosen, using a random sample of 100 documents while maintaining the same proportions across categories. This subset includes four distinct labels: "sci.space," "sci.electronics," "sci.crypt," and "sci.med.". Although only a small subset of documents is used for the clustering procedure, this subset is carefully sampled to preserve the original distribution of the selected macro-category. This makes it a meaningful and representative proxy for the full dataset. The choice of using a reduced sample is primarily motivated by computational efficiency. However, the methodology remains scalable: once the clusters are identified based on the sampled documents, the remaining documents can be assigned in a second phase using efficient similarity-based strategies (e.g., cosine similarity with cluster centroids). This two-step approach allows the method to be extended to much larger document collections without requiring the full dissimilarity matrix to be computed for all document pairs. Such sampling-based strategies are widely adopted in practice when working with large-scale text corpora. This approach ensures that the clustering process remains computationally feasible for large-scale document collections, as the initial sampling captures the essential distributional characteristics while the second phase efficiently extends the classification to the full dataset.

After applying basic preprocessing steps to this dataset, a lemmatizer from the NLTK library is chosen. Subsequently, words that appeared in less than 4% or more than 95% of the corpus are removed, leading to a final vocabulary of approximately 421 words, with a 90% sparsity rate. Note that during the preprocessing phases, unigrams, bigrams, and trigrams are all considered. In order to choose the optimal number of topics to employ in LDA, the coherence measures (the semantic similarity within top-ranked words per topic, and then an aggregate score) and the perplexity are used. From the analysis of Fig. 1, it is evident that the optimal number of topics is approximately 10. To compute the estimated p-values, 1000 bootstrap replicates are used and α is fixed to 0.05. As shown in Fig. 2(a), the p-value matrix exhibits a relatively low degree of sparsity, approximately 18%. Moreover, the p-values appear to capture relationships between documents more accurately than the usual similarity measures. In particular, by leveraging a topic significance degree measure [25], it is possible to examine the association between topics and clusters.

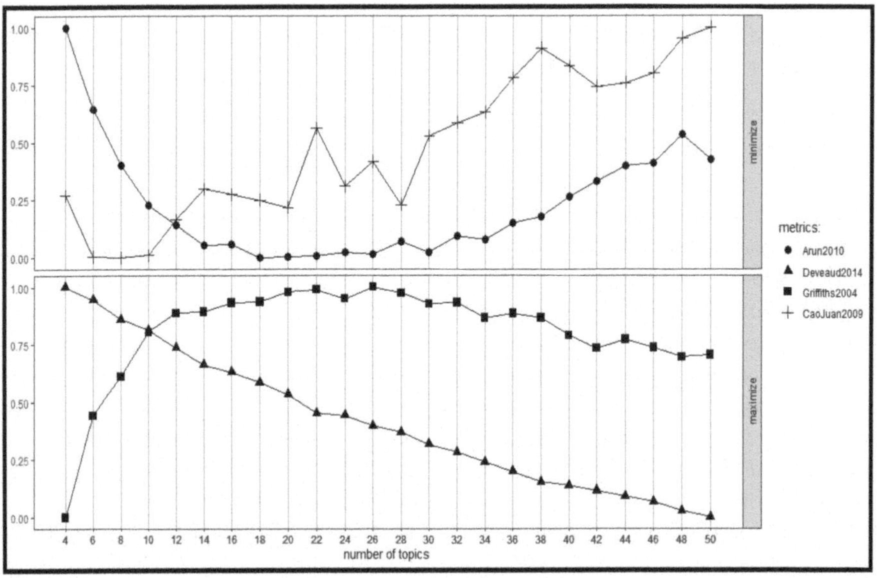

Fig. 1. Coherence scores for the reduced version of the 20 Newsgroups corpus.

Since the optimal number of clusters is known a priori, it is fixed at 4, and no further analysis is conducted regarding its selection. The resulting partition is consistent and coherent with the four known a priori categories. We also evaluated the performance of our model using the F_1 score and compared it against two well-established methods from the literature: Spectral Clustering with the Boundrange string kernel, and K-Means with cosine similarity. On this particular dataset, our method achieved an improvement of 5.5% over Spectral Clustering and 7.43% over K-Means.

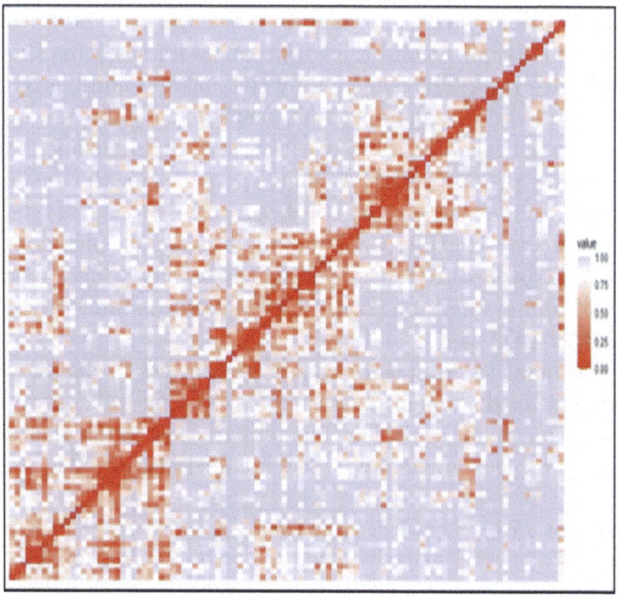

Fig. 2. Dissimilarity matrix obtained using $D(\mathbf{d}_i, \mathbf{d}_{i^*}) = 1 - \widehat{p}_{ii^*}$ (a) for the reduced version of the 20 Newsgroups corpus.

In conclusion, this study presents a small example of how the proposed method works, demonstrating generally good results. More specifically, the newly defined dissimilarity measure showcases its full flexibility and interpretability in this particular scenario, suggesting that it can be effectively employed in this setup. Additionally, it may be of further interest to explore its application with other clustering methods.

5 Concluding Remarks

A novel approach to document clustering is explored, leveraging the p-value associated with the homogeneity of topic distributions within documents. Comparing documents based on their topics offers a fresh perspective, moving away from traditional methodologies in this framework. This approach provides improved explanations and interpretations of the clustering results. However, it should be noted that the newly defined dissimilarity measure is not compatible with all clustering algorithms, particularly those that require a metric space [13]. Nevertheless, it can still be applied in practice to several well-known algorithms, such as Partitioning Around Medoids (PAM) or more in general Hierarchical Clustering methods.

As evident, the clustering procedure depends primarily on the number of bootstrap replicates used to construct the dissimilarity matrix and on the number of topics, while it is only marginally influenced by the choice of the confidence level. A commonly adopted rule of thumb is to use approximately 1000 bootstrap replicates, although optimization-based approaches for selecting this number will be explored in future work. For determining the optimal number of topics, established metrics such as topic coherence and perplexity can be employed. The role of the confidence level has already been discussed, highlighting that it does not affect the clustering process itself. As such, it can typically be fixed within a reasonable range, such as 0.01 to 0.1.

Time is not directly considered in the analysis, but topics may change over time. Therefore, the procedure for estimating the p-value (affecting the clustering procedure) could be modified to account for time, for instance, by employing the dynamic version of LDA [17].

Lastly, it is crucial to emphasize that a new internal cluster validity index needs to be defined for document clustering, as existing indices are not perfectly suitable for selecting the number of clusters in this particular domain. This remains an open problem in the field, as evidenced also by [20].

Acknowledgement. We acknowledge financial support under the PRIN 2022 project by the Italian Ministry of University and Research (MUR) entitled "Latent variable models and dimensionality reduction methods for complex data" – Project No. 20224CRB9E , CUP B53C24006310006.

References

1. Aggarwal, C.C., Hinneburg, A., Keim, D.A.: On the surprising behavior of distance metrics in high dimensional space. In: Proceedings of the 8th International Conference on Database Theory, ICDT '01, pp. 420–434 (2001)
2. Assent, I.: Clustering high dimensional data. WIREs Data Min. Knowl. Discov. **2**(4), 340–350 (2012)
3. Baghel, R., Dhir, R.: A frequent concepts based document clustering algorithm. Int. J. Comput. Appl. **4**(5), 6–12 (2010)
4. Blei, D.M., Ng, A.Y., Jordan, M.I.: Latent Dirichlet allocation. J. Mach. Learn. Res. **3**, 993–1022 (2003)
5. Creţulescu, R.G., Morariu, D.I., Breazu, M., Volovici, D.: DBSCAN algorithm for document clustering. Int. J. Adv. Stat. IT&C Econ. Life Sci. **9**(1), 58–66 (2019)
6. Cozzolino, I., Ferraro, M.B.: Document clustering. WIREs Comput. Stat. **14**(6), e1588 (2022)
7. Cutting, D.R., Karger, D.R., Pedersen, J.O., Tukey, J.W.: Scatter/gather: a cluster-based approach to browsing large document collections. In: Proceedings of the 15th Annual International ACM SIGIR Conference on Research and Development in Information Retrieval, SIGIR '92, pp. 318–329. Association for Computing Machinery (1992)
8. Davé, R.N., Sen, S.: Robust fuzzy clustering of relational data. IEEE Trans. Fuzzy Syst. **10**(6), 713–727 (2002)
9. Dhillon, I.S., Guan, Y., Kulis, B.: Kernel k-means: spectral clustering and normalized cuts. In: Proceedings of the Tenth ACM SIGKDD International Conference on Knowledge Discovery and Data Mining, pp. 551–556 (2004)

10. Ding, C.H.Q., He, X., Zha, H., Gu, M., Simon, H.D.: A min-max cut algorithm for graph partitioning and data clustering. In: Proceedings 2001 IEEE International Conference on Data Mining, pp. 107–114 (2001)
11. Efron, B.: Bootstrap methods: another look at the jackknife. Ann. Stat. **7**(1), 1–26 (1979)
12. Fraley, C., Raftery, A.E.: Model-based clustering, discriminant analysis, and density estimation. J. Am. Stat. Assoc. **97**(458), 611–631 (2002)
13. Kaufman, L., Rousseeuw, P.J.: Finding Groups in Data: An Introduction to Cluster Analysis. Wiley, Cham (1990)
14. Knittel, J., Koch, S., Ertl, T.: Efficient sparse spherical k-means for document clustering. In: Proceedings of the 21st ACM Symposium on Document Engineering, pp. 1–4 (2021)
15. Koller, D., Sahami, M.: Hierarchically classifying documents using very few words. In: Proceedings of the Fourteenth International Conference on Machine Learning, ICML '97, pp. 170–178. Morgan Kaufmann Publishers Inc., San Francisco, CA, USA (1997)
16. Kontoghiorghes, L., Colubi, A.: Testing the homogeneity of topic distribution between documents of a corpus. In: García-Escudero, L.A., et al. (eds.) Building Bridges between Soft and Statistical Methodologies for Data Science, pp. 248–254 (2022)
17. Lafferty, J., Blei, D.: Correlated topic models. In: Weiss, Y., Schölkopf, B., Platt, J. (eds.) Advances in Neural Information Processing Systems, vol. 18. MIT Press (2005)
18. Mehta, V., Bawa, S., Singh, J.: WEClustering: word embeddings based text clustering technique for large datasets. Complex Intell. Syst. **7**(6), 3211–3224 (2021). https://doi.org/10.1007/s40747-021-00512-9
19. Miao, D., Duan, Q., Zhang, H., Jiao, N.: Rough set based hybrid algorithm for text classification. Expert Syst. Appl. **36**(5), 9168–9174 (2009)
20. Misuraca, M., Spano, M., Balbi, S.: BMS: an improved Dunn index for document clustering validation. Commun. Stat. - Theory Methods **48**(20), 5036–5049 (2018)
21. Mustakim, Indah, R.G.N., et al.: DBSCAN algorithm: twitter text clustering of trend topic Pilkada Pekanbaru. J. Phys. Conf. Ser. **1363**(1), 012001 (2019)
22. Parsons, L., Haque, E., Liu, H.: Subspace clustering for high dimensional data: a review. ACM SIGKDD Explora. Newsl. **6**(1), 90–105 (2004)
23. Salton, G.: Introduction to Modern Information Retrieval. McGraw-Hill (1983)
24. Yin, J., Wang, J.: A Dirichlet multinomial mixture model-based approach for short text clustering. In: Proceedings of the 20th ACM SIGKDD International Conference on Knowledge Discovery and Data Mining, KDD '14, pp. 233–242 (2014)
25. Ma, Y., Wang, Y., Jin, B.: A three-phase approach to document clustering based on topic significance degree. Expert Syst. Appl. **41**(18), 8203–8210 (2014)

From Vectors to Networks: Comparing Conventional and Graph-Based Approaches to Unsupervised Text Categorisation

Maria Spano[1], Michelangelo Misuraca[2(✉)], and Luigi Celardo[3]

[1] Department of Economics and Statistics, University of Naples Federico II, I-80126, Naples, Italy

[2] Department of Management and Innovation Systems, University of Salerno, 84084 Fisciano, (SA), Italy
mmisuraca@unisa.it

[3] National Research Council, Institute of Polymers, Composites and Biomaterials, 80125 Naples, Italy

Abstract. One of the primary tasks of text mining is to organise a large number of unlabeled documents into a smaller set of meaningful and coherent clusters that are similar in content. Clustering algorithms typically operate on *document* × *term* matrices, where each document is represented as a vector in an algebraic format. Alternatively, a collection of documents can be represented using a *documents* × *documents* structure, which can be viewed as an adjacency matrix and graphically depicted as a graph. In network analysis, *community detection* is used on these graphs to identify groups of nodes that share common characteristics and perform similar functions. This paper aims to evaluate different data structures and grouping criteria, showing the effectiveness of various alternatives in a text categorisation strategy. We conduct a comparative study involving classical text clustering methods and community detection approaches, examining and discussing their performances.

Keywords: unsupervised classification · community detection · comparative study

1 Introduction

When we aim to automatically organise a collection of documents, extracting the wide range of information contained in the text is critical. Due to its specific encoding, the quantitative processing of this information can be complex. Among the different text mining tasks, clustering encompasses techniques that enable grouping documents using only textual content. Clustering algorithms are typically used on *document* × *term* matrices, encoding the documents in the analysed collection as vectors. This algebraic representation, known as the *vector*

A. D'Ambrosio et al. (Eds.): CLADAG-VOC 2025, SCDAKO, pp. 304–315, 2025.
https://doi.org/10.1007/978-3-032-03042-9_27

space model [29], is the most common method for transforming documents into structured data, disregarding grammatical and syntactical functions of terms within the text. The strategies for grouping documents correspond to those used in classical data analysis, based on optimising a criterion function. The primary approaches to this task rely on hierarchical and partitive methods.

Alternatively, data can be structured via a *document* × *document* matrix, interpreted as an adjacency matrix and visualised as a graph. Each document is a node, and edges represent semantic similarity via shared terms. Various similarity measures assess the proximity between documents. From a network analysis standpoint, community detection techniques reveal groups of nodes with shared properties or roles. These methods conceptually align with cluster analysis.

This paper compares classical vector space-based and graph-based clustering to assess how data representation and grouping strategies affect text categorisation. We apply both clustering and community detection techniques, examining the influence of different term-weighting schemes and similarity measures on document proximity. The paper is structured as follows. Section 2 outlines the theoretical basis for textual data representation and clustering methods, reviewing major contributions. Section 3 reports a comparative study on a news document collection, evaluating the performance of various clustering techniques. Section 4 concludes by summarising findings, discussing limitations, and suggesting future research directions.

2 Theoretical Background

Diverse methods can be used to model documents in quantitative analysis. Most clustering algorithms rely on a bag-of-words scheme, considering a document as an unordered set of terms, disregarding grammatical and syntactic roles. This model remains widely used due to its structural simplicity and interpretability, making data representation more transparent and computationally efficient. Given a *corpus* of n documents d_i (with $i = 1, \ldots, n$), each document can be represented as a vector in a space defined by the p terms of the corpus *vocabulary*:

$$d_i = (t_{i1}, \ldots, t_{im}, \ldots, t_{ip}) \tag{1}$$

where t_{im} represents the importance of the m-th term in the document d_i. Typically, term frequency measures this importance, indicating the number of times terms occur in the document. However, other weighting schemes can also be applied (for a broader discussion on the choice of t_{im}, see [2]).

By juxtaposing the document vectors, we can construct a *documents* × *terms* matrix $\mathbf{T}$. It is possible to derive a *terms* × *terms* matrix from this structure or, alternatively, a *documents* × *documents* matrix. In the first matrix, each element represents the number of times two terms co-occur within the document collection. In the second matrix, each element counts the number of terms shared between two documents. In both cases, more complex weighting methods can be considered to express the relationships between pairs of terms [39] or pairs of documents [13]. These matrices provide a foundation for clustering and network-based approaches, allowing for extracting meaningful patterns from text.

It is possible to use supervised and unsupervised methods, assuming that a natural grouping into M classes exists, to organise the set of n documents into K distinct subsets. The key difference is that supervised methods, or *text classification*, employ prior information related to themes or topics to define the categories, often relying on expert knowledge. In contrast, unsupervised methods extract knowledge directly from documents, leveraging the underlying semantic structures without additional information. The latter approach, known as clustering, is essential for several real-world applications, including organising and browsing document collections [6,36], performing text summarisation [1,42], and improving supervised strategies' effectiveness [9,22]. Document clustering assumes that documents from different categories exhibit distinct term distributions, although some overlap may occur. The diverse methods can be broadly classified into hierarchical and partitional approaches [14,38]. Hierarchical clustering produces a nested structure without requiring a predefined number of clusters. It can be *divisive*, iteratively splitting clusters, or *agglomerative*, merging documents into progressively larger clusters [18]. While widely used, agglomerative clustering is computationally expensive and irreversible, as document assignments remain fixed after merging [37]. Partitional clustering, or *centre-based*, assigns documents into a predefined number of groups, optimising a criterion function iteratively. Despite being more scalable, partitional clustering is sensitive to initialisation and may converge to local optima [33,41]. Both approaches can be applied to *documents* $\times$ *terms* or *document* $\times$ *document* matrices, though the latter allows similarity-based formulations without explicitly computing distances.

An alternative approach is network-based clustering, where the *document* $\times$ *document* matrix is interpreted as an adjacency matrix and represented as a graph [15]. A graph $\mathsf{G} = \{\mathscr{V}, \mathscr{E}, \varphi\}$ consists of a set of n vertices (documents), l edges (document similarities), and an incidence function φ. Weights on edges encode document proximity with a non-metric representation of textual data [26]. In a network analysis perspective, *community detection* aims to identify cohesive document groups by maximising intra-group similarity while minimising inter-group connections [10]. These methods align with clustering strategies but operate in a graph-theoretic perspective, leveraging network topology to partition data [30,35]. Various studies explore community detection in text mining, particularly on term graphs [16,20] and hybrid document-term networks [5]. However, applications focused on document graphs remain relatively scarce [19]. Network-based clustering offers an alternative to traditional methods. It captures complex relationships between documents and enhances text organisation through graph structures.

In this study, we illustrate the differences in applying the diverse approaches, compare the respective performances, and show the influence of different encoding schemes on the resulting data structure obtained through clustering.

3 Empirical Evaluation

As claimed, clustering and community detection share the same logic to achieve equivalent results in a text categorisation framework. In the analysis presented in the following, we used a dataset from Kaggle[1] containing 2,225 complete news articles published on the BBC website between 2004 and 2005. These articles report stories categorised into five topical areas [11]. The type-token ratio of the whole collection is 0.031, which supports the use of clustering techniques. Table 1 shows the distribution of documents across the topics, along with some descriptive statistics.

Table 1. Main characteristics of BBC news collection.

Topic	# of doc	% of doc	Avg terms per doc	Avg sentences per doc
Business	510	22.92	374.25	15.67
Entertainment	386	17.35	375.88	16.30
Politics	417	18.74	511.80	20.88
Sports	511	22.97	374.30	16.90
Tech	401	18.02	563.37	24.04

Before the analysis, we removed numbers, dates, punctuation, and URLs from the original documents. Text was normalised by eliminating special characters and non-space separators. The cleaned corpus was then lemmatised, English stop words were excluded, and only terms occurring at least twice and in a minimum of two documents were retained.

3.1 Study Set-up

Following the pre-processing phase, we constructed a matrix *documents* × *terms* **T** consisting of 2,225 rows and 14,511 columns. Each element t_{ij} represents the frequency of the j-th term in the i-th document. The corpus was treated by converting text to lowercase, removing punctuation, stopwords, and numerical values, and applying lemmatisation. From this term-frequency matrix **T**, according to [21], we derived a Boolean matrix $\mathbf{T_b}$, and a TF-IDF matrix $\mathbf{T_{td}}$. The Boolean encoding scheme merely accounts for the presence (1) or absence (0) of a term in each document. The TF-IDF scheme, on the other hand, provides a continuous weight based on a term's frequency in the document and its rarity across the corpus, balancing local and global importance. The whole framework used in the study is summarised in Table 2, which outlines the evaluated weighting schemes, clustering algorithms, and community detection methods.

[1] https://www.kaggle.com/pariza/bbc-news-summary.

Table 2. Comparative study set-up.

Weighting schemes	*Boolean* *Term Frequency* *Term Frequency – Inverse Document Frequency* [28]
Clustering algorithms	*UPGMA* [31] *K-means* [17] *Spherical K-means* [8]
Community detection	*Louvain algorithm* [3] Proximity measures: *Jaccard similarity, cosine similarity, matching coefficient*

To evaluate the performance of these unsupervised methods, we applied three clustering algorithms—UPGMA [31], K-means [17], and Spherical K-means [8]—to each weighting scheme. These algorithms are widely adopted in text categorisation tasks [20]. Since UPGMA produces a hierarchical structure, we extracted a five-cluster solution to match the number of topical categories in the BBC dataset. Likewise, we fixed the number of clusters to five for the centroid-based methods (K-means and Spherical K-means).

We then extended the analysis to include community detection based on document similarity matrices derived from each representation. These matrices were built using Jaccard similarity (for Boolean weighting), cosine similarity (for all schemes), and a matching coefficient based on the dot product of document vectors. In the Boolean case, this coefficient measures shared term presence; with TF-IDF, it also captures term specificity across the corpus. These matrices were interpreted as adjacency matrices and used to apply the Louvain algorithm [3], known for its efficiency and firm performance in large-scale networks [40]. Unlike clustering methods, Louvain does not require a predefined number of communities, allowing for data-driven structure discovery that may or may not align with the original five-category division.

3.2 Main Results

The different approaches were evaluated based on the accuracy and runtime of the categorisation process. We used two standard external criteria to compare the performance of the alternative strategies: *purity* [18] and *normalised mutual information* [34]. These criteria allowed us to assess the solutions obtained against the gold standard, representing the "true" classification of the BBC news collection. These metrics evaluate the quality of a solution by using the ground truth class labels in the dataset and the clustering results generated by a given procedure. The advantage of using purity and NMI over other well-known external validation measures – such as *precision*, *recall* and *F-measure* [32] – relies on the fact that they do not require a one-to-one correspondence between clusters and classes. Purity measures the coherence of class labels within clusters on a scale from 0 to 1, assessing the fraction of correctly labelled documents in each cluster. The purity is calculated as:

$$purity(\Omega, \mathbb{C}) = \frac{1}{n} \sum_{k=1}^{K} \max_{m} |\omega_k \cap c_m| \tag{2}$$

A purity value close to 0 indicates poor clustering, while a value of 1 denotes that all documents in each cluster share the same class label. Higher purity values indicate better clustering. However, achieving high purity is more straightforward with more clusters, making it hard to balance quality and number of clusters [18]. NMI offers a purity measure independent of the number of clusters. It builds on *mutual information* (MI) to quantify the shared statistical information between two distributions. Normalising MI allows for the comparison of partitions of different sizes. Unlike purity, sensitive to unbalanced classes, NMI robustly indicates the agreement level between a clustering solution and the gold standard [24]. Formally:

$$NMI(\Omega, \mathbb{C}) = \frac{I(\Omega; \mathbb{C})}{\sqrt{H(\Omega) + H(\mathbb{C})}} = \\ = \frac{\sum_{k=1}^{K} \sum_{m=1}^{M} |\omega_k \cap c_m| \log \frac{n|\omega_k \cap c_m|}{|\omega_k||c_m|}}{\sqrt{\sum_{k=1}^{K} |\omega_k| \log \frac{|\omega_k|}{n} \sum_{m=1}^{M} |c_m| \log \frac{|c_m|}{n}}} \tag{3}$$

where $I(\Omega; \mathbb{C})$ represents the mutual information, while $H(\Omega)$ and $H(\mathbb{C})$ denote the corresponding entropies. The metric is 0 when clusters are random concerning class membership and 1 for a perfect match between Ω and $\mathbb{C}$. Notably, the running times depend on the technical characteristics of the computer used to perform the analyses. The results presented in the following were obtained on a MacBook Pro 2.8 GHz Intel Core i7, 16 GB 1133 MHz DDR3 with macOS Sierra ver. 10.12.6.

Table 3. Comparison of clustering results on BBC news collection.

	Boolean			Term Frequency			TF-IDF		
	Purity	*NMI*	*Time*	*Purity*	*NMI*	*Time*	*Purity*	*NMI*	*Time*
UPGMA	0.240	0.006	136.026	0.231	0.000	137.069	0.237	0.006	127.045
Kmeans	**0.746**	**0.595**	43.266	**0.643**	**0.419**	33.380	**0.757**	**0.419**	15.920
Sph Kmeans	**0.948**	**0.842**	15.486	**0.888**	**0.721**	17.314	**0.949**	**0.846**	18.194
Louvain									
jaccard	0.230	0.000	8.891	-	-	-	-	-	-
cosine	0.082	0.169	3.011	0.082	0.169	3.015	0.134	0.047	4.005
matching coeff	**0.790**	**0.662**	3.080	**0.740**	**0.559**	3.559	**0.962**	**0.880**	5.782

Table 3 offers each algorithm's validation measures and run times. The best-performing algorithms are highlighted in bold font.

Focusing first on the Boolean weighting scheme, it is noticeable that Spherical K-means achieves the highest Purity (0.948) and NMI (0.842) among all methods reported. Although Louvain with matching coefficient yields upright performance (Purity = 0.790, NMI = 0.662), it does not surpass Spherical K-means regarding clustering quality. A consideration, however, is computational time. While Spherical K-means takes about 15.486 s, the Louvain algorithm with matching coefficient is notably faster (3.080 s) but yields a lower-quality partition. Turning to the term frequency scheme, Spherical K-means again provides the highest Purity (0.888) and NMI (0.721), indicating superior clustering performance relative to K-means and Louvain under the matching coefficient. Here, the Louvain algorithm is still very fast (3.559 s versus Spherical K-means's 17.314 s), but it reaches only moderate levels of Purity (0.740) and NMI (0.559). Hence, if the clustering quality is the primary objective, Spherical K-means outperforms the other methods in this scenario. When we move to the TF-IDF weighting, the situation changes: Louvain with matching coefficient attains the highest Purity (0.962) and NMI (0.880) among all configurations tested, and it also maintains a relatively low run time (5.782 s). Spherical K-means is still quite strong (Purity = 0.949, NMI = 0.846) but runs more slowly (18.194 s), and the slight difference in quality (0.949 vs. 0.962 Purity; 0.846 vs. 0.880 NMI) favours the Louvain approach under TF-IDF. Consequently, the Louvain plus matching coefficient offers excellent clustering accuracy and comparatively efficient computation for TF-IDF weighting. If the goal is to select a single "best" solution from the entire table, Louvain with matching coefficient under TF-IDF stands out. It achieves the highest Purity (0.962) and NMI (0.880) of all tested configurations and remains efficient at under six seconds, faster than many of the other top-accuracy methods.

This conclusion is further supported by the confusion matrices presented in Fig. 1, which illustrate the best clustering solutions (marked in bold in Table 3). Several patterns visually confirm the quantitative findings. Each cell's colour corresponds to the number of documents in a given class-cluster pairing, with greener shades indicating higher counts. A strong diagonal (one prominent green cell per row) suggests that an algorithm groups most documents belonging to a particular "true" category (the five-cluster solution is based on the categories in the BBC dataset).

Focusing first on the Boolean weighting (top row), we see that K-means scatters certain classes (notably, "tech" is split between multiple clusters). In contrast, Spherical K-means consolidates categories more cleanly. There is a more pronounced diagonal band of green for "business", "entertainment", "sports" and "tech". Meanwhile, community detection under the matching coefficient also gives reasonable groupings, although it appears slightly less sharply diagonal than Boolean Spherical K-means. These observations align with the earlier table, where Boolean Spherical K-means yielded the best purity and NMI in that weighting scheme. A similar pattern holds for frequency weighting (middle row). K-means creates distinct clusters for "business" and "tech", but "entertainment" and "sports" are less clearly separated. Spherical K-means exhibits

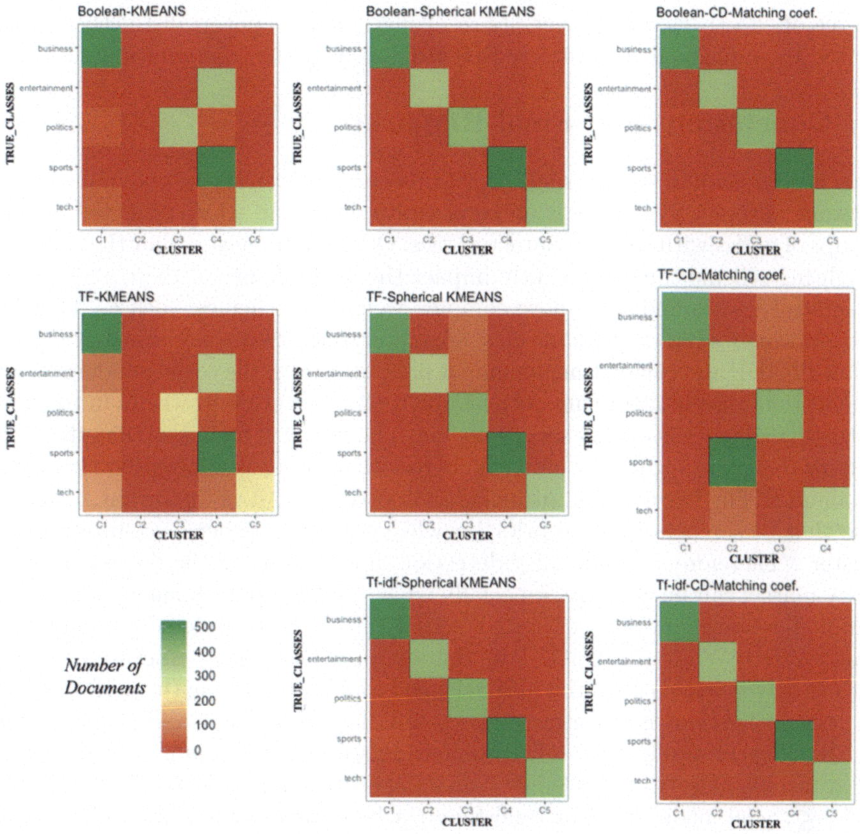

Fig. 1. Confusion matrices of the best clustering solutions.

stronger classcluster alignment: each row maps primarily to a single cluster column. Community detection with matching coefficient again yields an outcome between these extremes—strong on "tech" and "sports", but "entertainment" and "business" show a bit more dispersion than Spherical K-means. Numerically, this corresponds to the higher purity and NMI for TF-Spherical K-means observed in the table. Finally, the TF-IDF weighting (bottom row) shows the most visibly cohesive diagonal, particularly for the *CD-Matching coeff.* plot. Each row is concentrated in a single bright-green cell, indicating that most documents in each true class end up in a single cluster. This confirms why TF-IDF with community detection and matching coefficient achieved the highest overall purity and NMI in the preceding table: the confusion matrix shows minimal leakage of documents outside their correct clusters. TF-IDF Spherical K-means is also strong—there is a clear diagonal, though "politics" and "sports" appear slightly more scattered than with the community detection approach. Therefore, the clustering outcome obtained through Spherical K-means demonstrates a high degree of effectiveness; however, its resulting matrix is not as distinctly characterised by a diagonal structure as that produced by the community detection.

4 Conclusions and Final Remarks

This study compares clustering algorithms commonly used in text mining and network analysis. The focus is on community detection, which aims to discover groups of nodes with similar characteristics. Our results suggest that the choice of weighting scheme can significantly impact the performance of these algorithms. Specifically, we found that more complex weighting schemes, such as TF-IDF, outperformed simpler methods like binary or term frequency schemes. When using TF-IDF and calculating document proximity with the correspondence coefficient, the Louvain algorithm outperformed all alternative strategies in terms of both partition quality and execution time. The first key finding is that the community detection algorithm performs better than other unsupervised clustering techniques. Furthermore, in an unsupervised context, it is important to recognise that researchers have no prior knowledge of the number or composition of clusters. Therefore, a community detection approach using the document similarity matrix can be effective, especially when high computational efficiency and minimal assumptions are required. The findings can be considered a preliminary contribution. While the BBC News dataset provides a well-structured testbed with clearly labelled categories and balanced topics, confirming the performances on other collections is essential to assess findings' robustness and generalisability. Real-world applications of document clustering and community detection often involve noisy, imbalanced or domain-specific datasets (e.g. scientific publications, social media posts or legal texts), where topic boundaries are less clear and vocabulary distribution is more heterogeneous.

Future research will focus on replicating the comparative analysis using various textual collections, including unstructured or weakly labelled corpora, to evaluate how performance varies across different contexts. This step is particularly important for unsupervised methods, as they must cope with varying levels of intra-class similarity, thematic overlap and vocabulary sparsity. Evaluating clustering performance across multiple datasets can also help test the methods' scalability and stability, particularly regarding sensitivity to document length, class imbalance or domain-specific jargon [33]. One specific direction will involve applying the comparative framework to datasets characterised by heterogeneous document lengths, such as social media posts and news articles from online newspapers. The distinct challenges these two extremes pose – short, noisy, informal microtexts versus longer, structured, formal texts – have already been explored in clustering and community detection literature [12]. Sparsity and lexical brevity in microblogging data hinder conventional similarity-based approaches, necessitating dimensionality reduction and semantic enrichment strategies [27]. Conversely, longer content can lead to intra-document topic drift, necessitating strategies to segment or summarise documents before clustering. These methodological adaptations will be incorporated into future research to ensure fair comparisons and identify potential biases caused by the data structure itself. The extension can also involve broadening the range of possible algorithms. Alongside the Louvain algorithm, other well-established community detection methods can be considered. *Walktrap* [25], for example, showed good performances in several applicative domains. Including other techniques will enable us to evaluate the consistency of the results obtained using different network partitioning strategies and determine whether community detection offers systematic advantages over traditional clustering approaches in various scenarios.

Furthermore, an alternative approach to text representation has emerged in the form of word embedding methods, such as *fastText* [4], as well as contextual embeddings derived from transformer-based models like *BERT* [7]. Unlike the VSM, embeddings represent words and documents in a dense vector space, capturing semantic relationships beyond simple term co-occurrence. These models have shown great promise in improving performance across a wide range of natural language processing tasks, including text classification, summarisation, and retrieval [23]. Future research will investigate whether these vector-based representations can improve clustering performance, particularly in scenarios where document sparsity, informal language, or brevity pose significant limitations to traditional models.

References

1. Alguliyev, R.M., Aliguliyev, R.M., Isazade, R.M., Abdi, A., Idris, N.: COSUM: text summarization based on clustering and optimization. Expert. Syst. **36**(1), e12340 (2019)
2. Balbi, S., Misuraca, M.: Visualization techniques for non symmetrical relations. In: Sirmakessis, S. (Ed.) Knowledge Mining. Proceedings of the NEMIS Final Conference, 23–29. Springer (2005)
3. Blondel, V. D., Guillaume, J.-L., Lambiotte, R., Lefebvre, E.: Fast unfolding of communities in large networks. J. Stat. Mech.: Theory Exp. **2008**(10), p. P10008 (2008)
4. Bojanowski, P., Grave, E., Joulin, A., Mikolov, T.: Enriching word vectors with subword information. Trans. Assoc. Comput. Linguist. **5**, 135–146 (2017)
5. Celardo, L., Everett, M.G.: Network text analysis: a two-way classification approach. Int. J. Inf. Manag. **51**, 102009 (2020)
6. Cutting, D.R., Karger, D.R., Pedersen, J.O., Tukey, J.W.: Scatter/Gather: a cluster-based approach to browsing large document collections. In: Proceedings of the 15th Annual International ACM SIGIR Conference on Research and Development in Information Retrieval, 318–329 (1992). ACM
7. Devlin, J., Chang, M.-W., Lee, K., Toutanova, K.: BERT: Pre-training of deep bidirectional transformers for language understanding. In: Proceedings of NAACL-HLT, vol. 2019 (1), 4171–4186 (2019)
8. Dhillon, I.S., Modha, D.S.: Concept decompositions for large sparse text data using clustering. Mach. Learn. **42**, pp. 143–175 (2001). Springer
9. Dhillon, I.S., Mallela, S., Kumar, R.: A divisive information-theoretic feature clustering algorithm for text classification. J. Mach. Learn. Res. **3**, 1265–1287 (2003)
10. Fortunato, S., Hric, D.: Community detection in networks: a user guide. Phys. Rep. **659**, 1–44 (2016)

11. Greene, D., Cunningham, P.: Producing accurate interpretable clusters from high-dimensional data. In: European Conference on Principles of Data Mining and Knowledge Discovery, 486–494 (2005). Springer
12. Hong, L., Davison, B.D.: Empirical study of topic modeling in Twitter. In: Proceedings of the First Workshop on Social Media Analytics, 80–88 (2010). ACM
13. Huang, A.: Similarity measures for text document clustering. In: Holland, J., Nicholas, A., Brignoli, D. (Eds.) Proceedings of the 6th New Zealand Computer Science Research Student Conference, 49–56 (2008)
14. Jain, A.K., Murty, M.N., Flynn, P.J.: Data clustering: a review. ACM Comput. Surv. **31**(3), 264–323 (1999)
15. James, P.: Knowledge graphs. In: van der Riet, R.P., Meersman, R.A. (eds.) Linguistic Instruments in Knowledge Engineering, pp. 97–117. Elsevier (1992)
16. Jia, C., Carson, M.B., Wang, X., Yu, J.: Concept decompositions for short text clustering by identifying word communities. Pattern Recognit. **76**, 691–703 (2018)
17. MacQueen, J.: Some methods for classification and analysis of multivariate observations. In: Proceedings of the Fifth Berkeley Symposium on Mathematical Statistics and Probability, Volume 1: Statistics, 281–298 (1967). University of California Press
18. Manning, C.D., Raghavan, P., Schütze, H.: Introduction to Information Retrieval. Cambridge University Press (2008)
19. Mikhina, E.K., Trifalenkov, V.I.: Text clustering as graph community detection. Procedia Comput. Sci. **123**, 271–277 (2018)
20. Misuraca, M., Scepi, G., Spano, M.: A network-based concept extraction for managing customer requests in a social media care context. Int. J. Inf. Manag. **51**, 101956 (2020)
21. Misuraca, M., Spano, M.: Unsupervised analytic strategies to explore large document collections. In: Iezzi, D.F., Mayaffre, D., Misuraca, M. (eds.) Text Analytics: Advances and Challenges, pp. 17–28. Springer (2020)
22. Nigam, K., McCallum, A., Thrun, S., Mitchell, T.: Learning to classify text from labeled and unlabeled documents. In: Proceedings of the 15th National / 10th Conference on Artificial Intelligence / Innovative Applications of Artificial Intelligence, 792–799 (1998). AAAI
23. Petukhova, A., Matos-Carvalho, J.P., Fachada, N.: Text clustering with large language model embeddings. Int. J. Cogn. Comput. Eng. **6**, 100–108 (2025)
24. Pfitzner, D., Leibbrandt, R., Powers, D.: Characterization and evaluation of similarity measures for pairs of clusterings. Knowl. Inf. Syst. **19**(3), 361–394 (2009)
25. Pons, P., Latapy, M.: Computing communities in large networks using random walks. In: Yolum, I., Güngör, T., Gürgen, F., Özturan, C. (eds.) ISCIS 2005. LNCS, vol. 3733, pp. 284–293. Springer, Heidelberg (2005). https://doi.org/10.1007/11569596_31
26. Popping, R.: Knowledge graphs and network text analysis. Soc. Sci. Res. **42**(1), 91–106 (2003)
27. Rangrej, A., Kulkarni, S., Tendulkar, A.V.: Comparative study of clustering techniques for short text documents. In: Proceedings of the 20th International Conference Companion on World Wide Web, 111–112 (2011)
28. Salton, G., Buckley, C.: Term-weighting approaches in automatic text retrieval. Inf. Process. Manag. **24**(5), 513–523 (1988)
29. Salton, G., Wong, A., Yang, C.S.: A vector space model for automatic indexing. Commun. ACM **18**(11), 613–620 (1975)

30. Schaub, M.T., Delvenne, J.-C., Rosvall, M., Lambiotte, R.: The many facets of community detection in complex networks. Appl. Netw. Sci. **2**(1), 1–13 (2017). https://doi.org/10.1007/s41109-017-0023-6
31. Sokal, R.R., Michener, C.D.: A statistical method for evaluating systematic relationships. Univ. Kans. Sci. Bull. **38**(22), 1409–1438 (1958)
32. Sokolova, M., Lapalme, G.: A systematic analysis of performance measures for classification tasks. Inf. Process. Manag. **45**(4), 427–437 (2009)
33. Steinbach, M., Karypis, G., Kumar, V.: A comparison of document clustering techniques. Technical Report, Department of Computer Science and Engineering, University of Minnesota, TR 00-034 (2000)
34. Strehl, A., Ghosh, J.: Cluster ensembles - a knowledge reuse framework for combining multiple partitions. J. Mach. Learn. Res. **3**, 583–617 (2002)
35. Tang, L., Liu, H.: Graph mining applications to social network analysis. In: Aggarwal, C.C., Wang, H. (eds.) Managing and Mining Graph Data, pp. 487–513. Springer (2010)
36. Weisser, T., Sassmannshausen, T., Ohrndorf, D., Burggräf, P., Wagner, J.: A clustering approach for topic filtering within systematic literature reviews. MethodsX **7**, 100831 (2020)
37. Willett, P.: Recent trends in hierarchic document clustering: a critical review. Inf. Process. Manag. **24**(5), 577–597 (1988)
38. Xu, D., Tian, Y.: A comprehensive survey of clustering algorithms. Ann. Data Sci. **2**, 165–193 (2015)
39. Yan, X., Guo, J., Liu, S., Cheng, X., Wang, Y.: Learning topics in short texts by non-negative matrix factorization on term correlation matrix. In: Proceedings of the 2013 SIAM International Conference on Data Mining, 749–757. SIAM (2013)
40. Yang, Z., Algesheimer, R., Tessone, C.J.: A comparative analysis of community detection algorithms on artificial networks. Sci. Rep. **6**(1), 30750 (2016)
41. Zhao, Y., Karypis, G.: Empirical and theoretical comparisons of selected criterion functions for document clustering. Mach. Learn. **55**(3), 311–331 (2004)
42. Zheng, H.-T., Gong, S.-Q., Chen, H., Jiang, Y., Xia, S.-T.: Multi-document summarization based on sentence clustering. In: Loo, C.K., Yap, K.S., Wong, K.W., Teoh, A., Huang, K. (eds.) Neural Information Processing, pp. 429–436. Springer (2014)

A Note on Gradient-Based Parameter Estimation for Energy-Based Models

Luca Martino[1], Salvatore Ingrassia[1], Sofia Mangano[2(✉)], and Luca Scaffidi Domianello[1]

[1] Department of Economics and Business, University of Catania Corso Italia, 55 - 95129 Catania, Italy
{luca.martino,salvatore.ingrassia,luca.scaffidi}@unict.it
[2] Department of Mathematics and Computer Science, University of Catania, Viale Andrea Doria 6 - 95125 Catania, Italy
sofia.mangano@phd.unict.it

Abstract. Energy-based models (EBMs) are an important family of models where a piece of the likelihood is intractable, and hence unknown. For this reason, the parameter estimation in EBMs is a challenge for the standard estimation methods. In this paper, we present a critical discussion of gradient-based approaches for inference in energy-based models. We provide many details of different derivations, clarify connections and differences. We give practical suggestions for the application of the different schemes. Specifically, we focus on a suitable choice of the proposal/reference density that is crucial for the performance of the gradient-based procedures.

Keywords: Energy-based models · gradient descent · maximum likelihood

1 Introduction

An energy-based model (EBM) is a statistical model only specified up to the so-called partition function [8]. More specifically, considering empirical data coming from a sample space $\mathcal{X} \subseteq \mathbb{R}^d$, an EBM is defined as $p(\mathbf{x}|\boldsymbol{\theta}) = \frac{\phi(\mathbf{x}|\boldsymbol{\theta})}{Z(\boldsymbol{\theta})}$, depending on some parameter $\boldsymbol{\theta} \in \boldsymbol{\Theta} \subseteq \mathbb{R}^q$. The numerator $\phi(\mathbf{x}|\boldsymbol{\theta})$ is known and can be evaluated, whereas the denominator $Z(\boldsymbol{\theta}) = \int_{\mathcal{X}} \phi(\mathbf{x}|\boldsymbol{\theta}) d\mathbf{x}$ is analytically intractable and hence unknown. The intractable denominator $Z(\boldsymbol{\theta})$ is also referred to as the *partition function*. Moreover, the function $E(\mathbf{x}|\boldsymbol{\theta}) = -\log \phi(\mathbf{x}|\boldsymbol{\theta})$ is often called *energy* (this term comes from the statistical mechanics, see e.g. [14]). Smaller values of the latter correspond to larger values of $p(\mathbf{x}|\boldsymbol{\theta})$. We remark that traditional statistical models concerning classification, regression and density estimation can be reformulated in terms of energy-based models. In the Bayesian framework, energy-based models form the so-called double-intractable posteriors [1,7,13].

Even if nowadays the most common approaches avoid the normalization constant, like the *noise contrastive estimation* [3,4] and the *score matching* first

A. D'Ambrosio et al. (Eds.): CLADAG-VOC 2025, SCDAKO, pp. 316–328, 2025.
https://doi.org/10.1007/978-3-032-03042-9_28

proposed in [6], in this paper we revisit some earlier approaches, that conversely approximate the normalization constant, based on the *noisy* gradient descent of negative log-likelihood function [2,5] because we noticed that some important related statistical issues have been somehow overlooked (and/or missed) in the literature. We provide many details of different (complete) derivations, discussing connections and relationships and several practical suggestions for the choice of a proposal/reference density that is required in these gradient descent approaches. We remark also that to handle more flexible distributions, or data characterized by heterogeneous sub-groups, [12] quite recently proposed finite mixtures of energy-based models.

2 Energy-Based Models (EBMs)

Let $\phi(\mathbf{x}|\boldsymbol{\theta}) = e^{-E(\mathbf{x}|\boldsymbol{\theta})} \geq 0$ be a non-negative function defined on $\mathcal{X} \subseteq \mathbb{R}^d$, parametrized by a vector of parameters $\boldsymbol{\theta}$ taking values in $\boldsymbol{\Theta} \subseteq \mathbb{R}^q$. As we stated above, the non-negative function $E(\mathbf{x}|\boldsymbol{\theta}) = -\log \phi(\mathbf{x}|\boldsymbol{\theta})$ is often called energy function. We assume that $\phi(\mathbf{x}|\boldsymbol{\theta})$ is analytically known and we can evaluate it. Thus, an energy-based model $p(\mathbf{x}|\boldsymbol{\theta})$ is a parametrized family of density functions, defined for each $\boldsymbol{\theta}$ as

$$p(\mathbf{x}|\boldsymbol{\theta}) = \frac{\phi(\mathbf{x}|\boldsymbol{\theta})}{Z(\boldsymbol{\theta})}. \tag{1}$$

For given $\boldsymbol{\theta}$, generally the integral[1]

$$Z(\boldsymbol{\theta}) = \int_{\mathcal{X}} \phi(\mathbf{x}|\boldsymbol{\theta}) d\mathbf{x}, \tag{2}$$

is unknown because it cannot be solved analytically in closed form, and then the integral is said to be intractable.[2] Hence, the normalizing constant $Z(\boldsymbol{\theta})$ cannot be evaluated point-wise. This represents a challenge for a maximum likelihood estimation (MLE), as we discuss below.

3 MLE Approaches to Parameter Estimation in EBMs

Let $\underline{\mathbf{x}} = \{\mathbf{x}_1, \ldots, \mathbf{x}_N\} \in \mathcal{X}^N$ be an observed dataset that contains i.i.d. realizations distributed as the the EBM in Eq. 1 for a specific unknown vector of parameters $\boldsymbol{\theta}^*$, i.e., $\mathbf{x}_n \sim p(\mathbf{x}|\boldsymbol{\theta}^*)$ for all $n = 1, ..., N$. In order to estimate the parameter of the distribution, the likelihood function of $\boldsymbol{\theta}$ given $\underline{\mathbf{x}}$ is

$$L(\boldsymbol{\theta}|\underline{\mathbf{x}}) = p(\mathbf{x}_1, \mathbf{x}_2, \ldots, \mathbf{x}_N|\boldsymbol{\theta}) = \prod_{n=1}^{N} p(\mathbf{x}_n|\boldsymbol{\theta}) = \frac{1}{Z(\boldsymbol{\theta})^N} \prod_{n=1}^{N} \phi(\mathbf{x}_n|\boldsymbol{\theta}),$$

[1] All the integrals in this work are definite integrals. However, in the rest of the paper, for simplicity we avoid to write the integration domain.

[2] We assume that $\mathbf{x}$ be a continuous vector, although several considerations are also valid for the discrete case.

and the corresponding log-likelihood $\mathcal{L}(\boldsymbol{\theta}|\underline{\mathbf{x}})$ is

$$\mathcal{L}(\boldsymbol{\theta}|\underline{\mathbf{x}}) = \sum_{n=1}^{N} \log p(\mathbf{x}_n|\boldsymbol{\theta}) = \sum_{n=1}^{N} \log \phi(\mathbf{x}_n|\boldsymbol{\theta}) - N \log Z(\boldsymbol{\theta}). \tag{3}$$

Since $E(\mathbf{x}_n|\boldsymbol{\theta}) = -\log \phi(\mathbf{x}_n|\boldsymbol{\theta})$, the log-likelihood 3 becomes

$$\mathcal{L}(\boldsymbol{\theta}|\underline{\mathbf{x}}) = -\sum_{n=1}^{N} E(\mathbf{x}_n|\boldsymbol{\theta}) - N \log Z(\boldsymbol{\theta}). \tag{4}$$

The maximum likelihood estimation (MLE) of $\boldsymbol{\theta}$ is often reformulated as the minimization of cost function $J(\boldsymbol{\theta})$ defined as the negative log-likelihood function (NLL), i.e.,

$$J(\boldsymbol{\theta}) = \text{NLL}(\boldsymbol{\theta}|\underline{\mathbf{x}}) = -\mathcal{L}(\boldsymbol{\theta}|\underline{\mathbf{x}}) = \sum_{n=1}^{N} E(\mathbf{x}_n|\boldsymbol{\theta}) + N \log Z(\boldsymbol{\theta}), \tag{5}$$

so that $\widehat{\boldsymbol{\theta}} = \arg_{\boldsymbol{\theta}\in\boldsymbol{\Theta}} \max \mathcal{L}(\boldsymbol{\theta}|\underline{\mathbf{x}}) = \arg_{\boldsymbol{\theta}\in\boldsymbol{\Theta}} \min J(\boldsymbol{\theta})$. Generally, a standard widely-used optimization approach is based on the so called *gradient-descent*, i.e., computing a finite sequence of estimates $\{\boldsymbol{\theta}^{(t)}\}_{t=0}^{T}$, starting from some initial guess $\boldsymbol{\theta}^{(0)}$ (and suitable choices of the step value α_t) according to

$$\widehat{\boldsymbol{\theta}}_t = \widehat{\boldsymbol{\theta}}_{t-1} - \alpha_t \nabla_{\boldsymbol{\theta}} J(\widehat{\boldsymbol{\theta}}_{t-1}). \tag{6}$$

For a suitable large number T of iterations, we get the final estimate $\widehat{\boldsymbol{\theta}}_T \approx \widehat{\boldsymbol{\theta}} = \arg_{\boldsymbol{\theta}\in\boldsymbol{\Theta}} \min J(\boldsymbol{\theta})$, where $J(\boldsymbol{\theta})$ is given in Eq. 5. However, as we remarked above, $Z(\boldsymbol{\theta})$ is unknown for any $\boldsymbol{\theta}$, and cannot be computed in closed form. As a consequence, we cannot compute $\nabla_{\boldsymbol{\theta}} J(\boldsymbol{\theta})$ neither. Therefore, suitable strategies must be adopted for approximating $Z(\boldsymbol{\theta})$ (or its gradient).

3.1 The Geyer's Approach [2]

Theoretically speaking, the simplest approach relies on applications of some numerical method for computing $Z(\boldsymbol{\theta})$, and this approximation can be obtained by importance sampling (IS) [9]. It was proposed firstly in [2], and we also refer to it as *baseline* approach. Let $q(\cdot)$ be a known pdf with support $\mathcal{X}$ (which implies $\int q(\mathbf{y})d\mathbf{y} = 1$) and let $\underline{\mathbf{y}} = \{\mathbf{y}_1, \ldots, \mathbf{y}_M\}$, with $\mathbf{y} \in \mathcal{X} \subseteq \mathbb{R}^d$ $(m = 1, \ldots, M)$ be a sample of size M generated from $q(\mathbf{y})$, i.e. $\mathbf{y}_1, \ldots, \mathbf{y}_M \sim q(\cdot)$ that is chosen by the user; in this literature $q(\cdot)$ is called *proposal or reference density* [2]. The IS estimator $\widehat{Z}(\boldsymbol{\theta})$ of $Z(\boldsymbol{\theta})$ is given by

$$\widehat{Z}(\boldsymbol{\theta}) = \frac{1}{M} \sum_{m=1}^{M} \frac{\phi(\mathbf{y}_m|\boldsymbol{\theta})}{q(\mathbf{y}_m)} = \frac{1}{M} \sum_{m=1}^{M} w_m^{(\boldsymbol{\theta})}, \qquad \mathbf{y}_m \sim q(\mathbf{y}), \tag{7}$$

where $w_m^{(\boldsymbol{\theta})} = \frac{\phi(\mathbf{y}_m|\boldsymbol{\theta})}{q(\mathbf{y}_m)}$ are the (unnormalized) importance weights. It is easy to show that the estimator above is unbiased [9]. Hence, we can consider the following approximation of Eq. 5:

$$\begin{aligned} J_{\text{BL}}(\boldsymbol{\theta}) &= \sum_{n=1}^{N} E(\mathbf{x}_n|\boldsymbol{\theta}) + N \log \widehat{Z}(\boldsymbol{\theta}) \\ &= \sum_{n=1}^{N} E(\mathbf{x}_n|\boldsymbol{\theta}) + N \log \left[\frac{1}{M} \sum_{m=1}^{M} \frac{\phi(\mathbf{y}_m|\boldsymbol{\theta})}{q(\mathbf{y}_m)} \right], \quad \mathbf{y}_m \sim q(\mathbf{y}). \end{aligned} \tag{8}$$

Note that $J_{\text{BL}}(\boldsymbol{\theta})$ is a random variable depending on $\mathbf{y}_m$'s but, for fixed $\underline{\mathbf{y}}$, $J_{\text{BL}}(\boldsymbol{\theta})$ becomes a deterministic function of $\boldsymbol{\theta}$. In Fig. 1(a), we can see three different realizations of $J_{\text{BL}}(\boldsymbol{\theta})$. Then, the idea is to minimize this function [2],

$$\widehat{\boldsymbol{\theta}}_{\text{BL}} = \arg\min J_{\text{BL}}(\boldsymbol{\theta}). \tag{9}$$

We remark that the method requires *only one generation* of the artificial dataset, i.e., only M artificial data points $\mathbf{y}_1, \ldots, \mathbf{y}_M \sim q(\mathbf{y})$. After this generation and given $\{\mathbf{y}_1, \ldots, \mathbf{y}_M\}$, $J_{\text{BL}}(\boldsymbol{\theta})$ becomes a fixed, analytically known, and evaluable cost function. We point out also that there are two ways for reducing the variance of $J_{\text{BL}}(\boldsymbol{\theta})$:

- *increasing the sample size*: as $M \to \infty$, then $\widehat{Z}(\boldsymbol{\theta}) \to Z(\boldsymbol{\theta})$ and $J_{\text{BL}}(\boldsymbol{\theta}) \to J(\boldsymbol{\theta})$ as well.
- *select a good proposal density* $q(\mathbf{y})$ for fixed value of M; see the next section for more details.

3.2 On the Choice of the Proposal Density $q(\cdot)$

The proposal density must be satisfy the following two assumption:

1. The analytic form of the proposal density $q(\cdot)$ must be available, and we need to be able to evaluate it point-wise.
2. We need to be able to draw samples from $q(\cdot)$.

In order to reduce the variance of the estimator in 7, a better choice would be to choose a proposal density that depends on $\boldsymbol{\theta}$, i.e., $q(\mathbf{y}|\boldsymbol{\theta})$. Indeed, the optimal reference density in this scenario is

$$q_{\text{opt}}(\mathbf{y}|\boldsymbol{\theta}) = p(\mathbf{y}|\boldsymbol{\theta}) \propto \phi(\mathbf{y}|\boldsymbol{\theta}). \tag{10}$$

It can be shown that this choice minimizes the variance of $\widehat{Z}(\boldsymbol{\theta})$ [9,10]. We point out that, even though the choice $q_{\text{opt}}(\mathbf{y}|\boldsymbol{\theta}) = p(\mathbf{y}|\boldsymbol{\theta})$ minimizes the the variance of $\widehat{Z}(\boldsymbol{\theta})$, this choice presents one important drawback and an additional computational cost:

a. First of all, generally, we are not able to draw samples from the EBM model, i.e., from $q_{\texttt{opt}}(\mathbf{y}|\boldsymbol{\theta}) = p(\mathbf{y}|\boldsymbol{\theta})$. See Sect. 6 for further details.
b. Secondly, for any proposal density $q(\mathbf{y}|\boldsymbol{\theta})$ depending on $\boldsymbol{\theta}$, we should generate a set of artificial data for each $\boldsymbol{\theta}$, i.e., $\mathbf{y}_1^{(\boldsymbol{\theta})}, \ldots, \mathbf{y}_M^{(\boldsymbol{\theta})} \sim q(\mathbf{y}|\boldsymbol{\theta})$. Namely, for each evaluation of $J_{\text{BL}}(\boldsymbol{\theta})$ at some $\boldsymbol{\theta}$ we would require the generation of another set of artificial data. For instance, if we desire to evaluate $J_{\text{BL}}(\boldsymbol{\theta})$ in L different points $\boldsymbol{\theta}_1, ..., \boldsymbol{\theta}_L$, we would need to draw L artificial samples of size M, since for each $\boldsymbol{\theta}_l$, for $l = 1, ..., L$, we would have a different set $\{\mathbf{y}_1^{(\boldsymbol{\theta}_l)}, \ldots, \mathbf{y}_M^{(\boldsymbol{\theta}_l)}\}$. This fact could also increase variability in the evaluation of the function $J_{\text{BL}}(\boldsymbol{\theta})$.

In particular, this last consideration is shared by any proposal density $q(\mathbf{y}|\boldsymbol{\theta})$ depending on $\boldsymbol{\theta}$, and not just the optimal one.

Independent proposal. Choosing a proposal $q(\mathbf{y})$ independent from $\boldsymbol{\theta}$ allows us to generate only one set of artificial data $\{\mathbf{y}_1, \ldots, \mathbf{y}_M\}$, for all the processes. On the other hand, for some values of $\boldsymbol{\theta}$, the proposal $q(\mathbf{y})$ can be a non suitable choice, and the variance $\widehat{Z}(\boldsymbol{\theta})$ can vary drastically with $\boldsymbol{\theta}$. The experience with IS estimators suggests the use of a proposal $q(\mathbf{y})$ with great variance, bigger than the variance of $p(\mathbf{x}|\boldsymbol{\theta})$ for any $\boldsymbol{\theta}$, if possible [10].
Non-optimal but good choices. From a theoretical point of view, it is interesting to observe that a non-optimal but good choice of the proposal density is any density close to the true model generating the observed data, i.e.,

$$q(\mathbf{y}) \approx p(\mathbf{y}|\boldsymbol{\theta}^*) = \frac{\phi(\mathbf{y}|\boldsymbol{\theta}^*)}{Z(\boldsymbol{\theta}^*)}. \tag{11}$$

This proposal ensures to have small variance in the estimation of $Z(\boldsymbol{\theta})$ around of the true value $\boldsymbol{\theta}^*$. An even better (and more robust) choice of of $q(\mathbf{y})$ is a density that mimics the shape of $p(\mathbf{y}|\boldsymbol{\theta}^*)$ but with more variance (the area under $q(\mathbf{y})$ is more diffuse). In this scenario, we can avoid catastrophic behaviors of the IS estimator, as shown in the numerical experiments of [10]. See also the results depicted in Figs. 1 and 2 of Sect. 7. Clearly, we do not know $\boldsymbol{\theta}^*$ and we are not able to draw from $p(\mathbf{y}|\boldsymbol{\theta}^*)$. However, this consideration can drive the construction of a good proposal density.

4 Approximating the Gradient: a First Derivation

Noting that $\nabla_{\boldsymbol{\theta}} \log \widehat{Z}(\boldsymbol{\theta}) = \frac{1}{\widehat{Z}(\boldsymbol{\theta})} \nabla_{\boldsymbol{\theta}} \widehat{Z}(\boldsymbol{\theta})$, the gradient of the negative log-likelihood function in Eq. 5, scaled by a factor $1/N$, becomes:

$$\begin{aligned}
\frac{1}{N} \nabla_{\boldsymbol{\theta}} J_{\text{BL}}(\boldsymbol{\theta}) &= \frac{1}{N} \sum_{i=1}^{N} \nabla_{\boldsymbol{\theta}} E(\mathbf{x}_n|\boldsymbol{\theta}) + \frac{1}{\widehat{Z}(\boldsymbol{\theta})} \nabla_{\boldsymbol{\theta}} \widehat{Z}(\boldsymbol{\theta}) \\
&= \frac{1}{N} \sum_{i=1}^{N} \nabla_{\boldsymbol{\theta}} E(\mathbf{x}_n|\boldsymbol{\theta}) + \frac{1}{\frac{1}{M} \sum_{j=1}^{M} \frac{\phi(\mathbf{y}_j|\boldsymbol{\theta})}{q(\mathbf{y}_j)}} \frac{1}{M} \sum_{m=1}^{M} \frac{\nabla_{\boldsymbol{\theta}} \phi(\mathbf{y}_m|\boldsymbol{\theta})}{q(\mathbf{y}_m)}.
\end{aligned}$$

Moreover, recalling $\phi(\mathbf{x}|\boldsymbol{\theta}) = e^{-E(\mathbf{x}|\boldsymbol{\theta})}$, we have

$$\nabla_{\boldsymbol{\theta}}\phi(\mathbf{x}|\boldsymbol{\theta}) = -\nabla_{\boldsymbol{\theta}}E(\mathbf{x}|\boldsymbol{\theta})e^{-E(\mathbf{x}|\boldsymbol{\theta})} = -\nabla_{\boldsymbol{\theta}}E(\mathbf{x}|\boldsymbol{\theta})\phi(\mathbf{x}|\boldsymbol{\theta}).$$

Replacing above, we obtain

$$\begin{aligned}\frac{1}{N}\nabla_{\boldsymbol{\theta}}J_{\mathrm{BL}}(\boldsymbol{\theta}) &= \frac{1}{N}\sum_{i=1}^{N}\nabla_{\boldsymbol{\theta}}E(\mathbf{x}_n|\boldsymbol{\theta}) - \frac{1}{\sum_{j=1}^{M}\frac{\phi(\mathbf{y}_j|\boldsymbol{\theta})}{q(\mathbf{y}_j)}}\sum_{m=1}^{M}\frac{\nabla_{\boldsymbol{\theta}}E(\mathbf{y}_m|\boldsymbol{\theta})\phi(\mathbf{y}_m|\boldsymbol{\theta})}{q(\mathbf{y}_m)} \\ &= \frac{1}{N}\sum_{i=1}^{N}\nabla_{\boldsymbol{\theta}}E(\mathbf{x}_n|\boldsymbol{\theta}) - \sum_{m=1}^{M}\bar{w}_m^{(\boldsymbol{\theta})}\nabla_{\boldsymbol{\theta}}E(\mathbf{y}_m|\boldsymbol{\theta}),\end{aligned} \tag{12}$$

where we have defined the normalized importance weights as:

$$\bar{w}_m^{(\boldsymbol{\theta})} = \frac{\frac{\phi(\mathbf{y}_m|\boldsymbol{\theta})}{q(\mathbf{y}_m)}}{\sum_{j=1}^{M}\frac{\phi(\mathbf{y}_j|\boldsymbol{\theta})}{q(\mathbf{y}_j)}} = \frac{w_m^{(\boldsymbol{\theta})}}{\sum_{j=1}^{M}w_j^{(\boldsymbol{\theta})}}, \tag{13}$$

and $w_m^{(\boldsymbol{\theta})} = \frac{\phi(\mathbf{y}_m|\boldsymbol{\theta})}{q(\mathbf{y}_m)}$ are the unnormalized importance weights.

Remark 1. If we are able to draw samples from the model, i.e., $\mathbf{y}_m^{(\boldsymbol{\theta})} \sim q_{\mathrm{opt}}(\mathbf{y}|\boldsymbol{\theta}) = p(\mathbf{y}|\boldsymbol{\theta})$, using the optimal proposal density, the normalized IS weights become $\bar{w}_m^{(\boldsymbol{\theta})} = \frac{1}{M}$ for each m, and Eq. 12 becomes

$$\frac{1}{N}\nabla_{\boldsymbol{\theta}}J_{\mathrm{BL}}(\boldsymbol{\theta}) = \frac{1}{N}\sum_{i=1}^{N}\nabla_{\boldsymbol{\theta}}E(\mathbf{x}_n|\boldsymbol{\theta}) - \frac{1}{M}\sum_{m=1}^{M}\nabla_{\boldsymbol{\theta}}E(\mathbf{y}_m^{(\boldsymbol{\theta})}|\boldsymbol{\theta}). \tag{14}$$

However, recall that we have a new generation of artificial data $\mathbf{y}_m^{(\boldsymbol{\theta})}$ for each $\boldsymbol{\theta}$.

Remark 2. Assume again that we are able to draw samples from the model, i.e., $\mathbf{y}_m^{(\boldsymbol{\theta})} \sim q_{\mathrm{opt}}(\mathbf{y}|\boldsymbol{\theta}) = p(\mathbf{y}|\boldsymbol{\theta})$. An additional issue is that we cannot evaluate completely $p(\mathbf{y}|\boldsymbol{\theta})$ since $Z(\boldsymbol{\theta})$ is unknown. Thus, we cannot evaluate the *unnormalized* IS weights

$$w_m^{(\boldsymbol{\theta})} = \frac{\phi(\mathbf{y}_m^{(\boldsymbol{\theta})}|\boldsymbol{\theta})}{q_{\mathrm{opt}}(\mathbf{y}_m^{(\boldsymbol{\theta})}|\boldsymbol{\theta})} = \frac{\phi(\mathbf{y}_m^{(\boldsymbol{\theta})}|\boldsymbol{\theta})}{p(\mathbf{y}_m^{(\boldsymbol{\theta})}|\boldsymbol{\theta})} = Z(\boldsymbol{\theta})\frac{\cancel{\phi(\mathbf{y}_m^{(\boldsymbol{\theta})}|\boldsymbol{\theta})}}{\cancel{\phi(\mathbf{y}_m^{(\boldsymbol{\theta})}|\boldsymbol{\theta})}} = Z(\boldsymbol{\theta}), \quad \forall\, m, \tag{15}$$

and, as a consequence, we cannot actually evaluate $J_{\mathrm{BL}}(\boldsymbol{\theta})$ in Eq. 8, but only its gradient $\nabla_{\boldsymbol{\theta}}J_{\mathrm{BL}}(\boldsymbol{\theta})$ in Eq. 14 (that only depends on the normalized weights, i.e., $\bar{w}_m^{(\boldsymbol{\theta})} = \frac{1}{M}$). However, if we are only interested in minimizing $J_{\mathrm{BL}}(\boldsymbol{\theta})$, evaluating its gradient $\nabla_{\boldsymbol{\theta}}J_{\mathrm{BL}}(\boldsymbol{\theta})$ is enough.

5 Approximating the Gradient: a Second Classical Derivation

Here, we follow a more classical derivation employed in different works about inference in EBMs [5]. Instead of approximating $Z(\boldsymbol{\theta})$, let us consider directly the computation of the gradient of the negative log-likelihood function 5 scaled by the factor $1/N$,

$$\frac{1}{N}\nabla_{\boldsymbol{\theta}} J(\boldsymbol{\theta}) = \frac{1}{N}\sum_{n=1}^{N} \nabla_{\boldsymbol{\theta}} E(\mathbf{x}_n|\boldsymbol{\theta}) + \frac{1}{Z(\boldsymbol{\theta})}\nabla_{\boldsymbol{\theta}} Z(\boldsymbol{\theta}).$$

Since $Z(\boldsymbol{\theta}) = \int \phi(\mathbf{x}|\boldsymbol{\theta})d\mathbf{x} = \int \exp(-E(\mathbf{x}|\boldsymbol{\theta}))d\mathbf{x}$, we have

$$\nabla_{\boldsymbol{\theta}} Z(\boldsymbol{\theta}) = -\int \nabla_{\boldsymbol{\theta}} E(\mathbf{x}|\boldsymbol{\theta}) \exp(-E(\mathbf{x}|\boldsymbol{\theta}))d\mathbf{x} = -\int \nabla_{\boldsymbol{\theta}} E(\mathbf{x}|\boldsymbol{\theta})\phi(\mathbf{x}|\boldsymbol{\theta})d\mathbf{x},$$

and replacing above we obtain

$$\begin{aligned}
\frac{1}{N}\nabla_{\boldsymbol{\theta}} J(\boldsymbol{\theta}) &= \frac{1}{N}\sum_{i=1}^{N} \nabla_{\boldsymbol{\theta}} E(\mathbf{x}_i|\boldsymbol{\theta}) - \frac{1}{Z(\boldsymbol{\theta})}\int \nabla_{\boldsymbol{\theta}} E(\mathbf{x}|\boldsymbol{\theta})\phi(\mathbf{x}|\boldsymbol{\theta})d\mathbf{x}, \\
&= \frac{1}{N}\sum_{i=1}^{N} \nabla_{\boldsymbol{\theta}} E(\mathbf{x}_i|\boldsymbol{\theta}) - \int \nabla_{\boldsymbol{\theta}} E(\mathbf{x}|\boldsymbol{\theta})p(\mathbf{x}|\boldsymbol{\theta})d\mathbf{x}, \qquad (16) \\
&\approx \frac{1}{N}\sum_{i=1}^{N} \nabla_{\boldsymbol{\theta}} E(\mathbf{x}_i|\boldsymbol{\theta}) - \frac{1}{M}\sum_{m=1}^{M} \nabla_{\boldsymbol{\theta}} E(\mathbf{y}_m^{(\boldsymbol{\theta})}|\boldsymbol{\theta}), \quad \mathbf{y}_m^{(\boldsymbol{\theta})} \sim p(\cdot|\boldsymbol{\theta}). \qquad (17)
\end{aligned}$$

Remark. Note that Eq. 17 coincides with Eq. 14 and we remark the following points (the first two have been previously discussed):

a. Generally, we are not able to draw samples from the EBM model, i.e., $p(\mathbf{y}|\boldsymbol{\theta})$. See Sect. 6 for more details.
b. We need to generate a different set of artificial data $\mathbf{y}_1^{(\boldsymbol{\theta})}, \dots, \mathbf{y}_M^{(\boldsymbol{\theta})}$ for each $\boldsymbol{\theta}$. This point is valid for any proposal density $q(\mathbf{y}|\boldsymbol{\theta})$ depending on $\boldsymbol{\theta}$.
c. Even if we are able to draw from $p(\mathbf{y}|\boldsymbol{\theta})$, we can approximate only the gradient $\nabla_{\boldsymbol{\theta}} J(\boldsymbol{\theta})$ but not $J(\boldsymbol{\theta})$ by Monte Carlo arguments. See last remark in Sect. 4 and Eq. 15. However, in order to minimize $J(\boldsymbol{\theta})$, the information of its gradient is enough. Moreover, other numerical integration methods could be employed for recovering $J(\boldsymbol{\theta})$ from its gradient, if required.

Another possibility. Here we describe an alternative procedure to avoid the generation of samples from the model $p(\mathbf{x}|\boldsymbol{\theta})$. We can use an IS approach for approximating the integral in Eq. 16, i.e., we can consider the following equality,

$$\begin{aligned}
\int \nabla_{\boldsymbol{\theta}} E(\mathbf{x}|\boldsymbol{\theta})p(\mathbf{x}|\boldsymbol{\theta})d\mathbf{x} &= \int \nabla_{\boldsymbol{\theta}} E(\mathbf{x}|\boldsymbol{\theta})\frac{p(\mathbf{x}|\boldsymbol{\theta})}{q(\mathbf{x})}q(\mathbf{x})d\mathbf{x}, \\
&= \frac{1}{Z(\boldsymbol{\theta})}\underbrace{\int \nabla_{\boldsymbol{\theta}} E(\mathbf{x}|\boldsymbol{\theta})\frac{\phi(\mathbf{x}|\boldsymbol{\theta})}{q(\mathbf{x})}q(\mathbf{x})d\mathbf{x}}_{(*)},
\end{aligned}$$

so that $Z(\boldsymbol{\theta})$ can be approximated by 7, the term $(*)$ can be approximated by similar IS arguments [9] and we get

$$\int \nabla_{\boldsymbol{\theta}} E(\mathbf{x}|\boldsymbol{\theta}) p(\mathbf{x}|\boldsymbol{\theta}) d\mathbf{x} \approx \frac{1}{\sum_{j=1}^{M} \frac{\phi(\mathbf{y}_j|\boldsymbol{\theta})}{q(\mathbf{y}_j)}} \sum_{m=1}^{M} \frac{\nabla_{\boldsymbol{\theta}} E(\mathbf{y}_m|\boldsymbol{\theta}) \phi(\mathbf{y}_m|\boldsymbol{\theta})}{q(\mathbf{y}_m)},$$

$$= \frac{1}{\sum_{j=1}^{M} w_j^{(\boldsymbol{\theta})}} \sum_{m=1}^{M} w_m^{(\boldsymbol{\theta})} \nabla_{\boldsymbol{\theta}} E(\mathbf{y}_m|\boldsymbol{\theta}),$$

$$= \sum_{m=1}^{M} \bar{w}_m^{(\boldsymbol{\theta})} \nabla_{\boldsymbol{\theta}} E(\mathbf{y}_m|\boldsymbol{\theta}), \qquad \mathbf{y}_m \sim q(\mathbf{x}), \tag{18}$$

where $\bar{w}_m^{(\boldsymbol{\theta})}$ are the normalized weights in Eq. 13. We solve the previous issues: (a) we are able to draw from $q(\cdot)$ (since we choose the proposal density $q(\cdot)$), and (b) we draw only once M artificial data. However, $q(\cdot)$ can be not suitable for some values of $\boldsymbol{\theta}$.

6 Generating Artificial Data from $p(\mathbf{X}|\boldsymbol{\theta})$

Generally, we are not able to draw artificial data from the EBM $p(\mathbf{x}|\boldsymbol{\theta})$ for a given $\boldsymbol{\theta}$. In this case, there are mainly two alternatives: apply an MCMC algorithm or an "IS - plus - resampling" scheme. Below, we describe the details of both. As an example of MCMC, we consider the Metropolis-Hastings (MH) algorithm.

MH algorithm. As an example of possible MCMC method, we provide a Metropolis-Hastings schemes with an independent proposal density $q(\cdot)$ (independent from the previous state of the chain) [11]. We consider the use of an independent proposal density to facilitate the comparison with the other schemes in the rest of the work:

1. Starting with an arbitrary initial vector $\mathbf{y}_0^{(\boldsymbol{\theta})}$.
2. For $m = 1, \ldots, M$:
   a. Draw $\mathbf{y}'$ from $q(\mathbf{y})$.
   b. Set $\mathbf{y}_m^{(\boldsymbol{\theta})} = \mathbf{y}'$ with probability,

$$\alpha = \min\left[1, \frac{p(\mathbf{y}'|\boldsymbol{\theta})}{p(\mathbf{y}_{m-1}^{(\boldsymbol{\theta})}|\boldsymbol{\theta})} \frac{q(\mathbf{y}_{m-1}^{(\boldsymbol{\theta})})}{q(\mathbf{y}')}\right] = \min\left[1, \frac{\phi(\mathbf{y}'|\boldsymbol{\theta})}{\phi(\mathbf{y}_{m-1}^{(\boldsymbol{\theta})}|\boldsymbol{\theta})} \frac{Z(\boldsymbol{\theta})}{Z(\boldsymbol{\theta})} \frac{q(\mathbf{y}_{m-1}^{(\boldsymbol{\theta})})}{q(\mathbf{y}')}\right],$$
$$\left[1, \frac{\phi(\mathbf{y}'|\boldsymbol{\theta})}{q(\mathbf{y}')} \frac{q(\mathbf{y}_{m-1}^{(\boldsymbol{\theta})})}{\phi(\mathbf{y}_{m-1}^{(\boldsymbol{\theta})}|\boldsymbol{\theta})}\right].$$

   Otherwise, set $\mathbf{y}_m^{(\boldsymbol{\theta})} = \mathbf{y}_{m-1}^{(\boldsymbol{\theta})}$ with probability $1 - \alpha$.
3. The output is $\{\mathbf{y}_1^{(\boldsymbol{\theta})}, \ldots, \mathbf{y}_M^{(\boldsymbol{\theta})}\}$.

IS plus resampling. Below, we present the details of the "IS plus resampling" scheme. Again, we need the use of a proposal density $q(\cdot)$ but the idea is get first a sample $\underline{\mathbf{z}} = \{\mathbf{z}_1, \ldots, \mathbf{z}_M\}$ with $\mathbf{z}_m \sim q(\cdot)$ $(m = 1, \ldots, M)$ and then consider resamples points $\{\mathbf{y}_1^{(\boldsymbol{\theta})}, \ldots, \mathbf{y}_M^{(\boldsymbol{\theta})}\}$ from $\underline{\mathbf{z}}$:

1. Draw $\mathbf{z}_1, \ldots, \mathbf{z}_M \sim q(\cdot)$.
2. Assign the weights

$$w_m^{(\boldsymbol{\theta})} = \frac{\phi(\mathbf{z}_m|\boldsymbol{\theta})}{q(\mathbf{z}_m)}, \qquad m = 1, \ldots, M. \tag{19}$$

3. Compute the normalized weights,

$$\bar{w}_m^{(\boldsymbol{\theta})} = \frac{w_m^{(\boldsymbol{\theta})}}{\sum_{i=1}^M w_i^{(\boldsymbol{\theta})}}, \qquad m = 1, \ldots, M. \tag{20}$$

4. Resample (bootstrap) M times with replacement, within $\{\mathbf{z}_1, \ldots, \mathbf{z}_M\}$ according to normalized weights $\bar{w}_m^{(\boldsymbol{\theta})}$, $m = 1, \ldots, M$. The M resampled samples will be denoted as $\{\mathbf{y}_1^{(\boldsymbol{\theta})}, \ldots, \mathbf{y}_M^{(\boldsymbol{\theta})}\}$. Note that $\mathbf{y}_m^{(\boldsymbol{\theta})} \in \{\mathbf{z}_1, \ldots, \mathbf{z}_M\}$ for any value of m. The outputs are the M resampled particles $\{\mathbf{y}_1^{(\boldsymbol{\theta})}, \ldots, \mathbf{y}_M^{(\boldsymbol{\theta})}\}$.

Note that, with respect to the derivation in Sect. 3.1, we have the additional step of the resampling. We finally highlight that in both cases, MCMC or IS, an *internal* proposal density is required. Then, generally, even if we would like to employ the model $p(\mathbf{x}|\boldsymbol{\theta})$ as a reference density, we need the choice and use of another proposal density $q(\mathbf{y})$, within the Monte Carlo sampling schemes.
Furthermore, in both cases, we obtain samples distributed (approximately) as the target density $p(\mathbf{y}|\boldsymbol{\theta})$. Since the target density $p(\mathbf{y}|\boldsymbol{\theta})$ changes with $\boldsymbol{\theta}$, then the obtained samples $\{\mathbf{y}_1^{(\boldsymbol{\theta})}, \ldots, \mathbf{y}_M^{(\boldsymbol{\theta})}\}$ depend also on the specific fixed $\boldsymbol{\theta}$. Therefore, we need to generate a different set of artificial data $\mathbf{y}_1^{(\boldsymbol{\theta})}, \ldots, \mathbf{y}_M^{(\boldsymbol{\theta})}$ for each $\boldsymbol{\theta}$, i.e., the use of the MCMC or "IS plus resampling" algorithm must be repeated for different $\boldsymbol{\theta}$.

7 Some Numerical Considerations

For the sake of simplicity, and to know the ground-truth, in order to evaluate the different performance, we consider a one-dimensional Gaussian density with zero mean, $\mu = 0$, as observation model, i.e.,

$$p(x|\theta) = \mathcal{N}(x|0, \theta) = \frac{1}{\sqrt{2\pi\theta^2}} \exp\left(-\frac{x^2}{2\theta^2}\right), \tag{21}$$

where

$$\phi(x|\theta) = \exp\left(-\frac{x^2}{2\theta^2}\right), \quad E(x|\theta) = \frac{x^2}{2\theta^2}, \quad Z(\theta) = \sqrt{2\pi\theta^2}. \tag{22}$$

We assume that $Z(\theta)$ is unknown and apply the gradient approaches described in this work. We set $\theta^* = 2$ and $N = 100$, so that the observed data are drawn as

$$x_n \sim p(x|\theta^*) = \frac{1}{\sqrt{8\pi}} \exp\left(-\frac{x^2}{8}\right), \tag{23}$$

for $n = 1, \ldots, N$. We consider two possible Gaussian reference/proposal densities: the first one is an independent Gaussian proposal with $\mu_p = 2$ and $\sigma_p = 2$, whereas the second one is a Gaussian proposal (which depends on θ) with $\mu_p = 2$ and $\sigma_p = 2 + \theta$,

$$q_1(y) = \mathcal{N}(y|2, 2) = \frac{1}{\sqrt{8\pi}} \exp\left(-\frac{(y-2)^2}{8}\right), \tag{24}$$

$$q_2(y|\theta) = \mathcal{N}(y|2, 2+\theta) = \frac{1}{\sqrt{2\pi(2+\theta)^2}} \exp\left(-\frac{(y-2)^2}{2(2+\theta)^2}\right). \tag{25}$$

Namely, the artificial data are generated from $y_m \sim q_1(y)$ or $y_m^{(\theta)} \sim q_2(y|\theta)$, for $m = 1, \ldots, M$. We consider two values of $M \in \{100, 5000\}$. We test the results in 1000 independent runs to average the results. Figure 1 provides the results using $q_1(y)$. Figure 1(a) depicts three curves $J_{\text{BL}}(\theta)$ in Eq. 8 after generating three different realizations of artificial data $\{y_1, \ldots, y_M\} \sim q_1(y)$, with $M = 100$. In Figs. 1(b)-1(c) (corresponding to $M = 100$ and $M = 5000$, respectively) we show the 100% of variability of 1000 curve $J_{\text{BL}}(\theta)$ with a green area, the empirical mean of the curves $J_{\text{BL}}(\theta)$ with a dashed blue line, and the true negative log-likelihood $J(\theta)$ with a solid red line. Figure 2 depicts the same curves and results but considering the second proposal $q_2(y|\theta)$.

In Fig. 1, as $M \to \infty$ we can observe the converge of $J_{\text{BL}}(\theta)$ to $J(\theta)$ specially for values of $\theta < 2$, but the convergence struggles for values of $\theta > 2$, even with $M = 5000$. This is due to the choice of the proposal, which is not suitable for value of $\theta > 2$. Whereas, in Fig. 2, the green area (i.e., the variability of $J_{\text{BL}}(\theta)$) is much more smaller than in in Fig. 1, even with $M = 100$. For $M = 5000$, we have almost a perfect convergence. The reason is the $q_2(y|\theta)$ has a standard deviation, $2 + \theta$, always greater than the standard deviation of the model, that is 2 [10]. Therefore, $q_2(y|\theta)$ provides much better estimator $\widehat{\theta}$ in terms of mean square error, since it provides more efficient estimations of $\widehat{Z}(\theta)$.

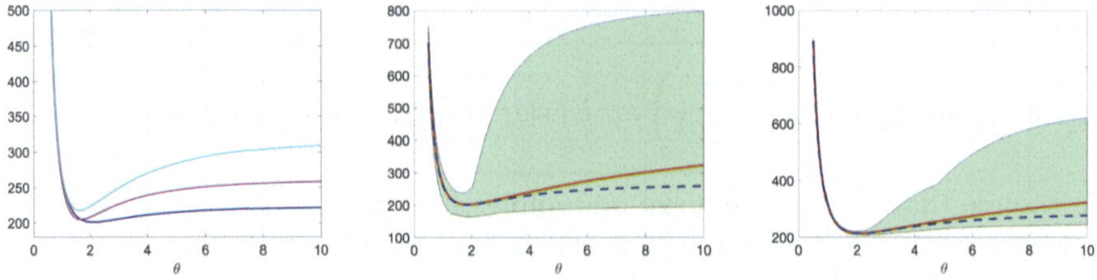

(a) $N = M = 100$ (b) $N = M = 100$ (c) $N = 100, M = 5000$

Fig. 1. (a) Approximation of the negative log-likelihood $J(\theta)$ (shown with a solid red line) using the first proposal with $\mu_p = 2$ and $\sigma_p = 2$. (b) and (c) The green area shows the 100% of variability of $J_{\text{BL}}(\theta)$. The empirical mean curve is depicted with a blue dashed line.

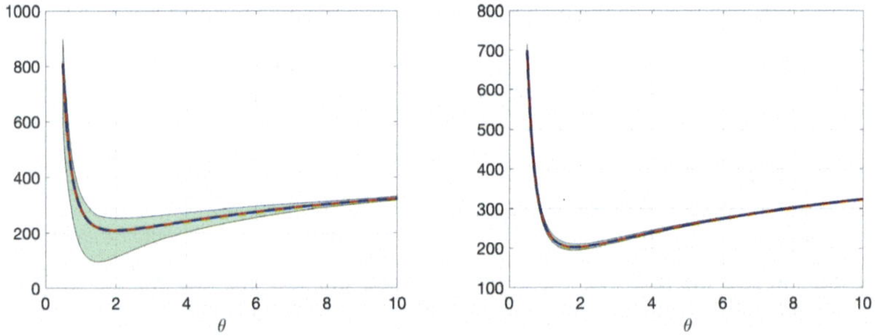

(a) $N = M = 100$ (b) $N = 100,\ M = 5000$

Fig. 2. Approximation of the negative log-likelihood $J(\theta)$ (shown with a solid red line) using the second proposal with $\mu_p = 2$ and $\sigma_p = 2 + \theta$ for $N = M = 100$ (a) and $N = M = 5000$ (b) . The green area shows the 100% of variability of $J_{\text{BL}}(\theta)$. The empirical mean curve is depicted with a blue dashed line.

8 Conclusions

We have provided a detailed description of the use of gradient-based approaches for parameter estimation in EBMs. Different complete derivations are described discussing connections and relationships. We have given several practical suggestions for the choice of a proposal/reference density in other to ensure a suitable approximation of the negative log-likelihood and its gradient. We have also remarked several possible relevant issues that have been in some way overlooked (or missed) in the literature. We remark also that similar issues hold also modeling data characterized by heterogeneous sub-groups. For this purpose, [12] propose a finite mixture of non-normalized densities, which is specified by

$$p(\cdot|\psi) = \sum_{g=1}^{G} \frac{\pi_g p_g(\cdot|\boldsymbol{\theta}_k)}{Z(\boldsymbol{\theta}_g)} = \sum_{g=1}^{G} \xi_g p_g(\cdot|\boldsymbol{\theta}_g), \tag{26}$$

where G is the number of components of the mixture, $\boldsymbol{\pi} = \{\pi_1, \dots, \pi_G\}$ are the mixing weights which sum to one, $\sum_{g=1}^{G} \pi_g = 1$.

Finally, we outline some ideas for future research. First of all, connections with other well-known approaches to estimate EBMs, such as the noise contrastive estimation and the score matching need to be investigated. Moreover, here we presented a simulation study concerning a simple Gaussian distribution to emphasize the behavior of the estimators; obviously, it would be interesting to test the performance of the estimators in more complex scenarios, for example using mixtures of non-normalized densities or with higher dimensional spaces. In the same framework, a deep investigation on the computational costs of the different gradient-based approaches could give additional insights into their applicability. Lastly, the usage of ulterior sampling techniques, such as inverse importance sampling, could deepen the relevance of the proposed approaches.

Acknowledgements. S. Ingrassia and L. Scaffidi Domianello acknowledge financial support from PNRR MUR project PE0000013-FAIR. L. Martino acknowledges financial support by the PIACERI Starting Grant BA-GRAPH (UPB 28722052144) of the University of Catania. The authors thank also Michael U. Gutmann for fruitful discussion during the preparation of the paper.

References

1. Caimo, A., Mira, A.: Efficient computational strategies for doubly intractable problems with applications to Bayesian social networks. Stat. Comput. **25**, 113–125 (2015)
2. Geyer, C.J.: On the convergence of Monte Carlo maximum likelihood calculations. J. Royal Stat. Soc. Ser. B (Methodological) **56**(1), 261–274 (1994)
3. Gutmann, M.U., Hyvärinen, A.: Noise contrastive estimation of unnormalized statistical models, with applications to natural image statistics. J. Mach. Learn. Res. **13**(2) (2012)
4. Gutmann, M.U., Kleinegesse, S., Rhodes, B.: Statistical applications of contrastive learning. Behaviormetrika **49**(2), 277–301 (2022)
5. Hinton, G.E.: Training products of experts by minimizing contrastive divergence. Neural Comput. **14**(8), 1771–1800 (2002)
6. Hyvärinen, A.: Estimation of non-normalized statistical models by score matching. J. Mach. Learn. Res. **6** (2005)
7. Liang, F.: A double metropolis-hastings sampler for spatial models with intractable normalizing constants. J. Stat. Comput. Simul. **80**(9), 1007–1022 (2009)
8. Le Cun, Y., Chopra, S., Hadsell, R., Ranzato, M., Huanh, G.J.: Energy-based models. In: Predicting Structured Data, pp. 191–246. MIT Press, Cambridge, Massachusetts (2007)
9. Llorente, F., Martino, L.: Optimality in importance sampling: a gentle survey (2025). arXiv:2502.07396
10. Llorente, F., Martino, L., Delgado, D., López-Santiago, J.: Marginal likelihood computation for model selection and hypothesis testing: An extensive review. SIAM Rev. **65**(1), 3–58 (2023)

11. Martino, L., Elvira, V.: Metropolis Sampling, pp. 1–18. John Wiley and Sons, Ltd (2017)
12. Matsuda, T., Hyvärinen, A.: Estimation of non-normalized mixture models. In: The 22nd International Conference on Artificial Intelligence and Statistics, pp. 2555-2563 (2019). PMLR.
13. Park, J., Haran, M.: Bayesian inference in the presence of intractable normalizing functions. J. Am. Stat. Assoc. **113**(523), 1372–1390 (2018)
14. Robert, C.P., Casella, G.: Monte Carlo Statistical Methods, Springer (2004)

Clustering Financial Time Series by Good and Bad Realized Volatility Decomposition

Raffaele Mattera[1,2](✉) and Germana Scepi[1,2]

[1] University of Campania, Caserta, Italy
raffaele.mattera@unicampania.it, scepi@unina.it
[2] University of Naples "Federico II", Naples, Italy

Abstract. Financial time series are often clustered based on conditional volatility, estimated from GARCH models. However, realized measures based on high-frequency data provide a more accurate estimation of the latent volatility process. In this paper, we assess the similarity of realized volatility dynamics using an autoregressive metric and the decomposition of volatility into good and bad components. In particular, we introduce a novel weighted algorithm for improving the hierarchical clustering approach and apply it to the U.S. stocks traded in the Dow Jones Industrial Average (DJIA) index.

Keywords: cluster analysis · high-frequency data · weighted hierarchical clustering

1 Introduction

Clustering is essential for uncovering patterns and relationships in large datasets of financial time series. According to a financial principle called Hierarchical Risk Parity (Lopez de Prado, 2016), the identification of grouped structure across assets can be exploited to build financial portfolios (Raffinot, 2017; Iorio et al., 2018; Mattera et al., 2025) and to provide more accurate forecasts of prices (Mattera et al., 2024) and volatility (Mattera and Otto, 2024). Therefore, clustering is in general useful for both portfolio management and risk assessment. Clustering of financial time series can be achieved in several ways. All the proposed approaches agree that in order to derive reliable clustering in the financial domain, the empirical regularities of the assets (i.e. the stylised facts) must be taken into account in the calculation of dissimilarities. Examples of interesting features that have been considered by previous research are: returns' correlations (Mantegna, 1999), long memory (Lahmiri, 2016), tail connectedness and spillover effects (Cerqueti et al., 2023; De Luca and Zuccolotto, 2021), spline coefficients (Iorio et al., 2018), and distributional features (e.g. see Cerqueti et al. 2022). Other approaches are based on a combination of static features characterizing the financial time series (e.g. see Bastos and Caiado (2021)). However,

A. D'Ambrosio et al. (Eds.): CLADAG-VOC 2025, SCDAKO, pp. 329–340, 2025.
https://doi.org/10.1007/978-3-032-03042-9_29

volatility dynamics remains the most important feature for clustering purposes. In this regard, Otranto (2008) proposed a distance between two GARCH processes, while Caiado and Crato (2010) and L'ucio and Caiado (2022) proposed alternative approaches based on estimated parameters from GARCH models. An important problem when using volatility-based clustering techniques is that the conditional variance is latent and therefore not directly observable. The GARCH models estimate volatility from daily squared returns, which is a noisy but unbiased measure of volatility. Andersen and Bollerslev (1998) showed that daily volatility is best measured by aggregating high-frequency squared returns rather than by daily squared returns. It is nowadays widely accepted that this approach to measuring volatility based on high-frequency data, called Realized Volatility (RV), is the most accurate. Following previous research (e.g. see Otranto and Gargano (2015)), our paper starts from the idea that we get more accurate clustering quality with more accurate estimation of volatility. As a main novelty, we introduce a clustering approach based on a key property of RV, that is, its decomposition into Good and Bad components. We enrich the existing volatility-based hierachical clustering algorithms with additional information improving the interpretability of the results in applications where the difference between the two components is relevant. The used demposition is based on the sign of intraday returns, capturing the asymmetric impact of positive and negative price movements on volatility. Good volatility arises from positive returns and it is often associated with stable market conditions, while Bad volatility arises from negative returns and it is related to market stress. This decomposition enriches the standard realized volatility measure by distinguishing between volatility due to positive versus negative returns, offering insight into whether market turbulence is driven by optimism or distress. To assess the similarity of financial time series in terms of these dynamics, we apply the autoregressive metric of Piccolo (1990) which captures temporal dependencies in realized volatility patterns. To partition the assets, we propose to use a weighted hierarchical clustering approach that accounts for the relative contributions of Good and Bad volatility.

The rest of the paper is structured as follows. Section 2 presents the proposed methodology, discussing firstly how realized volatility and the autoregressive metric are computed and, secondly, the proposed weighted clustering approach. Section 3 shows the application of the proposed approach to study the clustering structure of U.S. stocks included in the Dow Jones Industrial Average (DJIA) index. Section 4 concludes with final remarks and future research directions.

2 Methodology

2.1 Realized Volatility Estimation and the Autoregressive Metric

The measurement of daily realized volatility relies on high-frequency data, such as prices observed at 1-minute intervals. Let $t = 1, \dots, T$ be the temporal index for a generic day t, and let $m = 1, 2, \dots, M$ be the intraday time index. We call the high-frequency price of a generic i-th financial time series as $y_{i,t,m}$. To compute the Realized Volatility (RV), we first consider the intraday log-returns

$$r_{i,t,m} = \log(y_{i,t,m}) - \log(y_{i,t,m-1}). \tag{1}$$

We assume that intraday prices $y_{i,t,m}$ are observed at equally spaced time intervals (e.g., every minute). For $m = 1$, the lagged price $y_{i,t,m-1}$ corresponds to the opening price of day t. Given these intraday returns, the standard estimator for RV is given by the realized variance (Patton and Sheppard 2015)

$$RV_{i,t} = \sum_{m=1}^{M} r_{i,t,m}^2. \tag{2}$$

To further improve this measure, Patton and Sheppard (2015) introduced estimators that decompose the RV into components driven by positive and negative returns:

$$RV_{i,t} = RV_{i,t}^{+} + RV_{i,t}^{-}. \tag{3}$$

The Good component , measured as follows,

$$RV_{i,t}^{+} = \sum_{m=1}^{M} r_{i,t,m}^2 \cdot I(r_{i,t,m} \geq 0), \tag{4}$$

captures the contribution to total realized variance from periods when returns are non-negative, while the Bad component

$$RV_{i,t}^{-} = \sum_{m=1}^{M} r_{i,t,m}^2 \cdot I(r_{i,t,m} < 0), \tag{5}$$

isolates the impact of negative returns, which are often linked to market downturns, uncertainty, or panic-driven trading. We calculate the distance between two RV processes i and j in terms of their autoregressive metric. More specifically, consider the following collection of realized volatility time series

$$\boldsymbol{X} = \{RV_{i,t} : i = 1, \ldots, N; t = 1, \ldots, T\}. \tag{6}$$

where the generic element $RV_{i,t}$ is the RV for the i-th stock at day t. The matrix $\boldsymbol{X}$ has therefore dimensions $N \times T$, where each row corresponds to a stock and each column represents a specific daily observation, which is obtained according to the aggregation of intra-daily values. We compute the AR(∞) representation of each row of the matrix and calculate the Euclidean distance of the resulting π-weights, that is,

$$d_{i,j} = \sum_{p=1}^{P} (\pi_{i,p} - \pi_{j,p})^2, \tag{7}$$

where $\pi_{i,p}$ and $\pi_{j,p}$ are the set of P autoregressive coefficients of the AR(∞) representation for the i-th and j-th realized volatility series, respectively. For each i-th asset we select the best ARMA model order with the BIC criterion (Hyndman and Khandakar, 2008) and adopt a simple truncation approach using the first ten π-weights in (7). The distance $d_{i,j}$ is a well-defined measure because of the absolute convergence of the sequences of processes belonging to the admissible ARIMA class, and satisfies the properties of a metric Piccolo (1990). Given the decomposition in (3), we apply the autoregressive metric to both components, thus obtaining two alternative distance measures $d_{i,j,Good}$ and $d_{i,j,Bad}$. We note that using the autoregressive metric to define dissimilarity is consistent with the literature on volatility clustering. Several GARCH-based distances that are frequently employed in the analysis of volatility dynamics rely on the autoregressive representation of the GARCH process (D'Urso et al., 2013; Otranto, 2008). Differently, the proposed approach focuses on the RV, which is directly measurable. This distinction is important because GARCH models inherently assume an unobservable volatility component, whereas our metric operates in a setting where volatility is observable. For this reason, alternative dissimilarity measures can be applied, such as correlation-based measures (Mantegna, 1999) or periodogram-based measures (Caiado et al., 2006). However, they lack a clear financial interpretation in the context of volatility. For instance, although the concept of correlation between returns is well understood, the meaning of correlation between volatility processes is less so.

2.2 The Proposed Clustering Approach

The starting point of the proposed approach is the hierarchical clustering model of Ward. Recent research extended this method to the spatio-temporal setting (see Morelli et al. (2025)), where both temporal patterns and spatial dependencies are incorporated into the clustering process. We apply this framework to a novel setting by considering the decomposition of realized volatility into its Good and Bad components. In general, the hierarchical clustering algorithm starts with N singleton clusters, where each unit belongs to a separate cluster. At each iteration step, the two most similar clusters are merged based on a dissimilarity measure. The process continues until all units are assigned to a cluster. Compared to other clustering methods such as k-means or mixture models, the hierarchical approach shows the advantage that the number of clusters does not need to be pre-specified, still ensuring scalability for a moderate number of units. Let us define $P_K = (C_1, \ldots, C_k, \ldots, C_K)$ as a partition of the dataset into K clusters, with C_k representing the k-th cluster ($k = 1, \ldots, K$). The Ward hierarchical algorithm merges clusters at each step to minimize the total inertia, given by

$$W(P_K) = \sum_{k=1}^{K} I(C_k), \tag{8}$$

where $I(C_k)$ is the within-cluster inertia, that is,

$$I(C_k) = \sum_{i \in C_k} \sum_{j \in C_k} \frac{w_i w_j}{2 \sum_{i \in C_k} w_i} d_{i,j}^2, \tag{9}$$

where w_i and w_j are weights associated with the units i and j and $d_{i,j}$ is the pairwise dissimilarity, such that a lower value of $d_{i,j}$ indicates higher similarity between units i and j. In the context of realized volatility, the units i and j are two time series, and the pairwise dissimilarity $d_{i,j}$ can be computed using the autoregressive metric of Piccolo (1990). To incorporate the informational content of Good and Bad dynamics into (10), we consider a weighted inertia formulation, which extend the algorithm proposed in Chavent et al. (2018) to the temporal domain. Instead of weighting a temporal inertia with a spatial one, we propose to weight the inertia based on the similarity of Good and Bad realized volatility as follows:

$$I(C_k) = (1-\alpha) \sum_{i \in C_k} \sum_{j \in C_k} \frac{w_i w_j}{2 \sum_{i \in C_k} w_i} d^2_{i,j,Good} + \alpha \frac{w_i w_j}{2 \sum_{i \in C_k} w_i} d^2_{i,j,Bad}, \tag{10}$$

where $d_{i,j,Good}$ represents the squared distance between the Good realized volatility series of financial assets i and j, computed using the ARMA-based metric, and $d_{i,j,Bad}$ represents the squared distance between the Bad realized volatility series of financial assets i and j. The parameter $\alpha \in [0, 1]$ allows us to balance the relative importance of clustering based on upside or downside volatility dynamics. When $\alpha = 0$, clustering is based only on Good volatility, while with $\alpha = 1$ only Bad volatility is considered. The clustering algorithm follows the standard Ward-like procedure, but is based on weighted inertia, so it iteratively merges clusters to minimize the total weighted inertia. The main challenge, similar to spatio-temporal clustering, is the data-driven selection of the optimal value of α, which determines the relative weight given to Good and Bad volatility in the clustering process. For this aim, we adopt the heuristic approach proposed in Chavent et al. (2018), which works as follows. First, we define the total within-cluster inertia based on the distances in terms of Good component as $W_{\text{good}}(\mathcal{C})$, and the total within-cluster inertia due to Bad component as $W_{\text{bad}}(\mathcal{C})$. Additionally, we define $W_{\text{good}}(\mathcal{P}_1)$ and $W_{\text{bad}}(\mathcal{P}_1)$ as the total inertia associated with Good and Bad RV dissimilarities, respectively. Following Chavent et al. (2018), we compare the proportion of total inertia explained by the partition obtained using good volatility information

$$W_{\text{good}}(\mathcal{C}) = 1 - \frac{W_{\text{good}}(\mathcal{C})}{W_{\text{good}}(\mathcal{P}_1)}. \tag{11}$$

with that explained by bad volatility dynamics

$$W_{\text{bad}}(\mathcal{C}) = 1 - \frac{W_{\text{bad}}(\mathcal{C})}{W_{\text{bad}}(\mathcal{P}_1)}. \tag{12}$$

By calculating these quantities over a grid of α values, we can analyse the difference between the two criteria over all values of α. The best compromise is given by the value α associated with the lowest difference between the two curves. This heuristic approach provides a relatively simple criterion for selecting the optimal mixing parameter. To select the number of clusters, we adopt an automatic elbow criterion. We finally remark that the computational complexity of the proposed algorithm is the same as in the standard Ward (1963) hierarchical clustering.

3 Application to US Stock Market

3.1 Data and Estimated Models

We consider the daily realized volatility measures, computed using 1-minute returns, from March 2003 to March 2024 for the $N = 30$ stocks included in the Dow Jones Industrial Average (DJIA) index[1]. The list of stocks and the corresponding ARIMA model orders for realized volatility (2), Good volatility (4), and Bad volatility (5) are shown in Table 1. The estimated ARIMA model orders for RV, Good RV, and Bad RV reveal important differences in how these components evolve over time. While some stocks exhibit symmetry in their volatility dynamics, others show significant divergences between Good and Bad RV. More precisely, we find a group of stocks, including CSCO, CVX, DOW, JNJ, JPM, MMM, PG, TRV, and VZ, which display identical ARIMA models across all three measures. This suggests that for these stocks, volatility dynamics do not significantly differentiate between Good and Bad components, implying a more balanced response to market fluctuations. In contrast, stocks such as DIS, GS, HON, and MRK show identical ARIMA models for Good and Bad RV but differ from the total RV model. This pattern suggests that aggregating positive and negative volatility may mask underlying asymmetries in their individual structures. Other stocks, such as AAPL, IBM, MSFT, and UNH, exhibit cases where one of the Good or Bad components closely aligns with the RV model, while the other deviates. This indicates that realized volatility dynamics in these stocks are partially asymmetric, with one component behaving more similarly to the overall measure. The most striking differences arise in stocks like BA, HD, INTC, NKE, V, and WMT, where all three ARIMA models differ. This suggests strong asymmetry in the persistence and structure of good and bad volatility. In these cases, positive and negative volatility follow distinct autoregressive processes, highlighting how markets react differently to upward and downward movements. These findings underscore the need to model Good and Bad components separately, rather than relying only on aggregate realized volatility. The divergence in ARIMA structures across stocks highlights how the underlying drivers of volatility can vary significantly, with some stocks exhibiting symmetric dynamics while others show clear asymmetries.

3.2 Results

The first aspect to be determined in the weighted clustering approach presented in the previous section is the weight α. For this purpose, we adopt the approach of Chavent et al. (2018), explained in Sect. 2, which consists in choosing the value of α finding the best compromise between the loss in mixed inertia due to the good component and that due to the bad component of RV. Figure 1 shows the compromise, in the y-axis, between $W_{\text{good}}(\mathcal{C})$ (red line) and $W_{\text{bad}}(\mathcal{C})$

[1] We thank and acknowledge the CaPiRe research project for providing the aggregated data. Further details can be found at the following link: https://capire.stat.unipd.it/.

Table 1. Estimated ARIMA model orders (Hyndman and Khandakar, 2008) by type of realized volatility estimate. The ARMA distance is computed between the estimated processes.

Stock	RV (Eq. 2)	Good (Eq. 4)	Bad (Eq. 5)
AAPL	ARIMA(3,0,5)	ARIMA(5,0,5)	ARIMA(3,0,3)
AMGN	ARIMA(5,0,1)	ARIMA(4,0,5)	ARIMA(3,0,3)
AMZN	ARIMA(5,0,1)	ARIMA(5,0,3)	ARIMA(5,0,1)
AXP	ARIMA(5,0,3)	ARIMA(5,0,5)	ARIMA(3,0,5)
BA	ARIMA(5,0,4)	ARIMA(5,0,3)	ARIMA(4,0,2)
CAT	ARIMA(4,0,3)	ARIMA(5,0,3)	ARIMA(3,0,3)
CRM	ARIMA(4,0,4)	ARIMA(5,0,3)	ARIMA(3,0,1)
CSCO	ARIMA(5,0,1)	ARIMA(5,0,1)	ARIMA(5,0,1)
CVX	ARIMA(5,0,5)	ARIMA(5,0,5)	ARIMA(5,0,5)
DIS	ARIMA(5,0,5)	ARIMA(5,0,5)	ARIMA(5,0,4)
DOW	ARIMA(5,0,5)	ARIMA(5,0,5)	ARIMA(5,0,5)
GS	ARIMA(5,0,5)	ARIMA(5,0,5)	ARIMA(5,0,4)
HD	ARIMA(3,0,1)	ARIMA(3,0,1)	ARIMA(1,0,1)
HON	ARIMA(5,0,1)	ARIMA(5,0,1)	ARIMA(5,0,1)
IBM	ARIMA(3,0,5)	ARIMA(5,0,1)	ARIMA(3,0,5)
INTC	ARIMA(5,0,3)	ARIMA(4,0,1)	ARIMA(5,0,1)
JNJ	ARIMA(1,0,1)	ARIMA(1,0,1)	ARIMA(1,0,1)
JPM	ARIMA(5,0,0)	ARIMA(5,0,0)	ARIMA(5,0,0)
KO	ARIMA(5,0,5)	ARIMA(5,0,3)	ARIMA(5,0,5)
MCD	ARIMA(5,0,3)	ARIMA(5,0,5)	ARIMA(5,0,5)
MMM	ARIMA(4,0,4)	ARIMA(4,0,4)	ARIMA(4,0,4)
MRK	ARIMA(5,0,3)	ARIMA(5,0,3)	ARIMA(3,0,1)
MSFT	ARIMA(5,0,2)	ARIMA(5,0,5)	ARIMA(3,0,5)
NKE	ARIMA(2,0,2)	ARIMA(3,0,3)	ARIMA(3,0,5)
PG	ARIMA(1,0,1)	ARIMA(1,0,1)	ARIMA(1,0,1)
TRV	ARIMA(5,0,5)	ARIMA(5,0,5)	ARIMA(5,0,5)
UNH	ARIMA(2,0,3)	ARIMA(5,0,1)	ARIMA(5,0,3)
V	ARIMA(3,0,1)	ARIMA(3,0,3)	ARIMA(2,0,2)
VZ	ARIMA(1,0,1)	ARIMA(1,0,1)	ARIMA(1,0,1)
WMT	ARIMA(4,0,0)	ARIMA(4,0,5)	ARIMA(1,0,5)

(black line) for a grid of α values ranging from 0 to 1 with an increment of 0.05, in the x-axis. The optimal α is given by the intersection of the two lines. By applying the heuristic procedure to our dataset, we find a value of $\alpha = 0.65$, thus meaning that the Good component of RV is weighted more than the Bad component. This result suggests that stocks dissimilarities are more pronounced during periods of positive returns, represented by Good volatility, while periods of negative returns are characterized by more homogenous behaviors in the market. As a consequence, Good volatility is more informative for clustering than Bad volatility. This difference cannot be highlighted with standard clustering approaches.

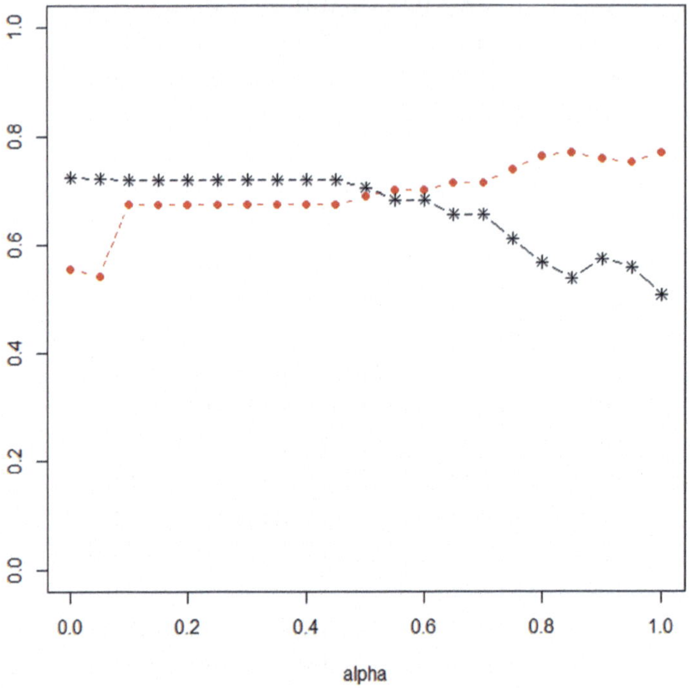

Fig. 1. Selection of the mixing parameter α with the heuristic procedure described in Sect. 2.

We then choose the number of clusters according to an automatic elbow criterion, which suggests a partition into $K = 8$ clusters. The obtained clusters are shown, using multidimensional scaling (MDS) technique on the weighted Good-Bad RV distance, in Fig. 2. The stocks' distribution in the MDS map offers a clear visual validation of the clustering, as compact groupings correspond to similar assets in terms of RV, while outliers and peripheral stocks highlight differentiated risk profiles which are not aligned with the market.

In particular, Fig. 2 highlights the existence of four singletons: BA, DOW, GS and TRV, signaling idiosyncratic volatility patterns likely driven by sector-specific (e.g., aerospace, investment banking) risk factors. Cluster 1 is the largest cluster and includes large-cap technology and healthcare stocks (e.g., AAPL, JNJ, MSFT), reflecting stable co-movements typical of growth and defensive sectors.

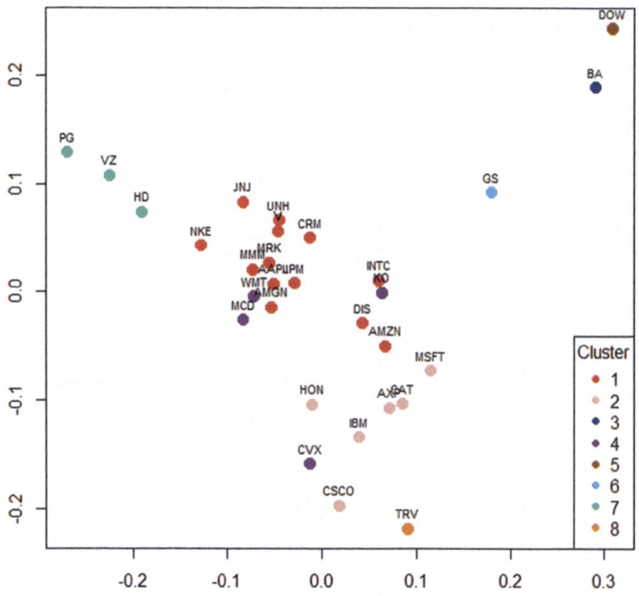

Fig. 2. Visualization of the clustering results obtained with the proposed approach.

The anaysis of π-weight patterns is useful to provide a characterization of the obtained clusters. Figure 3 shows the π-weight profiles associated with good volatility for different clusters. In terms of π-weights, we find that stocks in Cluster 1 exhibit negative and fast-decaying coefficients, which is resonable for growth or defensive assets. For these stocks, economic fundamentals and investor expectations tend to be stable, which may lead the system to quickly absorbe volatility shocks. Interestingly, stocks in Cluster 3 show flat π-weights, indicating assets with a persistent memory in the volatility process. Finally, Fig. 3 suggests that the main difference between stocks in Cluster 2 and Cluster 4 is that the second have a generally faster decay of the weights to zero than the first. Moreover, while the π weights of stocks in Cluster 4 fluctuate substantially around zero, those of the assets in Cluster 2 are more stable. Another advantage of incorporating the decomposition of realized volatility into Good and Bad components for clustering lies in the definition of more compact and well-separated clusters compared to using total realized volatility alone. To assess this advantage, we compare the partitions obtained using the Ward hierarchical clustering method based on the autoregressive metric of RV with those derived from the proposed weighted clustering approach, as shown in Fig. 2. The comparison is conducted using the Dunn Index, which is defined as the ratio of the smallest inter-cluster distance to the largest intra-cluster distance. The Dunn Index ranges from zero to infinity, with higher values indicating better clustering quality. Our proposed weighted clustering approach achieves a Dunn Index value of 0.4, whereas the clustering based on total realized volatility yields a value of 0.32. This indicates that the weighted approach produces an increase in the cluster validity index of about 25%.

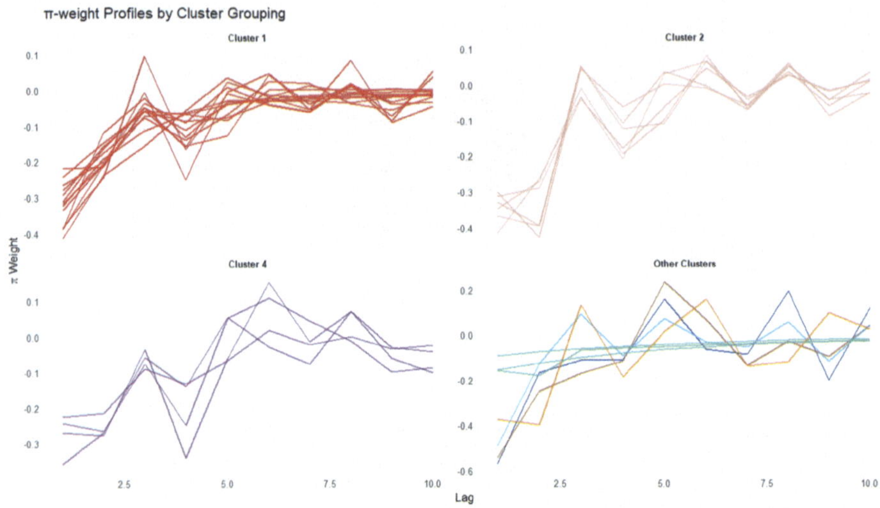

Fig. 3. Visualization of the π-weights for each cluster.

Conclusions

This paper proposes a novel methodology for clustering financial time series by decomposing realized volatility into good and bad components, thus accounting for the asymmetric impact of positive and negative returns. Following the rationale of existing volatility-based clustering methods, we propose an approach based on the autoregressive metric, which is applied separately to each component. Combined with a weighted hierarchical clustering procedure, this framework captures richer volatility dynamics and provide different information on the clusters formation compared to existing approaches. Our empirical analysis to Dow Jones Industrial Average (DJIA) index constituents reveals that good volatility helps more in discriminating stocks compared to bad volatility. Moreover, we found interesting clusters aligning with some known market segments such as growth, defensive, or sector-specific stocks. We find that large-cap technology and healthcare stocks cluster together due to their fast-decaying good-volatility profiles, while idiosyncratic stocks such as Boeing or Goldman Sachs, exhibit different patterns in volatility persistency. These findings highlight the role of clustering in enhancing decision-making processes. By identifying groups of assets that exhibit similar volatility behaviors, investors and risk managers can define appropriate strategies. Recognizing that certain clusters correspond to specific market segments, such as growth or defensive stocks, allows investors and risk managers for more informed asset allocation and risk assessment. Moreover, understanding the distinct volatility patterns, in terms of persistency, within each cluster helps in anticipating how different groups of assets may respond to market fluctuations. Future research could address the following paper limitations and further enhance the proposed methodology. While our approach is scalable to moderate cross-sections of stocks, its computational complexity may become relevant when applied to very large assets or ultra-high-frequency settings. In such cases, integrating more efficient clustering algorithms, such weighted k-means, may be needed to enhance compotuational performances of the method. To assess the robustness of the proposed approach, it would be interesting to perform a simulation study, considering also alternative distance definitions. Finally, the proposed clustering approach can be adopted for portfolio construction, by clustering assets with similar risk response profiles and designing allocation strategies accordingly.

References

Andersen, T.G., Bollerslev, T.: Deutsche mark-dollar volatility: intraday activity patterns, macroeconomic announcements, and longer run dependencies. J. Financ. **53**(1), 219–265 (1998)

Bastos, J., a. A., Caiado, J.: On the classification of financial data with domain agnostic features. Int. J. Approximate Reasoning **138**, 1–11 (2021)

Caiado, J., Crato, N.: Identifying common dynamic features in stock returns. Quant. Financ. **10**(7), 797–807 (2010)

Caiado, J., Crato, N., Peña, D.: A periodogram-based metric for time series classification. Comput. Stat. Data Anal. **50**(10), 2668–2684 (2006)

Cerqueti, R., D'Urso, P., De Giovanni, L., Giacalone, M., Mattera, R.: Weighted score-driven fuzzy clustering of time series with a financial application. Expert Syst. Appl. **198**, 116752 (2022)

Cerqueti, R., D'Urso, P., De Giovanni, L., Mattera, R., Vitale, V.: Fuzzy clustering of financial time series based on volatility spillovers. Ann. Oper. Res. 1–20 (2023)

Chavent, M., Kuentz-Simonet, V., Labenne, A., Saracco, J.: Clustgeo: an r package for hierarchical clustering with spatial constraints. Comput. Stat. **33**(4), 1799–1822 (2018)

De Luca, G., Zuccolotto, P.: Hierarchical time series clustering on tail dependence with linkage based on a multivariate copula approach. Int. J. Approximate Reasoning **139**, 88–103 (2021)

D'Urso, P., Cappelli, C., Di Lallo, D., Massari, R.: Clustering of financial time series. XXPhys. A **392**(9), 2114–2129 (2013)

Hyndman, R.J., Khandakar, Y.: Automatic time series forecasting: the forecast package for r. J. Stat. Softw. **27**, 1–22 (2008)

Iorio, C., Frasso, G., D'Ambrosio, A., Siciliano, R.: A P-spline based clustering approach for portfolio selection. Expert Syst. Appl. **95**, 88–103 (2018)

Lahmiri, S.: Clustering of Casablanca stock market based on Hurst exponent estimates. XXPhys. A **456**, 310–318 (2016)

Lopez de Prado, M.: Building diversified portfolios that outperform out-of-sample. J. Portfolio Manage. (2016)

L'ucio, F. and Caiado, J.: Covid-19 and stock market volatility: a clustering approach for s&p 500 industry indices. Financ. Res. Lett. **49**, 103141 (2022)
Mantegna, R.N.: Hierarchical structure in financial markets. Eur. Phy. J. B **11**(1), 193–197 (1999)
Mattera, R., Athanasopoulos, G., Hyndman, R.J.: Improving out-of-sample forecasts of stock price indexes with forecast reconciliation and clustering. Quant. Financ. **24**(11), 1641–1667 (2024)
Mattera, R., Otto, P.: Network log-arch models for forecasting stock market volatility. Int. J. Forecast. **40**(4), 1539–1555 (2024)
Mattera, R., Scepi, G., Kaur, P.: Time series clustering for high-dimensional portfolio selection: a comparative study. Soft Comput. (2025)
Morelli, C., Boccaletti, S., Maranzano, P., Otto, P.: Multidimensional spatiotemporal clustering-an application to environmental sustainability scores in Europe. Environmetrics **36**(2), e2893 (2025)
Otranto, E.: Clustering heteroskedastic time series by model-based procedures. Comput. Stat. Data Anal. **52**(10), 4685–4698 (2008)
Otranto, E., Gargano, R.: Financial clustering in presence of dominant markets. Adv. Data Anal. Classif. **9**(3), 315–339 (2015)
Patton, A.J., Sheppard, K.: Good volatility, bad volatility: signed jumps and the persistence of volatility. Rev. Econ. Stat. **97**(3), 683–697 (2015)
Piccolo, D.: A distance measure for classifying Arima models. J. Time Ser. Anal. **11**(2), 153–164 (1990)
Raffinot, T.: Hierarchical clustering-based asset allocation. J. Portfolio Manage. **44**(2), 89–99 (2017)
Ward, J.H.: Hierarchical grouping to optimize an objective function. J. Am. Stat. Assoc. **58**(301), 236–244 (1963)

Enhancing Statistical Inference in Mixed-Effect Three-Tree Model: A Data-Carving Estimation Strategy with an Application on Amyotrophic Lateral Sclerosis Data

Giulia Vannucci[1](✉), Roberta Siciliano[1], Valentina Iuzzolino[2], Gianmaria Senerchia[2], and Raffaele Dubbioso[2]

[1] Department of Electrical Engineering and Information Technology, University of Naples Federico II, Naples, Italy
giulia.vannucci@unina.it

[2] Department of Neuroscience, Reproductive Sciences and Odontostomatology, University of Naples Federico II, Naples, Italy

Abstract. In the framework of mixed-effects models, this paper explores the Three-Tree Mixed-Effects Model for longitudinal data. This model is a semi-parametric extension of the linear mixed-effects model, comprising a linear component and three tree-based components. This approach results in a model capable of handling interactions and nonlinearities while ensuring interpretability. Moreover, we propose an algorithm for estimating model parameters based on the data-carving post-selection inference procedure. The performance of the proposed algorithm is evaluated through a Monte Carlo study. The proposed methodology is applied to a real case study on Amyotrophic Lateral Sclerosis (ALS).

Keywords: mixed three-tree model · post-selection inference · data carving

1 Introduction

Mixed-effects models [1,2] provide a powerful statistical framework to deal with hierarchical data. In general, hierarchical data consist of unit-level observations nested within group-level clusters, such as individual cross-sectional data or repeated measures. These models are flexible tools used across various research fields. However, their flexibility comes with increased complexity, making careful model specification essential to avoid incorrect inferences.

In medical research, analyzing longitudinal data with repeated measures is crucial for understanding disease progression, treatment effects, and patient-specific variations. Longitudinal studies involve collecting data from the same subjects over multiple time points, allowing researchers to capture within-subject correlations and assess temporal trends. However, traditional statistical methods

A. D'Ambrosio et al. (Eds.): CLADAG-VOC 2025, SCDAKO, pp. 341–352, 2025.
https://doi.org/10.1007/978-3-032-03042-9_30

often fail to account for the inherent dependencies in repeated measures data, leading to biased estimates and unriable inferences.

In this work, we employ the Three-Tree Mixed-Effect Model [3] for longitudinal data. This model integrates decision trees into a mixed-effects framework, allowing for a more interpretable, data-driven approach to modeling interactions and non linear dependencies. Furthermore, we propose a data-carving post-selection inference strategy [4] to enhance the statistical validity of model parameter estimates, addressing challenges commonly encountered in small-sample settings. Additionally, we apply our proposed methodology to a real-world case study on Amyotrophic Lateral Sclerosis (ALS). ALS is a progressive neurodegenerative disorder that primarily affects motor neurons, leading to severe functional impairments and ultimately respiratory failure. Understanding disease progression is crucial for optimizing patient management and evaluating treatment effects.

This paper is organized as follows: Sect. 2 provides a brief overview of the Three-Tree Mixed-Effects Model. Section 2.1 introduces a post-selection inference procedure using the data-carving approach for model parameter estimation, along with a minimal simulation study reported in Sect. 2.2. Section 3 presents a case study on the ALS data set. Finally, conclusions and future research directions are discussed in Sect. 4.

2 Mixed-Effect Models with Trees

Consider a quantitative response variable Y_{ij} measured on unit i, with $i = 1, \ldots, n_j$ belonging to cluster (or group) j, with $j = 1, \ldots, J$, where the total number of observations is $n = \sum_{j=1}^{J} n_j$. In [3] a mixed-effect model called *Three-tree mixed-effect model* is presented, which is formulated as:

$$Y_{ij} = \boldsymbol{X}_{ij}\boldsymbol{\beta} + \boldsymbol{Z}_j\boldsymbol{\gamma} + g(\boldsymbol{X}_{ij}, \boldsymbol{Z}_j) + u_j + \epsilon_{ij} \tag{1}$$

where $\boldsymbol{X}_{ij}$ represents the set of unit-level explanatory variables, $\boldsymbol{Z}_j$ represents the set of the cluster-level explanatory variables. The corresponding fixed-effect coefficients are denoted by $\boldsymbol{\beta}$ and $\boldsymbol{\gamma}$, respectively; ϵ_{ij} are assumed to be *iid* $N(0, \sigma_\varepsilon^2)$, and u_j are assumed to be *iid* $N(0, \sigma_u^2)$, with $\epsilon_{ij} \perp\!\!\!\perp u_j$. the authors propose incorporating a nonlinear function $g(\cdot)$ into the standard mixed-effects model to capture nonlinear fixed effects, which are assumed to not vary across clusters. Specifically, $g(\cdot)$ is assumed to be a nonparametric approximation based on three regression trees. Therefore, the model in (1) becomes:

$$Y_{ij} = \boldsymbol{X}_{ij}\boldsymbol{\beta} + \boldsymbol{Z}_j\boldsymbol{\gamma} + \sum_{t=1}^{3}\sum_{m=1}^{M} \mu_{tm}\mathbb{I}\{(\boldsymbol{H}_{tij}) \in R_{tm}\} + u_j + \epsilon_{ij} \tag{2}$$

where each tree T_t $(t = 1, 2, 3)$ is constructed over a subset of predictors: $\mathbf{H}_{1ij}$ is constructed using the unit-level predictors, $\mathbf{H}_{2ij}$ is constructed using the cluster-level predictors, and $\mathbf{H}_{3ij}$ is constructed using all predictors; $R_{t1}, \ldots, R_{tM_t}$ is the partition of the predictor space corresponding to the tree T_t. The region R_{tm} is determined by the product of all dummy variables associated with the binary splits along the path from the root node to each leaf in the tree T_t. In this perspective, each tree plays the role of a factor with M_t categories.

The model is capable of taking into account non-linearities and interactions among individual-level predictors, cluster-level predictors, and cross-levels within the framework of a semi-parametric approach preserving the interpretability. The model fitting is accomplished via a backfitting-like iterative procedure that involves the fit of three CART [5] trees on partial residuals, adjusting for a linear mixed-effects component. The parameters of the final model are obtained via maximum likelihood estimator for linear mixed-effect model on the response and the explanatory variables and the three factors corresponding to the trees selected at convergence of the iterative procedure. More details on the procedure can be found in [3].

2.1 Algorithm Based on Post Selection Inference: Data Carving

In [3], the authors addressed the problem of invalid classical tools of inference like p–values and confidence intervals in the contexts where the model is selected using the data with a post-selection inference procedure (see i.e. [6]). Specifically, they adopt the *split sampling* procedure [7].

The idea behind the split sampling is very simple yet powerful: the dataset is randomly divided into two parts, one for model selection, the selection set, and the other for inference, the estimation set. This approach enables valid post-selection tests and confidence intervals while mitigating post-selection inference issues. Furthermore, inference using sample splitting followed by bootstrapping yields robust, assumption-lean confidence intervals for the projection parameters [8].

However, applying this procedure to longitudinal data in small samples, that is a common scenario in clinical studies, is not feasible. When both the number of units within groups n_j and the number of groups J are not sufficiently large, the split sampling-based algorithm may fail to perform the estimation step. This occurs because model selection in the selection set may involve different levels of categorical variables which can be absent in the estimation set. In practice, conditioning on half of the sample in such cases is too restrictive and can lead to a loss of information on categorical variables (particularly those constructed iteratively by tree-based models) preventing algorithm estimation step due to the lack of categorical variable representation across the two validation and estimation sets.

Therefore, we propose to use a different procedure of post-selection inference: the *data carving*. This method was proposed by [4] in context of ordinary least square and Lasso-type estimators for *iid* data. Instead of conditioning half the data for the estimation set as is done in the split sampling, the idea is to leave a small fraction of the data out of the selection set for model choice, and use the whole dataset in the estimation set.

For multilevel data, the standard assumption of *iid* data applies at the cluster level. In the *data carving* approach, the dataset $\mathcal{D}$ is partitioned into two subsets: a *selection set* $\mathcal{D}_S$ and an *estimation set* $\mathcal{D}_E$. Unlike sample splitting, where the two subsets are disjoint, here we retain most of the data for estimation. Specifically, we randomly select a proportion p_{carve} of the J clusters, so we carve the data taking the p_{carve} proportion of the original data to form $\mathcal{D}_S$, which is used to select the Three-tree mixed model. For the choice of the proportion p_{carve}, in [4] a simulation study illustrate the trade-off between model selection accuracy and inference power in data carving. Interestingly, holding out even a very small fraction of the data (e.g., 12%) from the first stage can significantly improve the power of the second-stage inference. This counterintuitive finding highlights the importance of carefully selecting the carving proportion and suggests that further investigation into this trade-off is warranted. The estimation set $\mathcal{D}_E$ consists of the entire dataset $\mathcal{D}$, ensuring that all available information is utilized for inference on model parameters via standard linear mixed-effects procedures. However, inspired by the approach in [8], we suggest using bootstrap-based confidence intervals in the data carving framework as well, with the aim to reduce reliance on strong distributional assumptions and improve robustness, particularly in the context of complex models like mixed models. The proposed algorithm, called *PS-3T Data Carving*, is summarized in Algorithm 1.

Algorithm 1: PS-3T Data Carving Post-selection procedure based on data carving for Three-Tree model.

Input: $\mathcal{D} = \{Y_{ij}, \mathbf{X}_{ij}, \mathbf{Z}_j\}$, $i = 1, \ldots, n_j$, $j = 1, \ldots, J$, $\mathbf{H_1}_{ij}$, $\mathbf{H_2}_{ij}$, $\mathbf{H_3}_{ij}$ matrix definition.

Output: Estimates of Model (2) parameters with their standard errors

1 **Data carving:** Let $[J] := \{1, \ldots, J\}$. Randomly choose the subset $[J_S] \subset [J]$, of cardinality $\approx (p_{\text{carve}})J$, and define $[J_E] := [J]$.

2 **Selection step:** Select Three-Tree model to the subset of data $\mathcal{D}_S = \{y_{ij}, \mathbf{x}_{ij}, \mathbf{z}_j\}, \quad i = 1, \ldots, n_j, \quad j \in [J_S]$.

3 **Estimation step:** Compute parameter estimates for model $\widehat{\mathcal{M}}$ using the standard maximum likelihood estimator for mixed effect models on the entire dataset $\mathcal{D}_E = \{y_{ij}, \mathbf{x}_{ij}, \mathbf{z}_j\}, \quad i = 1, \ldots, n_j, \quad j \in [J_E]$.

2.2 Monte Carlo Study

We conduct a minimal simulation study to evaluate the inferential accuracy of Three-Tree model with data carving (PS-3T Data Carving) in comparison with Three-Tree model with split sampling (PS-3T Split Sampling) and a standard linear mixed model (LMM). For the three considered methods we computed the bootstrap confidence intervals. We consider the simulations setting of [3]. Specifically, we consider three scenarios with different data-generating processes. Each scenario includes two individual-level predictors, X_1 and X_2, having bivariate Normal distribution, $N_2(\mathbf{0}, \mathbf{\Sigma})$, and two cluster-level predictors, Z_1 and Z_2, with distribution $N_2(\mathbf{0}, \mathbf{\Gamma})$. We set the following values for the covariance matrices:$\mathbf{\Sigma} = \mathbf{\Gamma} = \begin{bmatrix} 1.0 & 0.4 \\ 0.4 & 1.0 \end{bmatrix}$. Additionally, $u_j \sim N(0, \sigma_u^2)$, with $\sigma_u^2 = 3$, and $\varepsilon_{ij} \sim N(0, \sigma_\varepsilon^2)$, with $\sigma_\varepsilon^2 = 1$. In Scenario 1,

$$f(\boldsymbol{X}, \boldsymbol{Z}) = \beta_0 + \beta_1 X_{1ij} + \beta_2 X_{2ij} + \gamma_1 Z_{1j},$$

with $\beta_0 = 5$, $\beta_1 = 1$, $\beta_2 = 0$, $\gamma_1 = 1$, and $\gamma_2 = 0$. In Scenario 2,

$$\begin{aligned} f(\boldsymbol{X}, \boldsymbol{Z}) =& \beta_0 + \beta_1 X_{1ij} + \beta_2 X_{2ij} + \gamma_1 Z_{1j} + \gamma_2 Z_{2j} + \mu_1 \mathbb{I}(X_{1ij} \geq 0) + \\ &+ \mu_2 \mathbb{I}(Z_{2j} < 0) + \mu_3 \mathbb{I}(X_{1ij} < 0) \mathbb{I}(Z_{2j} \geq 0), \end{aligned}$$

with $\beta_0 = 5$, $\beta_1 = 1$, $\beta_2 = 0$, $\gamma_1 = 1$, and $\gamma_2 = 0$. In Scenario 3,

$$f(\boldsymbol{X}, \boldsymbol{Z}) = \beta_0 + \beta_1 X_{1ij} + \beta_2 X_{2ij} + \gamma_1 Z_{1j} + \gamma_2 Z_{2j} + \mu_1 X_{2ij}^2 + \mu_2 Z_{1j} \ln |X_{1ij}|,$$

with $\beta_0 = 0$, $\beta_1 = 2$, $\beta_2 = 0$, $\gamma_1 = 4$ and $\gamma_2 = 0$, $\mu_1 = 2$, $\mu_2 = 2$. We set the number of clusters $J = 200$, with $n_j = 7$ observations per cluster, and we run 500 Monte Carlo replications. The Carving percentage for PS-3T Data Carving is set to 90%. A slightly modified version of the algorithm in [9] was used for the analysis. This modification has been implemented in R Software, version 4.2.3 (2023-03-15).

Table 1 reports the average lengths and Monte Carlo coverage of the 95% bootstrap confidence intervals computed as in [10] in Scenarios 1, 2, and 3 for β_1, β_2, γ_1, γ_2, and σ_u. The column Fail Prop, filled only for PS-3T Split Sampling, represents the proportion of Estimation Failures. PS-3T Data Carving produces narrower confidence intervals compared to both PS-3T Split Sampling and LMM, while maintaining good coverage rates in Scenario 1. Even in more complex settings (Scenarios 2 and 3), where interactions and nonlinearities are present, Data Carving remains a viable option, offering a balance between interval precision and interpretability. It is worth noticing that some coverage deterioration is observed for certain parameters, particularly in Scenario 3. However, this scenario assumes strong nonlinearities, suggesting that in highly complex models, further refinements could enhance its robustness. PS-3T Split Sampling method exhibits significant limitations in small-sample settings. While its confidence intervals are generally wider, leading to higher coverage rates, its lack of reliability is a major drawback: in approximately 30% of Monte Carlo runs, the algorithm fails to produce estimates. This suggests that in real-world scenarios, such as clinical studies where both the number of groups and individuals is low, PS-3T Split Sampling is not a feasible strategy for post-selection inference in the Three-Tree model. The standard Linear Mixed Model (LMM) approach demonstrates stable performance and good coverage. However, when the true data-generating process includes nonlinearities and interactions, LMM struggles, leading to biased inference and poor coverage for certain parameters.

3 Case-Study on ALS Dataset

Amyotrophic Lateral Sclerosis (ALS) is a progressive neurodegenerative disorder characterized by the selective loss of motor neurons in the primary motor cortex, brainstem motor nuclei, and the anterior horn of the spinal cord. This neuronal degeneration leads to a relentless decline in motor function, ultimately resulting in death, typically due to respiratory failure. ALS is classified as a

Table 1. Monte Carlo averages of 95% confidence interval lengths and Monte Carlo coverage in Scenarios 1, 2, and 3.

Algorithm	Parameters					Fail Prop
	β_1	β_2	γ_1	γ_2	σ_u	
	SCENARIO 1					
Monte Carlo mean of the interval length						
PS-3T Data Carving	0.200	0.206	0.747	0.728	0.336	-
PS-3T Split Sampling	0.315	0.321	1.189	1.137	0.484	0.274
LMM	0.124	0.123	0.542	0.542	0.364	-
Monte Carlo coverage						
PS-3T Data Carving	0.920	0.902	0.920	0.902	0.618	-
PS-3T Split Sample	0.940	0.934	0.942	0.942	0.710	0.274
LMM	0.924	0.948	0.946	0.950	0.948	-
	SCENARIO 2					
Monte Carlo mean of the interval length						
PS-3T Data Carving	0.646	0.158	0.635	1.418	0.355	-
PS-3T Split Sampling	0.847	0.246	1.059	1.929	0.509	0.284
LMM	0.202	0.202	0.690	0.693	0.694	-
Monte Carlo coverage						
PS-3T Data Carving	0.002	0.850	0.924	0.668	0.754	-
PS-3T Split Sampling	0.953	0.953	0.905	0.816	0.777	0.284
LMM	0.000	0.952	0.948	0.000	0.038	-
	SCENARIO 3					
Monte Carlo mean of the interval length						
PS-3T Data Carving	0.311	0.992	0.809	0.790	0.431	-
PS-3T Split Sample	0.500	1.332	1.314	1.231	0.629	0.328
LMM	0.449	0.450	0.681	0.684	0.585	-
Monte Carlo coverage						
PS-3T Data Carving	0.864	0.652	0.014	0.900	0.600	-
PS-3T Split Sampling	0.905	0.604	0.095	0.902	0.714	0.328
LMM	0.964	0.718	0.000	0.960	0.928	-

Table 2. Patient and visit-level variables considered for the analyses from the ALS data set.

VISIT-LEVEL VARIABLES	
ALSFRS-R (response)	ALS Functional Rating Scale-Revised assesses 12 domains, covering bulbar function, motor function, ambulation, and respiratory function, with each item scored from 4 (normal) to 0 (complete loss) (048)
FVC	Forced Vital Capacity, measured as a percentage of predicted value
Tracheostomy	Presence of tracheostomy (0 = no, 1 = yes)
Disease_duration	Time from symptom onset to clinical evaluation (months)
MRC UL	Upper limb muscle strength (Medical Research Council scale, 0–70)
MRC LL	Lower limb muscle strength (Medical Research Council scale, 0–60)
MRC Bulbar	Bulbar region muscle strength (Medical Research Council scale, 0–15)
PUMNS Bulbar	Upper Motor Neuron burden in the bulbar region (0–4)
PUMNS UL	Upper Motor Neuron burden in the upper limbs (0–28)
PUMNS LL	Upper Motor Neuron burden in the lower limbs (0–28)
CUT OFF LABILITY SCALE	Indicator of pseudobulbar affect (threshold score on CNS-Lability Scale)
PATIENT-LEVEL VARIABLES	
Sex	Patient gender (0 = male, 1 = female)
Diagnostic delay	Time from first symptom to ALS diagnosis (months)
Familiarity	Family history of ALS (0 = no, 1 = yes)
Therapy	Ongoing ALS-specific therapy (e.g., riluzole)
Age at onset	Age at first symptom appearance (years)
Onset form	Initial symptom presentation (multiple, bulbar, respiratory or spinal)

rare disease, with a relatively low incidence and prevalence. A recent systematic review reported a global incidence ranging from 0.26 per 100 000 person-years in Ecuador to 23.46 per 100 000 person-years in Japan. Prevalence values also show wide variability, from 1.57 per 100 000 in Iran to 11.80 per 100 000 in the United States. In Europe, crude incidence rates range from 1.11 per 100 000 person-years in Serbia to 5.55 per 100 000 person-years in Denmark, while point prevalence varies between 3.44 per 100 000 in Malta and 10.80 per 100 000 in Italy. Clinically, ALS manifests as progressive weakness, which typically begins in the limb muscles (spinal onset), more frequently affecting distal rather than

Table 3. Estimates and 95% bootstrap confidence intervals for a random intercept Three-Trees model using Algorithm 1 on ALS data set.

	Coefficient	L c.i.	U c.i.
FVC (40–60)	0.93	−0.16	1.98
FVC (60–80)	1.66	0.51	2.79
FVC (>80)	2.00	0.95	3.19
FVC (ND)	0.08	−1.08	1.13
Tracheostomy (yes)	−4.06	−5.61	−2.56
Disease_duration	−0.10	−0.12	−0.09
MRC_UL	0.17	0.15	0.20
MRC_LL	0.17	0.13	0.20
MRC_Bulbar	0.41	0.23	0.59
PUMNS_Bulbar	−0.25	−0.59	0.08
PUMNS_UL	−0.13	−0.25	0.00
PUMNS_LL	−0.02	−0.14	0.10
CUT_OFF_LABILITY_SCALE (1)	−0.37	−1.27	0.42
Sex (female)	1.28	0.17	2.28
Diagnostic_delay	0.12	0.07	0.17
Familiarity (yes)	−0.01	−0.97	0.93
Therapy (none)	1.76	−0.03	3.58
Therapy (Riluzolo)	1.50	−0.18	3.23
Therapy (Riluzolo and Edaravone)	−1.40	−3.61	0.89
Age_at_onset	−0.09	−0.14	−0.04
Onset_form (multiple)	−0.85	−6.86	5.66
Onset_form (respiratory)	0.66	−2.79	3.46
Onset_form (spinal)	2.99	1.71	4.23
Tree1:9 MRC_B>=0.5 & Dur>=12.5 & MRC_B>=9.5	1.26	0.11	2.45
Tree1:10 MRC_B>=0.5 & Dur<12.5 & Trach=0	2.71	1.40	4.07
Tree1:11 MRC_B>=0.5 & Dur<12.5 & Trach=1	6.65	4.45	8.99
Tree1:12 MRC_B<0.5 & Trach=1 & PUMNS_B>=0.5	−1.28	−4.09	1.48
Tree1:13 MRC_B<0.5 & Trach=1 & PUMNS_B<0.5	2.63	0.16	4.71
Tree1:14 MRC_B<0.5 & Trach=0 & PUMNS_B>=0.5	5.60	3.47	8.07
Tree1:15 MRC_B<0.5 & Trach=0 & PUMNS_B<0.5	11.86	10.06	14.09
Tree2:12 DiagD<88.5 & DiagD<8.5 & DiagD>=2.5	5.57	0.35	10.74
Tree2:13 DiagD<88.5 & DiagD<8.5 & DiagD<2.5	8.28	2.14	14.40
Tree2:14 DiagD<88.5 & DiagD>=8.5 & AgeO>=69.5	5.73	0.68	10.99
Tree2:15 DiagD<88.5 & DiagD>=8.5 & AgeO<69.5	7.19	2.55	11.66
Tree3:9 Dur<103.5 & MRC_LL<56.5 & MRC_LL<31.5	1.57	0.45	2.62
Tree3:10 Dur<103.5 & MRC_LL>=56.5 & DiagD>=17	0.16	−1.20	1.49
Tree3:11 Dur<103.5 & MRC_LL>=56.5 & DiagD<17	3.14	2.18	4.19
Tree3:6 Dur>=103.5 & MRC_LL>=22	2.71	0.32	5.26
Tree3:14 Dur>=103.5 & MRC_LL<22 & PUMNS_B<0.5	5.87	2.99	8.64
Tree3:15 Dur>=103.5 & MRC_LL<22 & PUMNS_B>=0.5	8.53	6.00	11.51
Constant	5.75	−0.94	12.49
σ_u	3.72	3.16	3.99
σ	3.80	3.59	3.92

Abbreviations (Tree splits only): MRC_B = MRC Bulbar, Dur = Disease duration, DiagD = Diagnostic delay, AgeO = Age at onset, PUMNS_B = PUMNS Bulbar, Trach = Tracheostomy.

proximal muscles. However, approximately 25% of patients present with bulbar-onset disease, which is associated with a worse prognosis. Other adverse prognostic factors include older age at onset, a more rapid disease progression rate, advanced clinical staging, and a reduced diagnostic delay, as earlier diagnosis often correlates with a more aggressive disease course.

We apply the Three-Tree model with Data Carving as post-selection inference procedure to the dataset released by the ALS Center of Federico II University Hospital of Naples, led by Prof. Raffaele Dubbioso, which we refer to as the ALS data set. The data has a multilevel structure, where individual-level units correspond to visits and cluster-level units correspond to patients. The ALS data set consists of 254 patients with a number of visits per patient ranging from 1 to 34 (mean number of visits = 5 , median number of visits = 4), for a total of 1 412 in visits. Each participant underwent a comprehensive clinical evaluation, which included the collection of demographic and medical history data. Additional assessments were performed to quantify disease severity and progression. Overall, the dataset consists of $p = 124$ variables, that can be summarized into three main groups: demographic and anamnesic variables, clinical variables, and disease progression variables. Demographic and anamnesic variables include patient-related information such as sex, birth date, height, weight, familial history, and age at onset. Clinical variables cover functional assessments (ALSFRS-R, MRC, PUMNS, MITOS, KINGS scores), respiratory measures (FVC, ventilation status, tracheostomy), and laboratory values (creatinine). Disease progression variables includes disease duration, rate of progression, diagnostic delay, and therapy details. We considered ALSFRS-R score as response variable and a subset of $p = 16$ explanatory variables. A brief description of the variables is listed in Table 2.

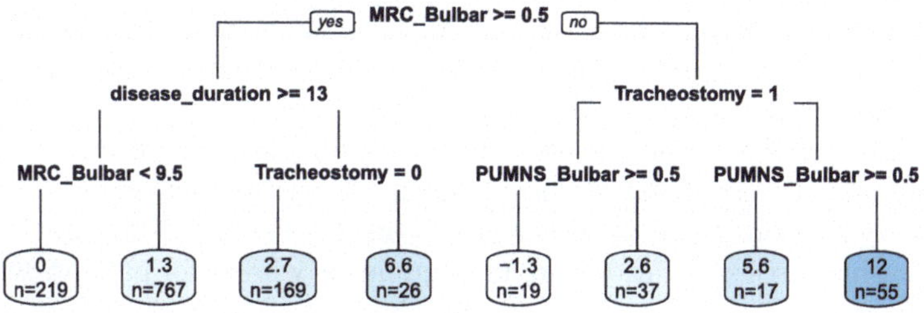

Fig. 1. Tree representing the regions selected on visit-level predictors of the Three-Trees model using Algorithm 1 for ALS data.

To make inference on coefficients of the predictors, we adopted the PS-3T data carving (Algorithm 1). The carving percentage was set to 90%. As suggested in [3], we keep the depth of the trees set to 3 to facilitate the interpretation of the results. The parameter estimates and their bootstrap confidence intervals for the model with ALSFRS-R as response variable and the set of explanatory

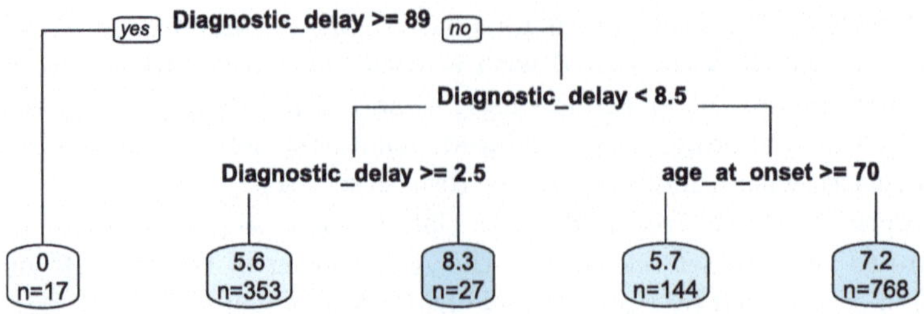

Fig. 2. Tree representing the regions selected on patient-level predictors of the Three-Trees model using Algorithm 1 for ALS data.

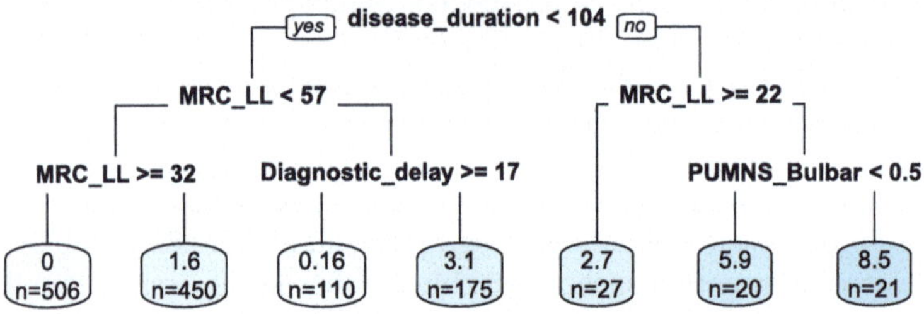

Fig. 3. Tree representing the regions selected both visit-level and patient level predictors of the Three-Trees model using Algorithm 1 for ALS data.

variables considered are reported in Table 3. The tree components of Three-tree model selected by the algorithm are depicted in Fig. 1, Fig. 2 and in Fig. 3. The algorithm represents the partition of the predictors of each tree as a factor with M_t labels. The factor is parameterized as $(M_t - 1)$ dummy variables for identifiability constraints, where the first region plays as the reference. The trees induced by the algorithm capture non-linearities and interactions among the response and the set of explanatory variables considered.

From Table 3, the parameters that have a significant main effect include Tracheostomy and disease duration, both showing negative coefficients. This indicates that, conditionally on the other predictors, the ALSFRS-R score decreases with increasing disease duration and it is significantly lower for patients with tracheostomy compared to those without. Furthermore, MRC scores for the lower limb, upper limb, and bulbar function have positive coefficients, suggesting that higher MRC scores are associated with a higher ALSFRS-R score. Other significant predictors include FVC (60–80 and > 80), Sex (female), Diagnostic delay, and Age at onset, which all show statistically significant effects. Regarding the tree components, the first tree depicted in Fig. 1 represents the regions selected on visit-level predictors. Each path of the tree can be interpreted as a *production rule* with a sequence of *if then* to yield at the mean of the response in the terminal node. For example, following the first path, the first partition consists of visits for patients with a MRC Blubar greater than 0.5 and less than 9.5, and a disease duration less than 13; following the second path, the second partition is composed of visits for patients with a MRC Blubar greater than 9.5 and a disease duration greater or equal than 13; and so on. Therefore, this tree suggests interactions and non-linearities among MRC Bulbar and disease duration, MRC Bulbar with disease duration and tracheostomy, and MRC Bulbar with PUMNS Bulbar and tracheostomy. The second tree depicted in Fig. 2 represents the regions selected on patient-level predictors. Here, interactions and non-linearities are observed between diagnostic delay and age at onset. The third tree depicted in Fig. 3 jointly considers all predictors: the interactions and non linearities suggested are among the disease duration and MRC lower limb, the disease duration with MRC lower limb and diagnostic delay, and disease duration with MRC lower limb and PUMS bulbar. The random effects standard deviation ($\sigma_u = 3.72$) indicates that there is substantial variability between patients. The residual standard deviation ($\sigma = 3.80$) represents the portion of the variability remains unexplained, indicating that additional factors might contribute to the ALSFRS-R score variability. One possibility could be to fit a random intercept and slope model, allowing slopes to vary between groups. For example, looking whether a random slope model for MRC Bulbar or PUMNS Bulbar better fits the data could be an interesting avenue for further research.

4 Conclusions and Future Research

In this paper, we employed the Three-Tree Mixed-Effect Model, a novel approach that combines mixed-effects modeling with decision trees to enhance the analysis of longitudinal data. This model enables the identification of non linearities and interactions among individual-level and cluster-level predictors while maintaining interpretability within a semi-parametric framework. To ensure valid inference after model selection, we proposed a data-carving post-selection inference strategy, which can be used in small-sample scenarios as an alternative to the sample-splitting procedure introduced in the original paper. The differences between the two strategies were explored in a simulation study, highlighting that data carving yields narrower confidence intervals while preserving quite good coverage properties compared to alternative methods. However, this minimal simulation study draw attention that further research is needed in small sample and longitudinal data analysis. Future research directions include exploring alternative tree-based partitioning strategies, exploring inferential accuracy with different settings of J number of clusters and n_j number of observations per cluster, while exploring the data-carving approach with different carving percentages. Moreover, a theoretical investigation of data carving in the context of maximum likelihood estimation for mixed models remains an open and important area of research, and further theoretical work is necessary to rigorously establish the properties of bootstrap-based inference within the data carving framework.

The case study on ALS represents an initial step in analyzing this type of data and disease. When applied to this dataset, the Three-Tree Mixed-Effect Model identified significant predictors of disease progression, including tracheostomy status, disease duration, and MRC scores, highlighting their complex interactions and non linearities through the tree components of the model. Further research is needed to examine alternative model structures, such as random slope models, to evaluate the predictive performance of the proposed approach for both new visits of existing patients and for new patients entering the study. Moreover, new research directions can be explored, such as utilizing tree-based results to define an MRC score categorization. Currently, based on clinical knowledge, the only established threshold is the maximum MRC score, which indicates the absence of pathology, while all lower scores imply pathology. With the proposed approach, one can investigate whether additional thresholds can be defined to differentiate levels of disease severity, offering a classification of patient disease status.

Notes and Comments. This work was supported by the Italian Ministry of Research, under the complementary actions to the NRRP "Fit4MedRob - Fit for Medical Robotics" Grant (PNC0000007). Data are available upon request to the authors.

References

1. Snijders, T.A.B. and Bosker, R.J.: Multilevel analysis: an introduction to basic and advanced multilevel modeling (2nd ed.). SAGE Publications Inc., Los Angeles (2012)
2. Pinheiro, J., Bates, D.: Mixed-effects models in S and S-PLUS, Springer, New York (2000). https://doi.org/10.1007/b98882
3. Gottard, A., Vannucci, G., Grilli, L., et al.: Mixed-effect models with trees. Adv. Data Anal. Classif. **17**, 431–461 (2023). https://doi.org/10.1007/s11634-022-00509-3
4. Fithian, W., Sun, D., Taylor, J.: Optimal inference after model selection. arXiv preprint (2014). https://doi.org/10.48550/arXiv.1410.2597
5. Breiman, L., Friedman, J., Olshen, R.A. et al.: Classification and Regression Trees (1st ed.). Chapman and Hall/CRC, Boca Raton (1984). https://doi.org/10.1201/9781315139470
6. Berk, R., Brown, L., Buja, A., et al.: Valid post-selection inference. Ann. Statist. **41**(2), 802–837 (2013). https://doi.org/10.1214/12-AOS1077
7. Cox, D.R.: A note on data-splitting for the evaluation of significance levels. Biometrika **62**(2), 441–444 (1975). https://doi.org/10.2307/2335385
8. Rinaldo, A., Wasserman, L., G'Sell, M.: Bootstrapping and sample splitting for high-dimensional, assumption-lean inference. Ann. Statist. **47**(6), 3438–3469 (2019). https://doi.org/10.1214/18-AOS1784
9. Gottard, A., Vannucci, G., Grilli, L. et al.: R package for 3Tree mixed effects models. (2023). https://github.com/agottard/Mix3Trees
10. Bates, D., Mächler, M., Bolker, B., Walker, S.: Fitting linear mixed-effects models using LME4. J. Stat. Softw. **67**(1), 1–48 (2015). https://doi.org/10.18637/jss.v067.i01

Author Index

A. D'Ambrosio et al. (Eds.): CLADAG-VOC 2025, SCDAKO, pp. 353–354, 2025.
https://doi.org/10.1007/978-3-032-03042-9

The manufacturer's authorised representative in the EU is Springer Nature Customer Service Centre GmbH, Europaplatz 3, 69115 Heidelberg, Germany. If you have any concerns regarding our products, please contact ProductSafety@springernature.com

Printed and bound by CPI Group (UK) Ltd, Croydon, CR0 4YY

01/10/2026

02225957-0010